AF505354

Sándor Szabó

Topics in Factorization of Abelian Groups

Birkhäuser Verlag
Basel · Boston · Berlin

Author:

Sándor Szabó
Institute of Mathematics and Informatics
University of Pecs
Ifjusag u. 6
7426 Pecs
Hungary
e-mail: szabos@ttk.pte.hu

2000 Mathematical Subject Classification 35J60, 35L45, 58J05, 58J45, 53C50, 83C05

A CIP catalogue record for this book is available from the
Library of Congress, Washington D.C., USA

Bibliografische Information Der Deutschen Bibliothek
Die Deutsche Bibliothek verzeichnet diese Publikation in der Deutschen Nationalbibliografie;
detaillierte bibliografische Daten sind im Internet über <http://dnb.ddb.de> abrufbar.

ISBN 3-7643-7158-7 Birkhäuser Verlag, Basel – Boston – Berlin

© 2004 Hindustan Book Agency (India)
Authorized edition by Birkhäuser Verlag, P.O. Box 133, CH-4010 Basel, Switzerland
for exclusive distribution worldwide except India
Part of Springer Science+Business Media
Cover design: Micha Lotrovsky, 4106 Therwil, Switzerland
Printed on acid-free paper produced from chlorine-free pulp. TCF ∞
Printed in Germany
ISBN 3-7643-7158-7

9 8 7 6 5 4 3 2 1   www.birkhauser.ch

Contents

Preface   **vii**

1 Hajós's and Rédei's theorems   **1**
   1.1   Introduction   1
   1.2   Simulated factorizations   7
   1.3   Hajós's theorem   13
   1.4   Rédei's theorem   20
   1.5   Keller's conjecture   26

2 Elementary arguments   **33**
   2.1   An axiomatic approach   33
   2.2   Direct product   39
   2.3   Nonperiodic factorizations   47

3 The machinery   **57**
   3.1   Vanishing sums   57
   3.2   Annihilators   66
   3.3   Zero divisors   78

4 Four characterization results   **89**
   4.1   Nonsubgroup factors   89
   4.2   Multiple factorizations   93
   4.3   Multiple cyclic factorizations   99
   4.4   Robinson's result   108

5 Applying the machinery   **123**
   5.1   Size of annihilators   123
   5.2   Cyclic type subsets   131
   5.3   Simulated subsets   138
   5.4   Periodic subsets   146

6 Cyclic prime component   **153**
   6.1 Cyclic prime component 153
   6.2 Cyclic groups 157
   6.3 Nonperiodic factorizations 161
   6.4 Simulated subsets 167

7 p-groups   **175**
   7.1 Groups of order p^4 175
   7.2 Factoring 2-groups 182
   7.3 Nonperiodic factorizations 195
   7.4 Elementary 2-component 201

8 The Hajós property   **215**
   8.1 Groups without Hajós property 215
   8.2 Groups with Hajós property 221

9 The Rédei property   **233**
   9.1 Latin squares 233
   9.2 The Rédei property 239
   9.3 Groups with Rédei property 252

10 Infinite groups   **263**
   10.1 Infinite constructions 263
   10.2 Infinite abelian groups 269

11 Further topics   **279**
   11.1 Some applications 279
   11.2 Miscellaneous results 287

A Simulated subsets   **297**

B Hajós's theorem   **301**

C Rédei's theorem   **307**

D Background of constructions   **311**

E Cyclotomic polynomials   **317**

References   **321**

Index   **332**

Preface

This volume contains an exposition of the factorization theory of abelian groups, some of its applications together with the necessary background materials. In 1941, G. Hajós solved a long standing geometric conjecture of H. Minkowski. This was the first, and indeed dramatic, illustration that considering a decomposition of an abelian group into a direct product of its subsets is a fruitful activity. In the ensuing decades the field grew to such an extent that it became imperative, in order to enhance further progress, to make accessible the most important techniques and results in a unified, simplified, and systematic manner.

Beginners can profitably read Chapters 1, and 2 and various later sections as a text book. These parts of the book can also be used as a reading material in a course on applications of abstract algebra. The experienced mathematician with interest in the applications of the factorization theory will find Chapters 3, and 11 particularly useful. The specialists of the field can use the volume as a reference book.

There are exercises throughout the book. These are mainly designed for beginners. The intent is to slow down the reader and ask for a few lines of computations, work out details in examples and fill in gaps in arguments.

The needs of the applications created and shaped the factorization theory in the past. Hopefully in the future it will be stimulated by newer and newer applications. However, we included some problems to try to point out directions in future research. Thus the problems are open problems.

Equations and tables are never referred from a different section than the one they appear. From this reason equations and tables are numbered by a single number. Other numbered items for example theorems are numbered by three numbers. The first one indicates the chapter, the second one refers to the section it appears.

Special thanks go to professors Keresztély Corrádi, Emil Molnár, Arthur D. Sands, and Sherman K. Stein. (The order is alphabetical.) I had the privilege and good fortune to learn from them.

Chapter 1

Hajós's and Rédei's theorems

1.1 Introduction

There is a temptation to think we know all about finite abelian groups because they are all direct products of cyclic groups.

Though we do understand their structure, many fascinating problems remain unsolved. After all, working in a direct product of cyclic groups there is no reason to expect that every interesting fact must be an immediate consequence of the way the group is constructed.

Let G be an abelian group of order v and let D be a subset of G with k elements. We say that D is a (v, k, λ) difference set if each nonidentity element g of G can be expressed precisely λ ways in the form $g = d_1 d_2^{-1}$, where $d_1, d_2 \in D$. One can check that if G is generated by g with $|g| = 7$, then $D = \{g, g^2, g^4\}$ is a $(7, 3, 1)$ difference set. (Here $|g|$ is the order of g.) Difference sets are valuable tools in constructing block designs. Symmetric block designs with parameters $(n^2 + n + 1, n + 1, 1)$ are finite projective planes. It is not known for which integers $n \geq 2$ there are projective planes. An n by n orthogonal matrix whose entries are $-1, +1$ is called a Hadamard matrix. If there is a block design with parameters $(4n-1, 2n-1, n-1)$, then there is a $4n$ by $4n$ Hadamard matrix. It is not known for which values of n there is an n by n Hadamard matrix. The existence of difference sets is a nontrivial and interesting problem.

Let G be an abelian group with basis elements $g_1, \ldots, g_n$ such that $|g_1| = \cdots = |g_n| = k+1$. If there is a code of length n over the alphabet $\{0, 1, \ldots, k\}$ that corrects one error, then G is a direct product of its subsets A, B such that $A \cap B$ contains the identity element and each nonidentity element a of A is uniquely expressible in the form $a = g_i^j$, where $1 \leq i \leq n$, $1 \leq j \leq k$. As an illustration we can consider the $n = 3$, $k = 1$ case with $A = \{e, g_1, g_2, g_3\}$, $B = \{e, g_1 g_2 g_3\}$. Here e is the identity element of G.

Let G be a finite abelian group of order n with identity element e. In 1961, P. Erdős, A. Ginsburg and A. Ziv showed that any sequence $a_1, \ldots, a_{2n-1}$ of the elements of G contains a subsequence $a_{i(1)}, \ldots, a_{i(n)}$ such that $a_{i(1)} \cdots a_{i(n)} = e$. There are many extensions of this theorem. In 1999, B. Bollobás and I. Leader proved that if $a_1, \ldots, a_{n+r}$ is a sequence of elements of G and $a_{i(1)} \cdots a_{i(n)} \neq e$ holds for each subsequence $a_{i(1)}, \ldots, a_{i(n)}$, then the products $a_{i(1)} \cdots a_{i(n)}$ vary over at least $r + 1$ distinct elements of G. Choosing here r to be $n - 1$ implies the earlier result.

For our next example we need a definition. Let f be a permutation of the elements of the finite abelian group G. If the elements $af(a)$ also form a permutation of the elements of G as a ranges over G, we say that f is a complete map of G. In 1947, L. J. Paige proved that G has a complete map unless G has only one element of order 2. The multiplication table of G forms a Latin square. A set of cells chosen from each row such that no two cells are from the same column is called a transversal. By Paige's result there is a transversal containing distinct entries if and only if G does not possess exactly one element of order 2. Complete maps and Latin squares will appear in Section 9.1 in another context.

We turn to number theory. Let A be a finite set of the integers Z. We say that translates of A tile Z if there is a subset C of Z such that each $z \in Z$ can be expressed uniquely in the form $a + c$, $a \in A$, $c \in C$. The problem considered here is to find necessary and sufficient conditions for A to tile Z. For the special case when the size of A is a prime power it was solved by D. J. Newman. For the case when the size of A is a product of two prime powers it was solved by E. M. Coven and A. Meyerowitz. Their solution is based on factoring finite abelian groups.

Factorization questions are relevant to the geometry of numbers. They are exceptionally relevant to tiling, packing and covering problems when the tile is a finite union of unit cubes. The fact that every finite abelian group is isomorphic to a factor group of the lattice of integer points with respect to an integer sub-lattice connects the vast field of tilings and abelian groups. For example, let $e_1, e_2, \ldots, e_n$ be the coordinate unit vectors of an n-dimensional orthogonal coordinate system. The union of the $nk + 1$ unit cubes whose centers are

$$je_i, \quad 1 \leq i \leq n, \quad 0 \leq j \leq k$$

is called an (n, k)-semicross. Here k refers to the length of the legs of the semicross, which is a star-body. The question of how thinly translated copies of a semicross by a lattice can cover n-space motivated A. Daddel in 1996 to ask the following question. Let G be a finite abelian group and suppose there are elements $g_1, g_2, \ldots, g_n$ in G such that each element in G can be represented in the form

$$g_i^j, \quad 1 \leq i \leq n, \quad 0 \leq j \leq k.$$

What is the maximum of the order of G in terms of k and n? Call this maximum order $f(n, k)$. Trivially, $f(n, k) \leq nk + 1$ and $f(2, k) = 2k + 1$. Daddel showed that $f(3, k) = 2k + 2$. He included results of computer calculations that suggest

$f(4, k) \leq 3k + 2$ and $f(5, k) \leq 3k + 5$. According to his data $f(4, k) \leq 3k + 3$ and $f(5, k) \leq 3k + 5$ hold for $k > 20$.

Combinatorial questions related to abelian groups vary widely. Those that are connected with expressing an abelian group as a direct product of its subsets are relevant for us. This book presents the results and develops the techniques of the factorization theory of abelian groups. Luckily many of the results are suitable for the broadest audience. The first six sections and also the sections on factorization constructions can be read with a modest background in abstract algebra.

Consider a family F of n-dimensional cubes in n-dimensional Euclidean space with the following properties.

(1) Each member of F is a translated copy of a fixed unit cube.

(2) The centers of the cubes in F form an n-dimensional lattice. In other words there are linearly independent vectors $v_1, \ldots, v_n$ such that the centers are linear combinations of $v_1, \ldots, v_n$ with integer coefficients.

(3) Each point of the space belongs to at least one member of F and if a point is not on the boundary of a cube, then the point belongs to only one cube.

A family of cubes F satisfying conditions (1), (2), (3) is called a lattice-like cube tiling.

About 1900, H. Minkowski conjectured that in an n-dimensional lattice-like cube tiling there are always two cubes sharing a complete $(n - 1)$-dimensional face. (Minkowski arrived at his conjecture working on simultaneous Diophantine approximation.) To simplify the terminology we will say that two cubes abutting along an entire $(n - 1)$-dimensional face form twins. If Minkowski's conjecture is correct, then an n-dimensional lattice-like cube tiling must have a well organized structure. If two cubes form a twin, then by the latticity there is an infinite column in the tiling formed by joining cubes. In addition the tiling is a union of translated copies of infinite columns. If we intersect the tiling by an $(n - 1)$-dimensional hyperplane perpendicular to the columns, we get an $(n - 1)$-dimensional lattice-like cube tiling for which Minkowski's conjecture applies.

In 1938, G. Hajós reformulated Minkowski's conjecture in terms of finite abelian groups. Let G be a finite abelian group written multiplicatively with identity element denoted by e. If $A_1, \ldots, A_n$ are subsets of G such that each element $g \in G$ is uniquely expressible in the form

$$g = a_1 \cdots a_n, \quad a_1 \in A_1, \ldots, a_n \in A_n,$$

then we say that G is a direct product of its subsets $A_1, \ldots, A_n$. We also will say that the equation $G = A_1 \cdots A_n$ is a factorization of G. A subset A of G with elements

$$e, a, a^2, \ldots, a^{r-1}$$

is called a cyclic subset. The group theoretical equivalent of Minkowski's conjecture now reads as follows.

If a finite abelian group G is a direct product of its cyclic subsets $A_1, \ldots, A_n$, then at least one of the factors is a subgroup of G. According to S. K. Stein the reformulation of Minkowski's geometric conjecture to an algebraic problem is nothing less remarkable than the metamorphosis of a caterpillar into a butterfly.

In 1941, G. Hajós proved Minkowski's conjecture in this group theoretical form. In the hands of subsequent workers the proof of Hajós's theorem underwent significant simplification. We are in a position now to present a fairly transparent proof.

Attempts were made to relax the cyclicity condition on the factors in Hajós's theorem. We devote Sections 2.1 and 5.2 to results that generalize Hajós's theorem by replacing the cyclic subsets with cyclic type subsets.

Before Hajós's proof in 1930, O. H. Keller conjectured that the conclusion of Minkowski's conjecture holds even if we drop the condition of latticity. Thus Keller conjectured that in any n-dimensional cube tiling there is a twin. Keller's conjecture also has a group theoretical equivalent. After more than 60 years, P. W. Shor and J. C. Lagarias refuted Keller's conjecture using the algebraic reformulation. Their constructions and a construction of J. Mackey can be found in Section 1.5.

In 1936, Ph. Furtwängler made another attempt to generalize Minkowski's conjecture. He replaced condition (3) by the following condition.

(3′) Each point of the space that is not on the boundary of any cube of F belongs to exactly k cubes.

A family of cubes F that satisfies (1), (2), (3′) is called a k-fold lattice-like cube tilin g. Furtwängler conjectured that in a k-fold lattice-like cube tiling there are always twins. This conjecture turned out to be incorrect. The following result is due to R. M. Robinson from 1979.

We say that the (n, k) pair is admissible if there is n-dimensional k-fold lattice-like cube tiling without twins. Robinson characterized the admissible pairs.

If $n \leq 3$, then there are no admissible pairs.

If $n = 4$, then $p^2 \mid k$ and p is an odd prime.

If $n = 5$, then $k = 3$ or $5 \leq k$.

If $6 \leq n$, then $2 \leq k$.

By a related result due to B. Gordon for each $n \geq 3$, $k \geq 2$ there is an n-dimensional k-fold (not necessarily lattice-like) cube tiling without twin cubes.

If a group is a direct product of cyclic groups of orders $t_1, \ldots, t_s$ we say that $t_1, \ldots, t_s$ are the invariants of G or that G is of type $(t_1, \ldots, t_s)$. Let $A_1, \ldots, A_n$ be subsets of a finite abelian group G. If each element g of G can be expressed exactly k ways in the form $g = a_1 \cdots a_n$, where $a_1 \in A_1, \ldots, a_n \in A_n$, then we say that G is a k-fold product of its subsets $A_1, \ldots, A_n$ or the product $A_1 \cdots A_n$ is a k-fold factorization of G. In the same way as 1-fold cube tilings correspond to 1-fold factorizations, k-fold cube tilings correspond to k-fold factorizations. Let $A_1, \ldots, A_n$ be subsets of a finite abelian group G. Let $A_1 \cdots A_n$ be a multiple factorization of the finite abelian group G, where the factors $A_1, \ldots, A_n$ are cyclic subsets. If for any choice of the factors it follows that at least one of the factors is always a subgroup of G, then G is either a cyclic group or G is of type $(p^\alpha, p, \ldots, p)$, where p is a prime. This result gives a characterization of those finite abelian groups for which Furtwängler's conjecture holds.

In 1965, L. Rédei proved a broad generalization of Hajós's theorem. Call a subset A of a finite abelian group normalized if $e \in A$. Call a factorization $G = A_1 \cdots A_n$ normalized if each factor is normalized. Rédei proved that if $G = A_1 \cdots A_n$ is a normalized factorization of the finite abelian group G and each A_i has a prime number of elements, then at least one of the factors must be a subgroup of G.

Let F be the family of groups whose type is one of the following or a subgroup of such a group.

$$(p^2), \qquad (p, p), \qquad (p, q),$$
$$(4, 2), \qquad (2, 2, 2, 2).$$

Here p, q are distinct primes. In other words F contains the groups with the above types and F is closed under the operation of forming subgroups. Let $G = A_1 \cdots A_n$ be a factorization of the finite abelian group G such that $e \in A_1, \ldots, e \in A_n$. If for any choice of the factors it follows that at least one of the factors is always a subgroup of G, then G must be a member of the F family. Conversely if $G \in F$, then in any factorization of $G = A_1 \cdots A_n$ with $e \in A_1 \cap \cdots \cap A_n$, at least one of the factors must be a subgroup of G.

Geometry is by no means the only field where factorizations have found essential applications. There are applications in combinatorics, coding theory, functional analysis, number theory and complexity of algorithms. We are not in position to do justice to the applications of the factorization theory for it could be the subject of another book. However it is our hope that the practitioners of the above mentioned fields will find the results the techniques and the examples presented here useful. Sections 1.2, 1.3, and 1.4 will present simplified proofs for Hajós's and Rédei's theorems suitable for a broad audience. The proofs are elementary in the sense that they do not rely either on group rings or on group characters. This alone was sufficient justification for writing on the subject for a more general audience than only the specialists in the subject. Chapter 3 will develop the machinery of the factorization theory: vanishing sums of roots of unity, characters, annihilators, and results on certain zero divisors of the group ring $Z(G)$. This will provide a platform to study results of a less elementary nature.

Motivated by Rédei's theorem, one main theme will emerge for the last part of the book. If $G = A_1 \cdots A_n$ is a factorization of the finite abelian group G such that the cardinalities of $A_1, \ldots, A_n$ are $q_1, \ldots, q_n$ respectively, then we call the n-tuple $(q_1, \ldots, q_n)$ the type of the factorization. A subset A of G is called periodic if there is an element $g \in G$ such that $g \neq e$ and $Ag = A$. By Rédei's theorem, if each component of $(q_1, \ldots, q_n)$ type of the factorization $G = A_1 \cdots A_n$ is a prime, then at least one of the factors $A_1, \ldots, A_n$ is periodic. We will be concerned with the following problem. Given a finite abelian group G and an n-tuple $(q_1, \ldots, q_n)$ such that the order of G is $q_1 \cdots q_n$, does it follow that in each factorization of $G = A_1 \cdots A_n$ with factorization type $(q_1, \ldots, q_n)$ at least one of the factors is always periodic? In other words is $(q_1, \ldots, q_n)$ a periodicity forcing factorization type for G?

There are groups for which each possible factorization type is periodicity forcing. Let U be the family of groups whose type is one of the following or is a subgroup of such a group.

$$
\begin{array}{llll}
(p^\alpha, q), & (p^2, q^2), & (p^2, q, r), & (p, q, r, s), \\
(p^3, 2, 2), & (p^2, 2, 2, 2), & (p, 2^2, 2), & (p, 2, 2, 2, 2), \\
(p, q, 2, 2), & (p, 3, 3), & (3^2, 3), & (2^\alpha, 2), \\
(2^2, 2^2), & (p, p).
\end{array}
$$

Here p, q, r, s are distinct primes the $p = 2$ and $p = 3$ cases are not excluded and $\alpha \geq 1$ is an integer. Plainly U contains the groups with the above types and U is closed under the operation of forming subgroups. If in each possible factorization of a group G one of the factors is always periodic, then G is a member of the U family and conversely if G belongs to U, then in each factorization $G = A_1 \cdots A_n$ at least one of the factors must be periodic. We could refer to U as the family of universally periodic groups since no matter how these groups are factored into subsets at least one of the factors always must be periodic.

In 1963, A. D. Sands asked if $G = BA_1 \cdots A_n$ is a factorization of a finite cyclic group G such that $|A_1|, \ldots, |A_n|$ are primes and $|B|$ is a product of two distinct primes, does it follow that one of the factors is periodic? The question was answered 32 years later in the affirmative by K. Corrádi and S. Szabó. In 2000, A. D. Sands proved the following more general result. If $G = A_1 \cdots A_n$ is a factorization of a finite cyclic group G such that each $|A_i|$ is either a prime power or a product of two distinct primes, then one of the factors is periodic. Section 6.2 will present the details. Sections 2.2, 7.1, and 7.4 are also motivated by Sands's original question in the sense that factorizations are considered where each but one factor has prime cardinality and the order of the exceptional factor is a product of two (not necessarily distinct) primes.

Consider a finite abelian group G and a factorization $G = AB$ of G. If for each choice of the factors it follows that either A or B is always periodic, then we say that G possesses the Hajós 2-property. Initially it was not obvious even whether there is any finite abelian group without the Hajós 2-property. It was G. Hajós who provided in 1949 the first examples for groups without Hajós 2-property. One might guess that there should be a Hajós n-property for various values of n. The guess is correct. However the Hajós n-property will appear only once in the book. If G has the Hajós 2-property, then G belongs to U. In fact, historically the list that determines U was compiled to describe the groups with the Hajós 2-property. It turned out later that if G has the Hajós 2-property, then in each factorization $G = A_1 \cdots A_n$ of G for any choice of n and the factors $A_1, \ldots, A_n$, at least one of the factors is always periodic.

The following problem is due to L. Rédei. Let p be a prime and let G be a group of type (p, p, p). From a factorization $G = AB$, does it follow that A or B is contained by a subgroup of order p^2? In other words does a factorization $G = AB$ imply that A or B does not span the whole of G? The same question was considered independently about elementary 2-groups in coding theory in connection with full rank tilings of the affine space $[\mathrm{GF}(2)]^n$. The scope can be broadened to all finite abelian groups. We say that the finite abelian group G has the Rédei property if from each factorization $G = AB$ it follows that A or B does not span the whole of G. In 1995, studying a problem in Fourier analysis R. Tijdeman conjectured that finite cyclic groups possess the Rédei property. We can see that the Rédei property appeared in three different branches of mathematics. In Section 9.2 it will be shown that only a tiny fraction of the finite abelian groups can have the Rédei property. In particular Tijdeman's conjecture will be disproved. Section 9.1 will describe how certain Latin squares are related to Rédei's problem on groups of type (p, p, p).

It looks useful to discuss factorizations of infinite abelian groups and the difference and similarity with the completely finite factorizations. Factorizations of infinite groups are treated in Chapter 10.

The reader will notice that the various chapters mostly deal with extensions of Hajós's and Rédei's theorems, or are occupied with developing a technical tool, or attempt to solve a characterization problem, or as a part of a characterization problem describe factorization constructions.

There is a phenomenon that can be stated such that each group we can test is too small. One converts a geometry (coding theory, number theory, functional analysis) problem to a problem of factoring finite abelian groups. Small groups do not admit a large variety of factorizations contrasted to larger groups. So inspecting small groups tends to suggest unsound conjectures. In fact groups whose possible factorizations can be checked exhaustively are too small to exhibit the typical behavior. The value of nontrivial factorization examples cannot easily be overstated. The factorization constructions presented in the book are considered to be very important contributions to the factorization theory of abelian groups and to its applications.

1.2 Simulated factorizations

In this section we consider a factorization $G = H_1 \cdots H_n$ of the finite abelian group G into subgroups $H_1, \ldots, H_n$ of G. Then we try to modify the factors by changing only one element in each subgroup $H_1, \ldots, H_n$. We will show that in the resulting factorization at least one of the factors must be a subgroup of G. In other words at least one of the factors, in fact, must remain unchanged.

The subset A of the finite abelian group G is said to be *simulated* if there is a subgroup H of G such that $|A| = |H| \geq 3$ and $|A \cap H| + 1 \geq |A|$. A simulated subset is either a subgroup of more than two elements or differ from such a subgroup in one element. ($|A|$ denotes the number of elements of A.)

Exercise 1.2.1. Let H be a finite abelian group with $|H| \geq 3$. Show that if $B \subset H$ such that $|B| > |H|/2$, then $\langle B \rangle = H$. Therefore a simulated subset uniquely determines its associated subgroup. (Here $\langle B \rangle$ stands for the smallest subgroup of H containing B.)

Let A be a simulated subset of G and let H be the corresponding subgroup of G such that $H \neq A$. There are elements $a \in A$ and $h \in H$ such that $a \notin H$ and $h \notin A$, that is, $A = (H \setminus \{h\}) \cup \{a\}$ and $H = (A \setminus \{a\}) \cup \{h\}$. Further there is an element $d \in G$ for which $a = hd$. Now $A = (H \setminus \{h\}) \cup \{hd\}$. If $d = e$, then $A = H$. The element d can be called the *distortion element* of A. The element h can be called the *distortion place* of A. Clearly, the subgroup H and the elements h and d determine the subset A.

A subset A is defined to be *periodic* if there exists an element g of $G \setminus \{e\}$ with $gA = A$. We refer to such elements g as *periods* of A. The empty set is not considered to be periodic and the identity element e is not considered to be a period. To a nonempty subset A of a finite abelian group G we assign the subset L such that

$$L = \bigcap_{a \in A} a^{-1} A.$$

Lemma 1.2.1. *Let A be a nonempty subset of a finite abelian group G. If for the subset L assigned to A it holds that $L \neq \{e\}$, then A is periodic.*

Proof. Let $A = \{a_1, \ldots, a_n\}$. Since A is not empty, $e \in L$. Suppose that $g \in L \setminus \{e\}$. Such g does exist since L is larger than $\{e\}$. As A is not empty, there are elements $b_1, \ldots, b_n \in A$ such that $g = b_1 a_1^{-1} = \cdots = b_n a_n^{-1}$. Note that $b_1, \ldots, b_n$ are pairwise distinct elements of A. Since $b_1, \ldots, b_n$ are different elements they are all the elements of A. Consequently,

$$\begin{aligned} gA &= \{ga_1, \ldots, ga_n\} \\ &= \{b_1 a_1^{-1} a_1, \ldots, b_n a_n^{-1} a_n\} \\ &= \{b_1, \ldots, b_n\} \\ &= A. \end{aligned}$$

This completes the proof.

Lemma 1.2.1 can be used to test the periodicity of a subset. We will refer to this as testing periodicity by translates.

Exercise 1.2.2. Let A be a subset of a finite abelian group G and let $g_1, \ldots, g_s$ be all the periods of A. Verify that $K = \{e, g_1, \ldots, g_s\}$ is a subgroup of G. The group G is a disjoint union of cosets modulo K. Show that A is a union of whole cosets, that is, A either contains a given coset or is disjoint to it. Prove that there is a subset B of G such that the product BK is direct and gives A. (Is this B unique?)

Exercise 1.2.3. Show that the subset L defined in Lemma 1.2.1 coincides with the subgroup K defined in Exercise 1.2.2.

If the finite abelian group G is a direct product of cyclic groups of order $t_1, \ldots, t_n$, then we say that the n-tuple $(t_1, \ldots, t_n)$ is the *type of G*. Be aware that a group might have different types. For example the cyclic group of order 6 has types $(2,3)$, $(3,2)$, (6). But a given type identifies a group without any ambiguity. We will use types of groups to identify groups. If p is a prime, then groups of type $(p, \ldots, p)$ are called *elementary p-groups*.

Exercise 1.2.4. Let G be an elementary 2-group of order 16 with basis elements x, y, u, v, that is, $G = \langle x, y, u, v \rangle$, where $|x| = |y| = |u| = |v| = 2$. Using Exercise 1.2.3 show that the sets

$$A = \{xy, xyu, xyv, yuv\},$$
$$B = \{xy, xyu, xyv, uv, xuv, yuv\},$$

are not periodic. ($|x|$ is the order of the element x.)

From the last part of Exercise 1.2.2 it follows that if A is a normalized subset of G and $|A|$ is a prime and in addition A is periodic, then A must be a subgroup of G. By the next lemma if A is a simulated subset and periodic, then A must be a subgroup.

Lemma 1.2.2. *Let A be a simulated subset of G and let H be the corresponding subgroup of G. If A is periodic, then $A = H$.*

Proof. Let $A = \big(H \setminus \{h\}\big) \cup \{hd\}$, where $h \in H$, $d \in G$. Let g be a period of A and suppose that $|g| = r$. We may assume that r is a prime. The permutation defined by

$$a \to ag, \quad a \in A$$

consists of cycles of length r. Consider the cycle which contains hd. If $r \geq 3$, then there must be at least two further elements a and b in this cycle contained by H. So $g^t a = b$ for some t, $0 \leq t \leq r - 1$. Therefore $g \in H$. Similarly, $g^s a = hd$ for some s, $0 \leq s \leq r - 1$ and so $d \in H$. If $r = 2$, then since $|A| \geq 3$ the permutation must contain at least one additional cycle. As before, using this second cycle we can see that $g \in H$ and using the first cycle we have that $d \in H$.

This completes the proof.

Tactically replacing a factor in a factorization is a powerful tool in studying factorizations. The proof of Theorem 1.2.1 will be the first illustration.

If A and A' are subsets of the finite abelian group G such that, for every subset B of G, if $G = AB$ is a factorization of G, then $G = A'B$ is also a factorization of G, then we shall say that A is *replaceable* by A'.

The next exercises describe elementary replacement results.

Exercise 1.2.5. Let $G = AB$ be a normalized factorization of the finite abelian group G. Show that $G = (a^{-1}A)B$ is also a normalized factorization of G for each $a \in A$. (Hint: Multiply both sides of the factorization $G = AB$ by a^{-1}.)

Exercise 1.2.6. Prove that if $G = AB$ is a factorization of the finite abelian group G, then $|G| = |A||B|$ and $A^{-1}A \cap BB^{-1} = \{e\}$. Here A^{-1} denotes $\{a^{-1} : a \in A\}$. Then show conversely if $|G| = |A||B|$ and $A^{-1}A \cap BB^{-1} = \{e\}$, then the product AB is direct and gives G. Using this observation verify that the factor A always can be replaced by A^{-1}.

Lemma 1.2.3. *Let $G = AB$ be a factorization of the finite abelian group G. Suppose that $A = \{h_1, \ldots, h_{s-1}, h_s d\}$ is a simulated subset of G and $H = \{h_1, \ldots, h_s\}$ is the corresponding subgroup of G. Then $dB = B$ and*

$$G = \{h_1, \ldots, h_{i-1}, h_i d^j, h_{i+1}, \ldots, h_s\}B$$

is also a factorization of G. In particular A is replaceable by H and if $d \neq e$, then B is periodic with period d.

Proof. Suppose that $h_1 = e$. The fact that $G = AB$ is a factorization of G is equivalent to the subsets

$$h_1 B, \ldots, h_{s-1} B, h_s dB$$

form a partition of G. We claim that $G = HB$ is a factorization of G, that is, the subsets

$$h_1 B, \ldots, h_{s-1} B, h_s B \tag{1}$$

forming a partition of G. To prove the claim assume the contrary, that the sets (1) do not form a partition of G, say

$$h_i B \cap h_s B \neq \emptyset$$

for some i, $1 \leq i \leq s - 1$. If $i \neq 1$, then we have the contradiction

$$h_i h_i^{-1} B \cap h_s h_i^{-1} B = h_1 B \cap h_j B \neq \emptyset$$

and $2 \leq j \leq s - 1$. If $i = 1$, then since $|H| = s \geq 3$ there is an $h_k \in H$ with $2 \leq k \leq s - 1$. Now we have the contradiction that

$$h_k h_1 B \cap h_k h_s B = h_k B \cap h_j B \neq \emptyset$$

and $1 \leq j, k \leq s - 1$, $j \neq k$.

Comparing the two partitions of G we get that $h_s dB = h_s B$ and so $dB = B$. It follows that $d^j B = B$ for each integer j. This gives that the sets

$$h_1 B, \ldots, h_{i-1} B, h_i d^j B, h_{i+1} B, \ldots, h_s B$$

form a partition of G.

This completes the proof.

The factorization $G = AB$ is clearly equivalent to the fact that the sets aB, $a \in A$ form a partition of G. It will be referred to as the partition reformulation of the factorization.

Theorem 1.2.1. *If $G = A_1 \cdots A_n$ is a factorization of the finite abelian group G, where the factors are simulated subsets of G, then at least one of the factors must be a subgroup of G, that is, there is an i, $1 \leq i \leq n$ such that $A_i = H_i$, where H_i is the subgroup associated with A_i.*

Proof. Let us start with the factorization

$$G = A_1 \cdots A_n, \tag{2}$$

where $A_1, \ldots, A_n$ are simulated subsets of the finite abelian group G. If $A_i = H_i$ for some i, $1 \leq i \leq n$, then the result holds. So we may suppose that $|A_i \cap H_i| = |A_i| - 1$, that is, the subset A_i is not a subgroup of G for any i, $1 \leq i \leq n$. Let $a_i \in A_i \setminus H_i$, $h_i \in H_i \setminus A_i$. There is a $d_i \in G$ such that $a_i = h_i d_i$. Clearly $d_i \neq e$. If $n = 1$, then the result is trivial. We assume that $n \geq 2$ and proceed by induction on n. By Lemma 1.2.3 in the factorization (2) the factor A_1 can be replaced by H_1 to give the factorization

$$G = H_1 A_2 \cdots A_n.$$

From this, considering the factor group G/H_1, we get the factorization

$$G/H_1 = (A_2 H_1)/H_1 \cdots (A_n H_1)/H_1$$

of the factor group G/H_1. Here $(A_i H_1)/H_1$ denotes the set of cosets $\{a_i H_1 : a_i \in A_i\}$. By the inductive assumption some factor $(A_i H_1)/H_1$ is a subgroup of G/H_1 and hence $H_1 A_i$ is a subgroup of G for some i, $2 \leq i \leq n$. We may assume that $i = 2$ because this is only a matter of indexing the factors $A_1, \ldots, A_n$ in the factorization $G = A_1 \cdots A_n$. We may consider a suitable factor group again to get a new factorization. Continuing in this way we conclude that

$$H_1 \subset H_1 A_2 \subset \cdots \subset H_1 A_2 \cdots A_n$$

are subgroups of G. More precisely we only proved that there is a permutation $B_2, \ldots, B_n$ of the factors $A_2, \ldots, A_n$ such that

$$H_1 \subset H_1 B_2 \subset \cdots \subset H_1 B_2 \cdots B_n$$

are subgroups of G. However, we may assume that the permutation is identical since this is only a matter of indexing the factors in (2).

Let $M = H_1 A_2 \cdots A_{n-1}$. Note that

$$M = H_1 A_2 \cdots A_{n-1}, \quad G = M A_n$$

are factorizations of M and G respectively. Let $a \in A_n$. From the factorizations

$$G = A_1 A_2 \cdots A_n,$$
$$G = H_1 A_2 \cdots A_n,$$
$$G = M A_n,$$

multiplying by a^{-1} we have the factorizations

$$G = A_1 A_2 \cdots A_{n-1}(a^{-1}A_n),$$
$$G = H_1 A_2 \cdots A_{n-1}(a^{-1}A_n),$$
$$G = M(a^{-1}A_n).$$

If $A_1 \subset M$, then $M = A_1 A_2 \cdots A_{n-1}$ is a factorization of M. (Exercise 1.2.7 after the proof is related to this step.) By the inductive assumption, $A_i = H_i$ for some i, $1 \le i \le n - 1$.

If $A_1 \not\subset M$, then from the factorization $G = M(a^{-1}A_n)$ it follows that $a^{-1}A_n$ is a complete set of representatives modulo M. Therefore, there exists an element c_a of $a^{-1}A_n$ such that the coset $c_a M$ contains the element $h_1^{-1}d_1^{-1}$, that is, for which $h_1 d_1 c_a \in M$. Let

$$\begin{aligned} C_a &= \left(H_1 \setminus \{h_1\}\right) \cup \{h_1 d_1 c_a\} \\ &= \left(A_1 \setminus \{a_1\}\right) \cup \{h_1 d_1 c_a\}. \end{aligned}$$

We claim that $M = C_a A_2 \cdots A_{n-1}$ is a factorization of M. To prove the claim note that $C_a \subset M$ and in addition products coming from $C_a A_2 \cdots A_{n-1}$ occur among the products coming from $A_1 A_2 \cdots A_{n-1}(a^{-1}A_n)$, and these latter are distinct since the product $A_1 \cdots A_{n-1}(a^{-1}A_n)$ is a factorization of G. To see this consider a product

$$x_1 x_2 \cdots x_{n-1}, \quad x_1 \in C_a, x_2 \in A_2, \ldots, x_{n-1} \in A_{n-1}.$$

If $x_1 \in A_1$, then

$$x_1 x_2 \cdots x_{n-1} \in A_1 A_2 \cdots A_{n-1}.$$

If $x_1 = h_1 d_1 c_a$, then as $h_1 d_1 \in A_1$ and $c_a \in (a^{-1}A_n)$ it follows that

$$x_1 x_2 \cdots x_{n-1} \in A_1 A_2 \cdots A_{n-1}(a^{-1}A_n).$$

Using the factorization $M = C_a A_2 \cdots A_{n-1}$ the inductive assumption gives that one of the factors $C_a, A_2, \ldots, A_{n-1}$ is a subgroup of M. We may assume that C_a is a subgroup of M since otherwise there is nothing to prove. In particular C_a is a periodic subset of M and so, by Lemma 1.2.2, $C_a = H_1$. Thus $h_1 d_1 c_a = h_1$, that is, $c_a = d_1^{-1}$ and so $d_1^{-1} \in a^{-1}A_n$. This gives

$$d_1^{-1} \in \bigcap_{a \in A_n} a^{-1}A_n.$$

The element d_1^{-1} is independent of the choice of a and $d_1^{-1} \ne e$. We can test the periodicity of A by its translates. By Lemma 1.2.1, A_n is periodic. By Lemma 1.2.2, A_n is a subgroup of G.

This completes the proof.

The maneuver in the above proof of defining the set C_a and showing that $M = C_a A_2 \cdots A_{n-1}$ is a factorization of M, will be used several times. We will refer to it as the Corrádi trick.

Exercise 1.2.7. Let $B_1, \ldots, B_m$ be subsets of the finite abelian group H such that $|H| = |B_1| \cdots |B_m|$ and the products

$$b_1 \cdots b_m, \qquad b_1 \in B_1, \ldots, b_m \in B_m$$

are distinct. Prove that the product $B_1 \cdots B_m$ is direct and gives H, that is, $H = B_1 \cdots B_m$ is a factorization of H.

The next exercises deal with applications of Theorem 1.2.1.

Exercise 1.2.8. Let G be a finite group of type $(2, \ldots, 2)$. Prove that if $G = A_1 \cdots A_n$ is a normalized factorization of G, where each $|A_i| = 4$, then at least one of the factors must be a subgroup of G. (Hint: Let $A = \{e, a, b, c\} \subset G$. Note that c can be written in the form $c = abd$, $d \in G$.)

A proof of Rédei's theorem for elementary 3-groups is the subject of the next exercise.

Exercise 1.2.9. Let G be a finite group of type $(3, \ldots, 3)$. Prove that if $G = A_1 \cdots A_n$ is a normalized factorization of G, where each $|A_i| = 3$, then at least one of the factors must be a subgroup of G. (Hint: Let $A = \{e, a, b\} \subset G$. Note that b can be written in the form $b = a^2 d$, $d \in G$.)

1.3   Hajós's theorem

A subset A of a finite abelian group G is defined to be *cyclic* if there is an element a of G and an integer r such that the elements $e, a, a^2, \ldots, a^{r-1}$ are distinct and give all the elements of A. In order to avoid trivial cases we will assume that $a \neq e$ and $r \geq 2$. The cyclic subset we just defined consists of the "first" r elements of the cyclic group generated by the element a.

Let $G = H_0$ be a finite abelian group. Let H_1 be a subgroup of G and let A_1 be a complete set of representatives modulo H_1. Now each element g of G is uniquely expressible in the form $g = a_1 h_1$, $a_1 \in A_1$, $h_1 \in H_1$. In other words, the product $A_1 H_1$ is direct and gives G. Similarly, if H_1 contains a subgroup H_2 and if A_2 is a complete set of representatives modulo H_2 in H_1, then $H_1 = A_2 H_2$ is a factorization of H_1. Hence $G = A_1 A_2 H_2$ is a factorization of G. Continuing in this way we get a factorization $G = A_1 \cdots A_{n-1} H_{n-1}$ of G. The factorization is constructed using the descending chain of subgroups

$$G = H_0 \supset H_1 \supset \cdots \supset H_{n-1} \supset H_n = \{e\}. \tag{1}$$

In order to illustrate the procedure let G be the cyclic group of order 32 generated by the element x, that is, $G = \langle x \rangle$, $|x| = 32$. Let the subgroup H_1 be $\langle x^2 \rangle$ and let the subset A_1 be $\{e, x^3\}$. This results in the factorization $G = A_1 H_1$ of G. As $H_2 = \langle x^8 \rangle$ is a subgroup of H_1 and $A_2 = \{e, x^{10}, x^{20}, x^{30}\}$ is a complete set of representatives in H_1 modulo H_2, $H_1 = A_2 H_2$ is a factorization of H_1.

Hence $G = A_1 A_2 H_2$ is a factorization of G. Finally let $H_3 = \langle x^{16} \rangle$ and $A_3 = \{e, x^{24}\}$. This gives the factorization $G = A_1 A_2 A_3 H_3$ of G. In the factorization $G = A_1 A_2 \cdots A_{n-1} H_{n-1}$ we can permute the factors to make the role of the chain of the subgroups $H_1, H_2, \ldots, H_{n-1}$ less apparent. In our example beside the factorization

$$G = \{e, x^3\}\{e, x^{10}, x^{20}, x^{30}\}\{e, x^{24}\}\{e, x^{16}\}$$

for instance

$$G = \{e, x^{10}, x^{20}, x^{30}\}\{e, x^{16}\}\{e, x^{24}\}\{e, x^3\}$$

is also a factorization of G.

Summing up our considerations we can say that there is a simple way to construct factorizations $G = A_1 \cdots A_n$ of the given finite abelian group G, where each factor is a cyclic subset of G. Namely, use a chain of subgroups (1), where the factor groups H_i/H_{i+1} are cyclic. By Hajós's theorem every factorization of G by cyclic subsets occurs as a result of this procedure. It turns out that to prove Hajós's theorem it is enough to consider only the factorizations $G = A_1 \cdots A_n$, where the cyclic factors are of prime cardinality. We formulate the observations leading to this reduction as exercises.

Exercise 1.3.1. Show that if $|a| = r$, then the cyclic subset

$$A = \{e, a, a^2, \ldots, a^{r-1}\}$$

is a subgroup. And conversely if A is a subgroup, then $|a| = r$.

Cyclic subsets of composite order factor into cyclic subsets of smaller order and in addition if one of the resulting factors is a subgroup so is the original one. This is the content of the next exercise.

Exercise 1.3.2. Verify that if A is a cyclic subset of G with composite cardinality, say

$$A = \{e, a, a^2, \ldots, a^{rs-1}\},$$

where $r \geq 2$ and $s \geq 2$, then A factors into two cyclic subsets

$$B = \{e, a, a^2, \ldots, a^{r-1}\},$$
$$C = \{e, a^r, a^{2r}, \ldots, a^{(s-1)r}\}.$$

Also A is the direct product of

$$B' = \{e, a, a^2, \ldots, a^{s-1}\},$$
$$C' = \{e, a^s, a^{2s}, \ldots, a^{(r-1)s}\},$$

that is, the decomposition of A into cyclic subsets is not necessarily unique.

Exercise 1.3.3. Prove that B cannot be a subgroup and that if C is a subgroup, then so is A. Similarly, prove that B' cannot be a subgroup and if C' is a subgroup, then so is A.

Thus in the factorization $G = A_1 \cdots A_n$ cyclic subsets of composite cardinality factor into cyclic subsets of prime cardinality. This leads to a factorization where all the cyclic factors are of prime cardinality. In addition, if there is a subgroup among the resulting factors, then there is one among the factors $A_1, \ldots, A_n$. Of course after such a transformation the number of the factors can increase.

Exercise 1.3.4. Verify Hajós's theorem for elementary p-groups, that is, for groups of type $(p, \ldots, p)$.

The following lemma is about replacing cyclic subsets of prime cardinality.

Lemma 1.3.1. *Let p be a prime and let t be an integer. Let G be an abelian group. The cyclic subset*

$$A = \{e, a, a^2, \ldots, a^{p-1}\}$$

of G can be replaced by

$$A' = \{e, a^t, a^{2t}, \ldots, a^{(p-1)t}\}$$

whenever t is relatively prime to p.

Proof. Let $G = AB$ be a factorization of G. By the partition reformulation of the factorization the sets

$$a^0 B, a^1 B, a^2 B, \ldots, a^{p-1} B \tag{2}$$

form a partition of G. We want to prove that the sets

$$a^{0t} B, a^{1t} B, a^{2t} B, \ldots, a^{(p-1)t} B \tag{3}$$

also form a partition of G, that is, $G = A'B$ is a factorization of G.

Multiply both sides of the factorization $G = AB$ by a to get $aG = (aA)B$. As $aG = G$ and $aA = \{a^1, a^2, \ldots, a^p\}$, the sets

$$a^1 B, a^2 B, \ldots, a^{p-1} B, a^p B \tag{4}$$

form a partition of G. Since sets (2) and (4) form a partition of G, we have $a^p B = B$. We may say that the multiplication by a permutes the sets (2) cyclically. Thus if i and j are integers (positive or negative) and they are congruent modulo p, then $a^i B = a^j B$. As t is relatively prime to p, (3) is a permutation of (2) and so $G = A'B$ is a factorization of G.

This completes the proof.

Let p be a prime and let G be a finite abelian group. If $p \mid |G|$, then G is a direct product of two subgroups H, K such that $|H|$ is a p power and $p \nmid |K|$. Each $g \in G$ can be written uniquely in the form $g = hk$, $h \in H$, $k \in K$. We call h the p-part of g and denote it by $g_{|p}$. We call k the p'-part of g and it will be denoted by $g_{|p'}$.

Exercise 1.3.5. Give an example to show that the conclusion of Lemma 1.3.1 does not necessarily hold if t is not relatively prime to p.

Exercise 1.3.6. Let p be a prime and consider the cyclic subset

$$A = \{e, a, a^2, \ldots, a^{p-1}\}.$$

Show that if A is a direct factor of a factorization of a finite abelian group, then the p-part of a cannot be e. (Hint: Assume the contrary that the p-part of a is e. Then apply Lemma 1.3.1 with $t = |a|$.)

Let $G = A_1 \cdots A_n$ be a factorization of the finite abelian p-group G, where A_i's are cyclic subsets

$$A_i = \{e, a_i, a_i^2, \ldots, a_i^{p-1}\}.$$

Any s of the elements $a_1, \ldots, a_n$ generate a subgroup of order at least p^s and the n elements $a_1, \ldots, a_n$ generate a subgroup of order p^n, that is, G itself.

Let p be a prime and let A be a subset of the finite abelian p-group G. If $\left|\langle A \rangle\right| = p^r$, then we denote r by $\gamma(A)$.

Lemma 1.3.2. *Let A be a subset of a finite abelian p-group such that $\gamma(B) \geq |B|$ for each $B \subset A$ and $\gamma(A) = |A|$. Then for each $a \in A$ there exists a power of p, say $s(a)$, such that*

$$\langle A \rangle = \prod_{a \in A} \left\{e, a^{s(a)}, a^{2s(a)}, \ldots, a^{(p-1)s(a)}\right\}$$

is a factorization of $\langle A \rangle$ and at least one of the factors is a subgroup of $\langle A \rangle$.

Proof. Suppose $|A| = 1$, that is, $A = \{a\}$. Let $|a| = p^r$. Now $r = \gamma(A) = |A| = 1$. Therefore the lemma holds when $|A| = 1$. We assume that $|A| \geq 2$ and proceed by induction on $|A|$. For a given $|A|$ consider the quantity

$$h(A) = \prod_{a \in A} |a|.$$

Note that if $e \in A$, then with the $B = \{e\}$ choice $0 = \gamma(B) < |B| = 1$, that is, A does not satisfy the hypothesis of the lemma. Thus $|a| \geq p$ for each $a \in A$.

Clearly, $h(A) \geq p^{|A|}$. If $h(A) = p^{|A|}$, then $|a| = p$ for each $a \in A$. Now $\langle A \rangle$ is an elementary p-group and the elements of A form a basis for $\langle A \rangle$. In this special case the lemma holds. We assume that $h(A) > p^{|A|}$ and for a given $|A|$ we proceed by induction on $h(A)$.

If $\gamma(B) > |B|$ holds for each $B \subset A$, $B \neq \emptyset$, $B \neq A$, then choose an element $c \in A$ and set $A' = \left(A \setminus \{c\}\right) \cup \{c^p\}$. (Observe that $|c| \geq p^2$ for each $c \in A$.) We will verify that the conditions of the lemma are satisfied for A'. Note that the index $\left|\langle A \rangle : \langle A' \rangle\right|$ is 1 or p, that is, $\gamma(A) = \gamma(A')$ or $\gamma(A) = \gamma(A') + 1$. Also note that $\gamma(A \setminus \{c\}) > |A| - 1$, that is, $\gamma(A \setminus \{c\}) \geq |A|$. Now

$$|A| = \gamma(A) \geq \gamma(A') \geq \gamma(A \setminus \{c\}) \geq |A|$$

gives that $\gamma(A') = |A| = |A'|$. Next we check that $\gamma(B) \geq |B|$ holds for each $B \subset A'$. If $c^p \notin B$, then $B \subset A$ and there is nothing to prove. If $c^p \in B$, then $B' = (B \setminus \{c^p\}) \cup \{c\}$ is a subset of A. Combining the fact that $\gamma(B') = \gamma(B)$ or $\gamma(B') = \gamma(B) + 1$ with $\gamma(B') > |B'|$ we get that $\gamma(B) \geq |B|$. Thus the conditions of the lemma hold for A'

As $h(A') < h(A)$ and $\langle A' \rangle = \langle A \rangle$, by induction on $h(A)$ we see that the lemma holds for A.

Now assume that $\gamma(B) = |B|$ holds for some $B \subset A$, $B \neq \emptyset$, $B \neq A$. Clearly, B satisfies the conditions of the lemma and $|B| < |A|$. By the inductive assumption on the cardinality of A, $\langle B \rangle$ has a desired factorization

$$\langle B \rangle = \prod_{b \in B} \left\{ e, b^{s(b)}, b^{2s(b)}, \ldots, b^{(p-1)s(b)} \right\}.$$

Similarly, the factor group $\langle A \rangle / \langle B \rangle$ has a desired factorization. Using the factorization of $\langle B \rangle$ and the factorization of the factor group $\langle A \rangle / \langle B \rangle$, we can construct a desired factorization for $\langle A \rangle$. (Exercise 1.3.7 after the proof helps to fill in the details.)

This completes the proof.

Exercise 1.3.7. Let $H = B_1 \cdots B_r$ be a factorization, where H is a subgroup of the finite abelian group G. Let $C_1, \ldots, C_s$ be subsets of G such that $(C_1 H)/H \cdots (C_s H)/H$ is a factorization of the factor group G/H. Prove that $B_1 \cdots B_r C_1 \cdots C_s$ is a factorization of G.

The next lemma proves Hajós's theorem for p-groups.

Lemma 1.3.3. *Let G be a finite abelian p-group and let*

$$G = A_1 \cdots A_n \tag{5}$$

be a factorization of G, where A_i's are cyclic subsets of form

$$A_i = \{e, a_i, a_i^2, \ldots, a_i^{p-1}\}.$$

Then at least one of the factors A_i is a subgroup of G.

Proof. Lemma 1.3.2 is applicable to $A = \{a_1, \ldots, a_n\}$. So for each i, $1 \leq i \leq n$ there is a power of p say, $s(i)$ and a cyclic subset

$$A_i' = \left\{ e, a_i^{s(i)}, a_i^{2s(i)}, \ldots, a_i^{(p-1)s(i)} \right\}$$

such that $G = A_1' \cdots A_n'$ is a factorization of G and at least one factor A_i' is a subgroup of G.

If $s(1) = \cdots = s(n) = 1$, then $A_i = A_i'$ for each i, $1 \leq i \leq n$ and we are done. So we may assume that

$$s(1) \neq 1, \ldots, s(m) \neq 1, s(m+1) = \cdots = s(n) = 1$$

since this is only a matter of indexing the factors. Now

$$G = A'_1 \cdots A'_m A_{m+1} \cdots A_n$$

is a factorization of G. Hence the element $a_1 \cdots a_m$ can be represented in the form

$$a_1 \cdots a_m = a_1^{s(1)t(1)} \cdots a_m^{s(m)t(m)} a_{m+1}^{t(m+1)} \cdots a_n^{t(n)},$$

where $0 \le t(i) \le p-1$. So

$$e = a_1^{s(1)t(1)-1} \cdots a_m^{s(m)t(m)-1} a_{m+1}^{t(m+1)} \cdots a_n^{t(n)}. \tag{6}$$

Remember that $s(i)$ is a power of p. Hence $s(i)t(i) - 1$ is relatively prime to p. By Lemma 1.3.1, in factorization (5) A_i can be replaced by

$$A_i^* = \left\{ e, a_i^{s(i)t(i)-1}, a_i^{2[s(i)t(i)-1]}, \ldots, a_i^{(p-1)[s(i)t(i)-1]} \right\}$$

for each i, $1 \le i \le m$ to get the factorization

$$G = A_1^* \cdots A_m^* A_{m+1} \cdots A_n.$$

Equation (6) violates this factorization. Therefore $s(1) = \cdots = s(n) = 1$.
This completes the proof.

Under certain conditions a cyclic subset can be replaced by a simulated one.

Lemma 1.3.4. *Let p be a prime and let $q \ge 2$ be an integer not divisible by p. If u and v are elements of order p and q respectively, then the cyclic subset*

$$A = \{e, uv, (uv)^2, \ldots, (uv)^{p-1}\}$$

can be replaced by the subset

$$A' = \{e, u, u^2, \ldots, u^{p-2}, u^{p-1}v\}.$$

Note that A' is a simulated subset if $p \ge 3$.

Proof. Consider a factorization $G = AB$. Using Lemma 1.3.1 with the choice $t = q$ replace A by $\langle u^q \rangle$. As $\langle u^q \rangle = \langle u \rangle$, A can be replaced by $A'' = \{e, u, u^2, \ldots, u^{p-1}\}$. By the partition reformulation of the factorization the subsets

$$u^0 B, u^1 B, u^2 B, \ldots, u^{p-1} B \tag{7}$$

form a partition of G.
For a given i, $0 \le i \le p-2$ choose an integer t such that

$$(uv)^t = u^{p-1-i}v.$$

Since p and q are distinct primes the system of congruences

$$\begin{aligned} t &\equiv p-1-i \qquad (\text{mod } p), \\ t &\equiv 1 \qquad\qquad\ (\text{mod } q) \end{aligned}$$

has a solution and so such a t exists. Clearly, t is relatively prime to p. Thus in the factorization $G = AB$ the factor A can be replaced by

$$C = \left\{ e, (uv)^t, (uv)^{2t}, \ldots, (uv)^{(p-1)t} \right\}.$$

Therefore the sets

$$B, (uv)^t B, (uv)^{2t} B, \ldots, (uv)^{(p-1)t} B$$

form a partition of G. The only information we need from this partition is that B and $u^{p-1-i}vB$ are disjoint for each i, $0 \le i \le p - 2$. We want to prove that the sets

$$B, uB, u^2 B, \ldots, u^{p-2} B, u^{p-1} vB$$

form a partition of G. Since sets (7) form a partition of G it is enough to show that $u^{p-1}vB$ and $u^i B$ are disjoint for each i, $0 \le i \le p - 2$. Suppose the contrary, that they are not disjoint. Multiplying by u^{-i} we have the contradiction that $u^{p-1-i}vB$ and B are not disjoint.

This completes the proof.

Now we are ready to prove Hajós's theorem. Further proofs are given in Appendix B.

Theorem 1.3.1. *If $G = A_1 \cdots A_n$ is a factorization of the finite abelian group G, where each A_i is either a simulated or a cyclic subset of G, then at least one of the factors must be a subgroup of G.*

Proof. Consider a factorization $G = A_1 \cdots A_n$, where each A_i is either cyclic or simulated. Decompose each cyclic subset into a product of cyclic subsets of prime cardinality. Let

$$G = B_1 \cdots B_m \tag{8}$$

be the resulting factorization, where the cyclic subsets are of prime cardinality. If $m = 1$, then $G = B_1$ and we are done. So we proceed by induction on m. If one of the factors is a subgroup of G we are done. So we suppose that none of the factors is a subgroup of G. Therefore $|B_i \cap H_i| = |B_i| - 1$ holds for the simulated factors.

First consider the case when a simulated factor occurs among the factors

$$B_1, \ldots, B_m.$$

We may assume that B_1 is simulated since this is only a matter of indexing the factors. Now using the Corrádi trick as in the proof of Theorem 1.2.1 we can conclude that one of the factors, say B_m, is periodic.

If B_m is a simulated subset of G, then by Lemma 1.2.2, it is a subgroup of G. If B_m is a cyclic subset of prime cardinality, then it is a subgroup of G since $e \in B_m$.

In the remaining part of the proof we may assume that each B_i is a cyclic subset of G of prime cardinality. The case when $|G|$ is a prime power is settled in Lemma 1.3.3. So we may assume that $|G|$ is not a prime power. Let p be the largest prime factor of $|G|$. Clearly, $p \geq 3$.

Consider a factor B_i of cardinality p. Let

$$B_i = \{e, b_i, b_i^2, \ldots, b_i^{p-1}\}$$

and let u and v be the p-part and the p'-part of b_i respectively. If $|u| \geq p^2$, then using Lemma 1.3.1 with the choice $t = |v|$ in factorization (8) replace B_i by

$$B_i' = \{e, u, u^2, \ldots, u^{p-1}\}.$$

Clearly, B_i' is not a subgroup of G. If this happens for each B_i with cardinality p, then the product of these factors provides a factorization of the p-component of G. The fact that none of the factors is a subgroup of G contradicts Lemma 1.3.3. So we may assume that $|u| = p$ for at least one factor B_i of cardinality p. As B_i is not a subgroup of G, $v \neq e$. Using Lemma 1.3.1 with the choice $t = |v|/q$ in factorization (8), replace B_i by

$$B_i' = \{e, uw, (uw)^2, \ldots, (uw)^{p-1}\},$$

where $|w| = q$ is a prime distinct from p. Using Lemma 1.3.4 replace B_i' by the nonsubgroup simulated subset

$$B_i'' = \{e, u, u^2, \ldots, u^{p-2}, u^{p-1}w\}.$$

Now the proof is reduced to the case when there is a simulated subset among the factors.

This completes the proof.

1.4 Rédei's theorem

We say that a factorization is *normalized* if each factor contains the identity element. There is a straightforward way to construct factorizations with normalized factors containing a prime number of elements. Consider a descending chain of subgroups

$$G = H_0 \supset H_1 \supset \cdots \supset H_{n-1} \supset H_n = \{e\}$$

of G. Let A_i be a complete set of representatives of H_{i-1} modulo H_i such that $e \in A_i$. If the indices $|H_{i-1} : H_i|$ are primes, then $G = A_1 \cdots A_{n-1}H_{n-1}$ is a normalized factorization of G with factors containing a prime number of elements.

By Rédei's theorem, if a finite abelian group is a direct product of its normalized subsets such that each factor has a prime number of elements, then at least one of the factors must be a subgroup. Thus the above procedure provides all possible normalized factorizations with factors of prime cardinalities.

Let $G = AB$ be a factorization of G. Then each $g \in G$ is uniquely expressible in the form $g = ab$, $a \in A$, $b \in B$. We call a the *A-part* of g and we denote it by $\alpha(g)$. Similarly, we call b the *B-part* of g and we denote it by $\beta(g)$. The A-part and B-part of g make sense only relative to the factorization $G = AB$.

Lemma 1.4.1. *Let $G = AB$ be a factorization of G and let $A = \{a_1, \ldots, a_n\}$. For each $g \in G$ the elements $\alpha(ga_1), \ldots, \alpha(ga_n)$ form a permutation of $a_1, \ldots, a_n$.*

Proof. Clearly, $\alpha(ga_i) \in A$. So we will show that $\alpha(ga_i) = \alpha(ga_j)$ implies $a_i = a_j$. From the equations

$$ga_i = \alpha(ga_i)\beta(ga_i), \quad ga_j = \alpha(ga_j)\beta(ga_j)$$

we get the equations

$$g = \alpha(ga_i)\beta(ga_i)a_i^{-1}, \quad g = \alpha(ga_j)\beta(ga_j)a_j^{-1}.$$

Then $\beta(ga_i)a_i^{-1} = \beta(ga_j)a_j^{-1}$ and $a_j\beta(ga_i) = a_i\beta(ga_j)$. Now as $a_i, a_j \in A$, $\beta(ga_i)$, $\beta(ga_j) \in B$ it follows that $a_i = a_j$.

This completes the proof.

If A is a subset of a finite abelian group G, then A^q will denote the set $\{a^q : a \in A\}$. We will refer to the next lemma as the power replacement lemma.

Lemma 1.4.2. *Let $G = AB$ be a factorization of G and let q be a prime such that $q \nmid |A|$. Then $G = A^q B$ is a factorization of G.*

Proof. Choose an $a \in A$, $g \in G$ and define T to be the set of all q tuples

$$(x_1, x_2, \ldots, x_q), \quad x_1, x_2, \ldots, x_q \in A$$

for which $\alpha(gx_1x_2 \cdots x_q) = a$. First note that $|T| = |A|^{q-1}$. Indeed, choose $x_1, x_2, \ldots, x_{q-1} \in A$ arbitrarily, then by Lemma 1.4.1, $\alpha[(gx_1x_2 \cdots x_{q-1})x_q] = a$ has a unique solution for x_q. Next note that if $(x_1, x_2, \ldots, x_q) \in T$, then $(x_2, \ldots, x_q, x_1) \in T$. We define a graph Γ. The vertices of Γ are the elements of T and we draw an arrow from the node $(x_1, x_2, \ldots, x_q)$ to the node $(x_2, \ldots, x_q, x_1)$. The graph Γ is a union of disjoint cycles. The cycles are of length 1 or of length q. When $x_1 = x_2 = \cdots = x_q$, then the node $(x_1, x_2, \ldots, x_q)$ is on a cycle of length 1. When $x_1, x_2, \ldots, x_q$ are not all equal, then the node $(x_1, x_2, \ldots, x_q)$ is on a cycle of length q. As $q \nmid |A|$ there must be a cycle of length 1 in Γ. In other words there is an $x_1 \in A$ such that $\alpha(gx_1^q) = a$. In addition x_1 is uniquely determined by a and g.

As the last step of the proof we claim that the product $A^q B$ is direct. Suppose that $a_1^q b_1 = a_2^q b_2$, $a_1, a_2 \in A$, $b_1, b_2 \in B$. Then $a_1^q b_2^{-1} = a_2 b_1^{-1}$. There are $a \in A$,

$b \in B$ such that $a_1^q b_2^{-1} = a_2^q b_1^{-1} = ab$. From the equation $b^{-1} a_1^q = ab_2$ we get that $\alpha(b^{-1} a_1^q) = a$. From the equation $b^{-1} a_2^q = ab_1$ we get $\alpha(b^{-1} a_2^q) = a$. From $\alpha(b^{-1} a_1^q) = \alpha(b^{-1} a_2^q) = a$ by Lemma 1.4.1 we get $a_1 = a_2$ which in turn implies $b_1 = b_2$.

This completes the proof.

Exercise 1.4.1. Let $G = AB$ be a factorization of G and let k be an integer relatively prime to $|A|$. Show that $G = A^k B$ is a factorization of G. (Hint: The $k = -1$ case is settled in Exercise 1.2.6. If k is positive, then k is a product of positive primes and we can apply Lemma 1.4.2 several times starting with the factorization $G = AB$. If k is negative, then $-k$ is positive and we can use a similar procedure starting with the factorization $G = A^{-1} B$.)

Lemma 1.4.3. *Let $G = AB$ be a factorization such that $e \in A$, $|A| = p$ is a prime. Then $G = A'B$ is a factorization of G, where $A' = \{e, a, a^2, \ldots, a^{p-1}\}$, $a \in A \backslash \{e\}$.*

Proof. By Exercise 1.4.1, $G = A^t B$ is a factorization of G whenever $p \nmid t$. Let $A = \{e, a_1, a_2, \ldots, a_{p-1}\}$. By the partition equivalent of the factorization $G = A^t B$ the sets

$$eB, a_1^t B, a_2^t B, \ldots, a_{p-1}^t B$$

form a partition of G. Similarly, the fact that $G = A'B$ is a factorization is equivalent to the sets

$$eB, a_k B, a_k^2 B, \ldots, a_k^{p-1} B$$

forming a partition of G. Here $A' = \{e, a_k, a_k^2, \ldots, a_k^{p-1}\}$. Since G is finite it is enough to show that $a_k^i B \cap a_k^j B = \emptyset$ for each i, j, $0 \leq i < j \leq p-1$. Assume the contrary that $a_k^i B \cap a_k^j B \neq \emptyset$. Multiplying by a_k^{-i} we get $eB \cap a_k^{j-i} B \neq \emptyset$. Set $t = j - i$. Clearly, $1 \leq t \leq p-1$ and so t is prime to p. Now $eB \cap a_k^t B \neq \emptyset$ contradicts the fact that $G = A^t B$ is a factorization of G.

This completes the proof.

Exercise 1.4.2. Let $G = AB$ be a factorization of G such that $|A| = p$ is a prime, $e \in A$. Further assume that A contains only (p, q)-elements. $A = \{e, a_1 b_1, a_2 b_2, \ldots, a_{p-1} b_{p-1}\}$, $|a_i| = p$, $|b_i| = 1$ or $|b_i| = q$ for each i, $1 \leq i \leq p-1$, $|b_1| = q$. Show that $G = A'B$ is a factorization of G, where $A' = \{e, a_1, a_1^2, \ldots, a_1^{p-2}, a_1^{p-1} b_1\}$. ($A'$ differs from the subgroup $\langle a_1 \rangle$ in one element.) (Hint: Replace A by the cyclic subset $\{e, a_1 b_1, (a_1 b_1)^2, \ldots, (a_1 b_1)^{p-1}\}$ and use Lemma 1.3.4.)

First we prove Rédei's theorem for groups of type (p, p), where p is a prime. We will use concepts from finite geometries. Let X be a subset of an affine plane. We say that X determines a direction if there are two points in X that span a line in this direction. The affine plane we need is $[\mathrm{GF}(p)]^2$. ($\mathrm{GF}(p)$ is isomorphic to the field of integers modulo p. We identify $\mathrm{GF}(p)$ with this field using $0, 1, \ldots, p-1$ as elements of the field.)

Lemma 1.4.4. *If X is a subset of the affine plane $[\mathrm{GF}(p)]^2$ such that $|X| = p$ is a prime and X determines at most $(p+1)/2$ directions, then X is a straight line.*

Proof. Choose a subset X of $[\mathrm{GF}(p)]^2$ such that $|X| = p$ is a prime and X determines at most $(p+1)/2$ directions. If X determines all $p+1$ directions on the plane, then it follows that $p+1 \geq (p+3)/2$ holds. This is not possible. So we may assume that X does not determine all directions. Consequently, we may introduce a coordinate system in such a way that the direction of the second coordinate axis is not determined by X. Hence X can be represented in the form

$$X = \{(k, b_k) : 0 \leq k \leq p - 1\},$$

where $b_0, \ldots, b_{p-1} \in \mathrm{GF}(p)$. It is convenient to record the directions determined by X with the slope of a representative straight line. We can do this since none of the directions determined by X is parallel to the second coordinate axis.
Set

$$U = \left\{ \frac{b_k - b_m}{k - m} : 0 \leq k, m \leq p - 1, k \neq m \right\}.$$

By the assumption of the lemma, $|U| \leq (p+1)/2$. Consider the polynomials

$$F_j(x) = \sum_{k=0}^{p-1} (b_k - kx)^j$$

in $\mathrm{GF}(p)[x]$ for each j, $1 \leq j \leq (p-1)/2$. From

$$\sum_{i=0}^{p-1} a_i^j = \begin{cases} 0, & \text{if } 0 \leq j \leq p - 2, \\ -1, & \text{if } j = p - 1, \end{cases} \tag{1}$$

it follows that $\deg F_j(x) \leq j - 1$. If $x \notin U$, then the elements $b_k - kx$ are all distinct as k varies over $\mathrm{GF}(p)$. So $x \notin U$ implies $F_j(x) = 0$. Since $\deg F_j(x) \leq j - 1$ it follows that if $j - 1 < p - |U|$, then F_j is the zero polynomial. In particular $F_j(x)$ is the zero polynomial when $1 \leq j \leq (p-1)/2$. Using the fact that every function from $\mathrm{GF}(p)$ to $\mathrm{GF}(p)$ is a polynomial of degree less than or equal to $p - 1$, we can represent b_k in the form

$$b_k = c_0 + c_1 k + c_2 k^2 + \cdots + c_{p-1} k^{p-1}.$$

We will show that $c_m = 0$ for each m, $2 \leq m \leq p - 1$. This will show that X is a straight line. Divide $p - 1$ by m with remainder.

$$p - 1 = ma + b, \quad 0 \leq b \leq m - 1.$$

Note that $1 \leq a$ and $a + b \leq (p-1)/2$ as $m \leq 2$. So $F_{a+b}(x)$ is the zero polynomial. Let us compute the coefficient of $(-x^b)$ in $F_{a+b}(x)$.

$$0 = \sum_{k=0}^{p-1} \binom{a + b}{b} b_k^a k^b = \binom{a + b}{b} \sum_{k=0}^{p-1} \left(c_m^a k^{am+b} + \sum_{j=b}^{p-2} d_j k^j \right)$$

with some $d_j \in \mathrm{GF}(p)$. Using (1) we get for this same coefficient

$$\binom{a+b}{b} c_m^a \sum_{k=0}^{p-1} k^{p-1} = -\binom{a+b}{b} c_m^a.$$

It follows that $c_m = 0$.

This completes the proof.

Lemma 1.4.5. *If $G = AB$ is a factorization of the group G of type (p, p) such that $e \in A \cap B$, $|A| = |B| = p$, then A or B is a subgroup of G.*

Proof. Let u, v be basis elements of G. The correspondence $u^i v^j \to (i, j)$ assigns points of the affine plane $[\mathrm{GF}(p)]^2$ to the elements of G. Subgroups of order p correspond to straight lines passing through the point $(0, 0)$. The $p + 1$ subgroups of order p of G correspond to the $p + 1$ directions available on the plane. Suppose that the elements $a_1, a_2 \in G$ correspond to the points $p_1, p_2 \in [\mathrm{GF}(p)]^2$. Then the direction determined by the points p_1, p_2 corresponds to the subgroup $\langle a_1 a_2^{-1} \rangle$ of G. Briefly, we will talk about the direction determined by a_1, a_2.

Next we will show that if $G = AB$ is a factorization, then the directions determined by the elements of A are distinct from the directions determined by the elements of B. Assume that there are $a_1, a_2 \in A$, $b_1, b_2 \in B$ such that $a_1 \neq a_2$, $b_1 \neq b_2$ and $\langle a_1 a_2^{-1} \rangle = \langle b_1 b_2^{-1} \rangle$. Multiply the factorization $G = AB$ by $a_2^{-1} b_2^{-1}$ to get the factorization $G = (Aa_2^{-1})(Bb_2^{-1})$. By Lemma 1.4.3, Aa_2^{-1}, Bb_2^{-1} can be replaced by $\langle a_1 a_2^{-1} \rangle$, $\langle b_1 b_2^{-1} \rangle$ respectively to get the factorization $G = \langle a_1 a_2^{-1} \rangle \langle b_1 b_2^{-1} \rangle$. But this is a contradiction as $\langle a_1 a_2^{-1} \rangle = \langle b_1 b_2^{-1} \rangle$. Since A and B determine distinct directions it follows that either A or B determines at most $(p + 1)/2$ directions. By the previous result A or B is a subgroup of G.

This completes the proof.

Next we prove Rédei's theorem for all finite abelian p-groups.

Lemma 1.4.6. *Rédei's theorem holds for any finite abelian p-group.*

Proof. Let G be an abelian group of order p^n and let $G = A_1 \cdots A_n$ be a factorization of G, where $|A_1| = \cdots = |A_n| = p$ and $e \in A_1, \ldots, e \in A_n$. We want to show that at least one of the factors $A_1, \ldots, A_n$ is a subgroup of G. The $n = 1$ case is trivial. We may assume that $n \geq 2$. By Lemma 1.4.3, every factor A_i can be replaced by a cyclic subset. If A_i contains an element of order at least p^2, then A_i can be replaced by a nonsubgroup cyclic subset. If each factor has an element of order at least p^2, then we can construct a factorization of G consisting of nonsubgroup cyclic subsets. By Lemma 1.3.3, it is not possible. So there is a factor, say A_1, whose nonidentity elements all have order p. We assume that $n \geq 2$ and start an induction on n.

By Lemma 1.4.3, the factor A_1 can be replaced by a subgroup H_1 in the factorization $G = A_1 A_2 \cdots A_n$ to get the factorization $G = H_1 A_2 \cdots A_n$. Considering the factor group G/H_1 we have the factorization

$$G/H_1 = (A_2 H_1)/H_1 \cdots (A_n H_1)/H_1.$$

By the inductive assumption there is a permutation $B_1, \ldots, B_n$ of the factors $H_1, A_2, \ldots, A_n$ such that

$$B_1, \; B_1 B_2, \ldots, B_1 B_2 \cdots B_n$$

is an ascending chain of subgroups of G and $B_1 = H_1$. For notational convenience we assume that $B_2 = A_2, \ldots, B_n = A_n$ since this is only a matter of reindexing the factors $A_2, \ldots, A_n$ in the factorization $G = A_1 A_2 \cdots A_n$. Consider the subgroup $M = H_1 A_2 \cdots A_{n-1}$. Clearly, each of the factors $H_1, A_2, \ldots, A_{n-1}$ is a subset of M. If $A_1 \subset M$, then $M = A_1 A_2 \cdots A_{n-1}$ is a factorization of M. By the inductive assumption at least one of the factors $A_1, \ldots, A_{n-1}$ is a subgroup of M and so is a subgroup of G.

For the remaining part of the proof we may assume that $A_1 \not\subset M$.

$$A_1 = \{e, x, x^2 d_2, \ldots, x^{p-1} d_{p-1}\},$$

where $d_2, \ldots, d_{p-1} \in G$. Clearly $\{d_2, \ldots, d_{p-1}\} \not\subset M$. The factor A_n can be replaced by $A'_n = \{e, y, y^2, \ldots, y^{p-1}\}$, for each $y \in A_n \setminus \{e\}$. Here we use the Corrádi trick again. Since $G = MA'_n$ is a factorization of G, A'_n is a complete set of representatives modulo M. There are $u_2, \ldots, u_{p-1} \in A'_n$ such that

$$(x^2 d_2)^{-1} \in u_2 M, \ldots, (x^{p-1} d_{p-1})^{-1} \in u_{p-1} M,$$

that is

$$x^2 d_2 u_2, \ldots, x^{p-1} d_{p-1} u_{p-1} \in M.$$

Set

$$B = \{e, x, (x^2 d_2) u_2, \ldots, (x^{p-1} d_{p-1}) u_{p-1}\}.$$

Note that $B \subset M$ and $BA_2 \cdots A_{n-1}$ is a factorization of M. Indeed, the products coming from $BA_2 \cdots A_{n-1}$ are among the products coming from $A_1 A_2 \cdots A_{n-1} A'_n$. From the factorization $M = BA_2 \cdots A_{n-1}$ the inductive assumption gives that B is a subgroup of M. In fact $B = \langle x \rangle = H_1$. For each i, $2 \le i \le p-1$, there is a j, $0 \le j \le p-1$ such that $x^i d_i u_i = x^j$. It follows that $u_i = d_i^{-1} x^{j-i} \in A'_n$ and so $|y| = p$. In addition each element in $A_n \setminus \{e\}$ has order p. Thus $d_i \in \langle x, y \rangle$. Hence $A_1 \subset \langle x, y \rangle$. If $A_n \subset \langle x, y \rangle$, then $A_1 A_2$ is a factorization of $\langle x, y \rangle$ and by Lemma 1.4.5, A_1 or A_2 is a subgroup. We may assume that $A_n \not\subset \langle x, y \rangle$. There is a $z \in A_n \setminus \{e\}$ such that $z \notin \langle x, y \rangle$. Replacing A_n by $A''_n = \{e, z, z^2, \ldots, z^{p-1}\}$ we get that $d_2, \ldots, d_{p-1} \in \langle x, z \rangle$. Therefore

$$d_2, \ldots, d_{p-1} \in \langle x, y \rangle \cap \langle x, z \rangle = \langle x \rangle$$

and so A_1 is a subgroup of G.

This completes the proof.

We are ready now to prove Rédei's theorem in full generality.

Theorem 1.4.1. *Let $G = A_1 \cdots A_n$ be a factorization of the finite abelian group G such that $e \in A_i$ and $|A_i|$ is a prime for each i, $1 \le i \le n$. Then at least one of the factors $A_1, \ldots, A_n$ is a subgroup of G.*

Proof. The theorem holds for $n = 1$. We start an induction on n and assume that $n \ge 2$. If $|G|$ is a power of 2, then by Lemma 1.4.6 at least one of the factors is a subgroup of G. So we may assume that $|G|$ has a prime factor p with $p \ge 3$ and suppose that $A_1, \ldots, A_t$ are all the factors among $A_1, \ldots, A_n$ with cardinality p. By Lemma 1.4.2 there is a factorization $G = A_1' \cdots A_t' A_{t+1} \cdots A_n$ of G such that A_i' contains only p-elements for each i, $1 \le i \le t$. Now $A_1' \cdots A_t'$ is a factorization of the p-component of G. By Lemma 1.4.6 at least one of the factors $A_1', \ldots, A_t'$ is a subgroup of the p-component of G and hence of G. Let A_1' be this factor. If A_1 is a subgroup of G, then there is nothing to prove. So we may assume that A_1 contains not only p-elements. Again using Lemma 1.4.2 if necessary we may assume that A_1 satisfies the conditions of Exercise 1.4.2. So there is a factorization $G = A_1^* A_2 \cdots A_n$ such that A_1^* is in the form

$$A_1^* = \{e, x, x^2, \ldots, x^{p-2}, x^{p-1}y\}.$$

Since A_1^* is simulated we can use the Corrádi trick in the way we have seen in the proof of Theorem 1.2.1 to conclude that one of the factors $A_2, \ldots, A_n$ is a subgroup of G.

This completes the proof.

1.5 Keller's conjecture

In 1930, O. H. Keller conjectured that in any tiling of Euclidean n-space by translates of a unit cube, some pair of cubes share a common $(n-1)$-dimensional face. We call two such cubes *twins*. In 1940, O. Perron verified Keller's conjecture for $n \le 6$.

In 1949 G. Hajós reformulated this geometric question in terms of factoring finite abelian groups. If there is a counterexample for Keller's conjecture in n dimensions, then there is a finite abelian group G a subset B of G and cyclic subsets $A_1, \ldots, A_n$ of G such that

$$A_i = \{e, a_i, a_i^2, \ldots, a_i^{r(i)-1}\},$$

$G = BA_1 \cdots A_n$ is a factorization of G and

$$BB^{-1} \cap \{a_1^{r(1)}, \ldots, a_n^{r(n)}\} = \emptyset.$$

Conversely, if there is a finite abelian group G such that $G = BA_1 \cdots A_n$ is a factorization of G and $BB^{-1} \cap \{a_1^{r(1)}, \ldots, a_n^{r(n)}\} = \emptyset$, then one can construct an

n-dimensional counterexample for Keller's conjecture. The following group theoretical assertion can be called the algebraic version of Keller's conjecture.

Conjecture 1.5.1. Let G be a finite abelian group. The factorization $G = BA_1 \cdots A_n$ implies $BB^{-1} \cap \{a_1^{r(1)}, \ldots, a_n^{r(n)}\} \neq \emptyset$.

In other words the factorization $G = BA_1 \cdots A_n$ implies that there is an i, $1 \leq i \leq n$ such that $a_i^{r(i)} \in BB^{-1}$. If $G = BA_1 \cdots A_n$ is a factorization of G such that $BB^{-1} \cap \{a_1^{r(1)}, \ldots, a_n^{r(n)}\} = \emptyset$, then we express this fact simply by saying that the factorization $G = BA_1 \cdots A_n$ is a counterexample for Keller's conjecture. The counterexamples can be standardized. The cyclic subsets can be factored into cyclic subsets of prime cardinalities. (After this procedure the dimension of the counterexample may change.) So we may assume that in a counterexample all $|A_1|, \ldots, |A_n|$ are primes. If a cyclic subset A_i is a subgroup of G, then $a_i^{r(i)} = e$ and so $a_i^{r(i)} \in BB^{-1}$. Thus in a counterexample a cyclic subset A_i cannot be a subgroup. Further we claim that the product of the cyclic subsets cannot be periodic in a counterexample.

In order to verify this claim let C be the product of the cyclic subsets, that is, let $C = A_1 \cdots A_n$ and assume that C is periodic. It is a fact (which we will prove in Section 5.4) that at least one of the cyclic subsets, say $A_i = \{e, a_i, a_i^2, \ldots, a_i^{r(i)-1}\}$ must be a subgroup of G. Now $a_i^{r(i)} = e$ and so $a_i^{r(i)} \in BB^{-1}$, that is, the factorization is not a counterexample for Keller's conjecture.

If there is a counterexample, then there is one for which $|G|$ is minimal. In a minimal counterexample B cannot be periodic since otherwise we would consider a factor group to get a counterexample with a smaller $|G|$. In a minimal counterexample the elements $a_1, \ldots, a_n$ must span G. In order to prove this claim set $H = \langle a_1, \ldots, a_n \rangle$. Restricting the factorization $G = BA_1 \cdots A_n$ to H gives the factorization $H = (B \cap H)(A_1 \cdots A_n)$ of H. If $H \neq G$, then we get a counterexample with a smaller $|G|$.

If the group G has the Hajós 2-property, then we get that one of the factors is periodic and so a group with the Hajós 2-property cannot provide a counterexample for Keller's conjecture. Inspecting the list of the groups with the Hajós 2-property we can see that the number of prime factors of $|G|$ is at most five except the cases when G is of type (p^α, q) or $(2^\alpha, 2)$. The number of the factors in a factorization of G with the Hajós 2-property whose type is not (p^α, q) or $(2^\alpha, 2)$ is at most five. In other words $n \leq 5$ must hold in these cases. Thus these cases are already covered by Perron's result.

Keller's conjecture holds for certain special classes of groups. We have just seen that Keller's conjecture holds for groups of types (p^α, q) or $(2^\alpha, 2)$. We mention some further results. Keller's conjecture holds for cyclic groups of order $p^\alpha q^\beta$. It was proved by A. D. Sands. For groups of invariants

$$(p^\alpha, p^\beta), \quad (p^\alpha, p, \ldots, p), \quad (p^\alpha, q, \ldots, q)$$

it was proved by K. Corrádi and S. Szabó and for groups of type

$$(p^\alpha, p^\beta, p, \dots, p), \quad (p^\alpha, p^\beta, q, \dots, q)$$

by D. Moews. Here p and q are distinct primes and in the last case $p \geq 5$; furthermore α and β are positive integers.

In 1986, S. Szabó pointed out that to decide Keller's conjecture we may restrict our attention to abelian groups of invariants $(4, \dots, 4)$. More precisely, if there is a counterexample for Keller's conjecture, then there is a finite abelian group G of type $(4, \dots, 4)$ with basis elements $a_1, \dots, a_n$, a subset B of G such that $G = B\{e, a_1\} \cdots \{e, a_n\}$ is a factorization of G and $BB^{-1} \cap \{a_1^2, \dots, a_n^2\} = \emptyset$. (We would like to stress that if there is an n-dimensional counterexample for Keller's conjecture, then this n is generally smaller than the number of generator elements in the group theoretical version.) Conversely, if there is a finite abelian group G of type $(4, \dots, 4)$ with basis elements $a_1, \dots, a_n$ such that $G = B\{e, a_1\} \cdots \{e, a_n\}$ is a factorization of G and $BB^{-1} \cap \{a_1^2, \dots, a_n^2\} = \emptyset$, then one can construct an n-dimensional counterexample for Keller's conjecture. (Here the number of the generator elements is equal to the dimension of the counterexample.) The following group theoretical assertion can also be viewed as an algebraic version of Keller's conjecture.

Conjecture 1.5.2. Let G be a finite abelian group of type $(4, \dots, 4)$ with basis elements $a_1, \dots, a_n$. If $G = B\{e, a_1\} \cdots \{e, a_n\}$ is a factorization of G, then there is an i, $1 \leq i \leq n$ such that $a_i^2 \in BB^{-1}$.

This algebraic form suggests a combinatorial approach. Let G be a finite abelian group of type $(4, \dots, 4)$ with basis elements $a_1, \dots, a_n$. Define a graph Γ_n on the elements of G. Let the elements a, b of G be adjacent if

$$ba^{-1} = a_1^{\alpha(1)} \cdots a_n^{\alpha(n)}$$

and there is an i, $1 \leq i \leq n$ such that

$$\alpha(i) \equiv 2 \pmod 4.$$

The edges of Γ_n are not directed. The edge $\{a, b\}$ of Γ_n is called a twin edge if $\alpha(j) \equiv 0 \pmod 4$ for each j, $1 \leq j \leq n$, $i \neq j$.

A subgraph of Γ_n is called a *clique* if it does not contain nonadjacent vertices. Let B be the set of vertices of a clique of size 2^n in Γ_n. Then $G = B\{e, a_1\} \cdots \{e, a_n\}$ is a factorization of G. In order to verify this claim set $C = \{e, a_1\} \cdots \{e, a_n\}$. Notice that if $c, c' \in C$, then in the representation

$$c'c^{-1} = a_1^{\gamma(1)} \cdots a_n^{\gamma(1)}$$

each $\gamma(i)$ is congruent to $-1, 0, 1$ modulo 4. Further if $b, b' \in B$ and $b \neq b'$, then in the representation

$$b'b^{-1} = a_1^{\beta(1)} \cdots a_n^{\beta(n)}$$

$n = 2$
00
12
20
32

$n = 3$
000
102
210
021
222
230
023
302

$n = 4$	
0000	0201
0012	2111
0120	0222
1320	2300
1032	2312
2130	2232
3032	2113
3320	0203

Tables 1 (a)-(c)

$n = 5$			
00000	02322	02112	23022
00012	21011	00213	22211
11132	21113	10211	23231
00332	20303	13020	23223
03132	21323	00230	12303
01020	23001	02220	21130
02100	23103	21331	02301
30211	33020	32303	31132

A		
0	0	0
1	0	2
2	1	0
0	2	1
2	2	2
2	3	0'
0'	2	3
3	0'	2

Table 1 (d) Table 3

there is a $\beta(i)$ congruent to 2 modulo 4. From this it follows that $BB^{-1} \cap CC^{-1} = \{e\}$ which means that the product BC is direct. On the other hand the cardinalities force that BC gives G.

Together with Γ_n it will be convenient to consider the graph Γ_n^*. The new graph Γ_n^* is constructed from Γ_n by erasing the twin edges. Thus if Γ_n^* has a clique of size 2^n, then there is an n-dimensional counterexample for Keller's conjecture.

By inspection K. Corrádi and S. Szabó found that each clique of size 2^n in Γ_n contains at least one twin edge for $n \leq 6$. (This comes as no surprise after Perron's result.)

A "typical" 2^n-clique in Γ_n is abundant in twin edges. This inspection brought the following "extreme" cliques to light. The cliques are listed in Tables 1 (a)-(d).

The tables are divided by a horizontal line. Each element below the line has an element above the line such that the two nodes are connected with a twin edge and these are all the twin edges in the clique. Therefore the elements above the lines form a clique in Γ_n^*. Table 2 summarizes the data.

In 1992, J. C. Lagarias and P. W. Shor refuted Keller's conjecture by a clever construction. Consider the 2^3-clique of Γ_3 listed in Table 1. Mark the three zeros

n	2	3	4	5	6
number of twins	2	3	4	4	4
clique size in Γ_n^*	2	5	12	28	60
2^n	4	8	16	32	64

Table 2

S_0	$S_{0'}$	S_2	S_1	$S_{1'}$	S_3
0000	1130	1132	1000	2130	2132
0012	1331	0211	1012	2331	1211
0332	1011	3020	1332	2011	0020
3132	1113	2303	0132	2113	3303
1020	0303		2020	1303	
2100	1323		3100	2323	
2112	3001		3112	0001	
0213	3103		1213	0103	
0230	3022		1230	0022	
2220	2211		3220	3211	
2301	3231		3301	0231	
2322	3223		3322	0223	

Table 4

by a prime in the last three rows and denote the resulting list by A. (See Table 3.) The 2^5-clique in Γ_5 listed in Table 1 has 12 vectors with value 0 in the first column, another 12 vectors with value 2 in the first column, and 4 vectors with value 1 in the first column. Deleting the first column the resulting sets of vectors are denoted by S_0, $S_{0'}$, S_2 respectively. Adding 1 modulo 4 to the first column in S_0, $S_{0'}$, S_2 we get S_1, $S_{1'}$, S_3.

Now construct a set of vectors B in the following way. If (a, b, c) is an element of A, then let

$$\big[(u_1, u_2, u_3, u_4),\ (v_1, v_2, v_3, v_4),\ (w_1, w_2, w_3, w_4)\big]$$

be an element of B, where

$$(u_1, u_2, u_3, u_4) \in S_a,\ (v_1, v_2, v_3, v_4) \in S_b,\ (w_1, w_2, w_3, w_4) \in S_c.$$

(This construction resembles the tensor product of matrices.) The claim is that B is a 2^{12}-clique in Γ_{12}^* and so Keller's conjecture fails in dimension 12.

Exercise 1.5.1. Verify that B has 2^{12} elements and that B is a clique in Γ_{12}^*.

J. C. Lagarias and P. W. Shor also constructed a counterexample in 10-space and consequently there are counterexamples for each dimension $n \geq 10$.

<table>
<tr><td colspan="2" align="center">C</td></tr>
<tr><td>0 0 0 0</td><td>0 0 1 2</td></tr>
<tr><td>0 3 3 2</td><td>2 1 0 0</td></tr>
<tr><td>2 1 1 2</td><td>0 2 3 0</td></tr>
<tr><td>2 2 2 0</td><td>2 3 2 2</td></tr>
<tr><td>3 1'3 2</td><td>1 1 3 2</td></tr>
<tr><td>1 0'2 0</td><td>3 0 2 0</td></tr>
<tr><td>0 2 1'3</td><td>0 2 1 1</td></tr>
<tr><td>2 3 0'1</td><td>2 3 0 3</td></tr>
</table>

<table>
<tr><td colspan="2" align="center">K</td></tr>
<tr><td>1 0 3 2</td><td>2 3 2 0</td></tr>
<tr><td>1 0 1 2</td><td>0 3 2 0</td></tr>
<tr><td>1 2 1 2</td><td>0 1 0 0</td></tr>
<tr><td>1 2 3 2</td><td>2 1 0 0</td></tr>
<tr><td>1 3 0 0</td><td>3 3 0 1</td></tr>
<tr><td>1 1 2 0</td><td>3 3 0 3</td></tr>
<tr><td>3 1 0 2</td><td>3 1 2 1</td></tr>
<tr><td>3 3 2 2</td><td>3 1 2 3</td></tr>
</table>

Table 5   Table 6

Exercise 1.5.2. Show that if there is a counterexample for Keller's conjecture in n-space, then there is a counterexample for Keller's conjecture in $(n+1)$-space.

The construction of the 10-dimensional counterexample starts with a 16-clique of Γ_4 listed in Table 5. There are four twin edges in the clique. These are the vectors appearing in one row below the line that divides the table. We marked certain 0s and 1s by a prime in the second and third columns to destroy these twins. The list is denoted by C. Then we construct a set of vectors in the following way. If (a, b, c, d) is an element of C, then let

$$\left[a, (u_1, u_2, u_3, u_4), (v_1, v_2, v_3, v_4), d\right]$$

be an element of D, where

$$(u_1, u_2, u_3, u_4) \in S_b, \quad (v_1, v_2, v_3, v_4) \in S_c.$$

The claim is that D is a 2^{10}-clique of Γ_{10}^* and consequently Keller's conjecture does not hold in 10-dimensional space.

In 2002, J. Mackey found a clique of size 2^8 in the graph Γ_8^*. Indeed he lists the 256 nodes of the clique. This proves that Keller's conjecture is false for $n = 8$. The construction of the 8-dimensional counterexample starts with a clique K of size 2^4 of Γ_4 listed in Table 6. Then define the blocks $S_{i,j}$, $0 \le i, j \le 3$ as indicated in Table 7. Next we construct a set of vectors L in the following way. If $(a, b, c, d) \in K$, then let

$$\left[(u_1, u_2, u_3, u_4), (v_1, v_2, v_3, v_4)\right]$$

be an element of L, where

$$(u_1, u_2, u_3, u_4) \in S_{a,b}, \quad (v_1, v_2, v_3, v_4) \in S_{c,d}.$$

One can verify that L is a clique of size 2^8 in Γ_8^*.

S_{00}	S_{02}	S_{21}	S_{23}	S_{12}	S_{10}	S_{33}	S_{31}
0211	2211	1011	1113	0000	0102	1210	3210
1132	1130	1331	1323	0230	0222	3302	1302
2303	0303	3103	3001	2112	2010	0023	0021
3020	3022	3223	3231	2322	2330	2131	2133
S_{20}	S_{22}	S_{01}	S_{03}	S_{32}	S_{30}	S_{13}	S_{11}
0213	2213	3111	3013	0012	0110	0131	0133
3132	3130	3321	3333	0332	0320	2023	2021
2301	0301	1003	1101	2100	2002	1212	3212
1020	1022	1233	1221	2220	2232	3300	1300

Table 7

The next exercise helps to construct new counterexamples starting with the Lagarias–Shor or Mackey counterexamples.

Exercise 1.5.3. Let G be a group of type $(4, \ldots, 4)$ with basis elements $a_1, \ldots, a_n$, a_{n+1} and let $H = \langle a_1, \ldots, a_n \rangle$. Suppose that $H = C\{e, a_1\} \cdots \{e, a_n\}$ is a factorization of H and $CC^{-1} \cap \{a_1^2, \ldots, a_n^2\} = \emptyset$. Set $B = C \cup Ca_1 a_{n+1}^2 = C\{e, a_1 a_{n+1}^2\}$. Show that $G = B\{e, a_1\} \cdots \{e, a_n\}\{e, a_{n+1}\}$ is a factorization of G and $BB^{-1} \cap \{a_1^2, \ldots, a_n^2, a_{n+1}^2\} = \emptyset$.

Problem 1.5.1. For which dimensions the geometric version of Keller's conjecture does hold? For which finite abelian groups the algebraic form of Keller's conjecture does hold?

Chapter 2

Elementary arguments

2.1   An axiomatic approach for Hajós's theorem

If a cyclic subset is periodic, then it is a subgroup. So Hajós's theorem can be reformulated such that if a finite abelian group is a direct product of cyclic subsets, then at least one of the factors is periodic. In this section we replace the cyclicity by an abstract property of the factors, which still guarantees that at least one of the factors is periodic.

Let G be a finite abelian group. Let A be a subset and let ϕ be a homomorphism of G into another group. We say that the subset A of G possesses the P property if for each homomorphism ϕ and subset B of $\phi(G)$ the factorization $\phi(G) = \phi(A)B$ implies that either $\phi(A)$ or B is periodic.

The main result of this section is the following. If the 2-component of a finite abelian group is cyclic and the group is a direct product of subsets possessing the P property, then at least one of the factors is periodic.

We list some types of subsets that do have the P property, so we do not study a vacuous property. We will see that cyclic subsets and simulated subsets possess the P property. When A is the cyclic subset $\{e, a, a^2, \ldots, a^{r-1}\}$, then A and aA differ mutually in at most one element, that is, there is at most one element of A which is not in aA and conversely there is at most one element of aA which is not in A. After L. Fuchs, the subset A of G is called *weakly periodic* if there is an element $a \in G \setminus \{e\}$ such that A and aA differ mutually in at most one element. We give another description of a weakly periodic set. Consider the map $f : G \to G$ defined by $f(g) = ga$ for each $g \in G$. Clearly f is a permutation of the elements of G and so it can be decomposed into disjoint cycles. Each cycle has the same length. This common length is $|a|$, the order of a. Elements belonging to the same cycle form a coset modulo $\langle a \rangle$, the subgroup generated by a. The elements of A are distributed over these cycles. Some cycles are completely filled by elements of A. There is at most one cycle which is not completely filled. The elements of this cycle form one connected arc. Thus a weakly periodic subset is in the form

$$C\langle a\rangle \cup b\{e, a, a^2, \ldots, a^{r-1}\},$$

where C is a subset, b is an element of G, further the union is disjoint and the products are direct. When C is empty A is a translated copy of a cyclic subset. When the translated cyclic part is missing, A is periodic.

We call the subset A a *distorted cyclic* subset if it is of form

$$\{e, a, a^2, \ldots, a^{i-1}, a^i d, a^{i+1}, \ldots, a^{r-1}\},$$

where $a, d \in G$. When $d = e$, then A is a cyclic subset. When $a^r = e$, that is, when $\{e, a, a^2, \ldots, a^{r-1}\}$ is a subgroup, then A is a simulated subset.

Let D be a periodic subset of G with period a. Thus the elements of D fill complete cosets modulo $\langle a\rangle$. Choose elements $d \in D$ and $c \in G$ and set $A = (D \setminus \{d\}) \cup \{c\}$. We call A a *distorted periodic* subset. If $c = d$, then $A = D$ and so A is periodic. If $c \in D \setminus \{d\}$, then $A = D \setminus \{d\}$ and so A is a weakly periodic subset. If $D = H$ is a subgroup of G, then A is a simulated subset.

In general the homomorphic image of a subset might be a multiset which contains elements with multiplicities distinct from 1 and 0. In order to see the details more sharply consider a homomorphism ϕ from the group G onto the group G' and let A be a subset of G. Let K be the kernel of ϕ. By the homomorphism theorem the factor group G/K is isomorphic to $\phi(G) = G'$. In the factor group G/K the homomorphic image $\phi(A)$ of A corresponds to the collection of cosets aK, where $a \in A$. If elements of A are incongruent modulo K, then the cosets aK are distinct and so $\phi(A)$ is a subset of $\phi(G) = G'$. In our considerations the only type of homomorphism ϕ occurring for which $\phi(G) = \phi(A)B$ is a factorization of $\phi(G)$. In these cases clearly $\phi(A)$ must be a subset and not merely a multiset of $\phi(G)$.

Consider the cyclic subset $A = \{e, a, a^2, \ldots, a^{r-1}\}$ of G and assume that elements of A are incongruent modulo K. Now $\phi(A)$ consists of the cosets

$$eK, aK, a^2K, \ldots, a^{r-1}K$$

and these cosets are distinct elements of the factor group G/K. Further as these cosets can be written in the forms

$$(aK)^0, (aK)^1, (aK)^2, \ldots, (aK)^{r-1},$$

the homomorphic image $\phi(A)$ is a cyclic subset of $\phi(G)$. Similar arguments yield that if A is a simulated, a weakly periodic or a distorted cyclic or a distorted periodic subset of G and the elements of A are incongruent modulo K, then $\phi(A)$ is again such a subset of $\phi(G)$.

The next lemma establishes that these subsets possess the P property.

Lemma 2.1.1. *Let $G = AB$ be a normalized factorization of the finite abelian group G, where A is cyclic or simulated or weakly periodic or distorted cyclic or distorted periodic subsets. Then either A or B is periodic.*

Proof. First assume that A is the cyclic subset $\{e, a, a^2, \ldots, a^{r-1}\}$. The partition reformulation of the factorization $G = AB$ shows that the sets

$$B, aB, a^2 B, \ldots, a^{r-1} B \tag{1}$$

form a partition of G. Multiplying the factorization $G = AB$ by a gives the factorization $G = Ga = (Aa)B$ and so the sets

$$aB, a^2 B, \ldots, a^{r-1} B, a^r B$$

also form a partition of G. If $a^i B \cap a^r B \neq \emptyset$ for some i, $1 \leq i \leq r-1$, then there is an element g of G such that $g \in a^i B \cap a^r B$. Then $a^{-1} g \in a^{i-1} B \cap a^{r-1} B$. But by (1) $a^{i-1} B \cap a^{r-1} B = \emptyset$. Consequently $a^r B = B$. From this it follows that either B is periodic or $a^r = e$ in which case A is periodic.

Suppose that A is a simulated subset and let H be the associated subgroup. Let us introduce the notation $H = \{h_1, \ldots, h_s\}$, where $h_1 = e$. We may assume that $A = \{h_1, h_2, \ldots, h_{s-1}, h_s d\}$ since this is only a matter of indexing the elements of H. If $|A| = 2$, then A is a cyclic subset and we are done. If $d = e$, then A is equal to the subgroup H which is clearly periodic. Thus we may assume that $|A| \geq 3$ and $d \neq e$. The factorization $G = AB$ gives that the subsets

$$h_1 B, h_2 B, \ldots, h_{s-1} B, h_s d B \tag{2}$$

form a partition of G. We show that in the factorization $G = AB$ the factor A can be replaced by the associated subgroup H, that is, the subsets

$$h_1 B, h_2 B, \ldots, h_{s-1} B, h_s B \tag{3}$$

form a partition of G. Comparing (2) and (3) gives that $h_s B = h_s d B$ and so d is a period of B. The only thing we should verify is that $h_s B \cap h_i B = \emptyset$ for each i, $1 \leq i \leq s - 1$. Assume the contrary that $x \in h_s B \cap h_i B$. If $i \neq 1$, then we have the contradiction

$$h_i^{-1} x \in h_i h_i^{-1} B \cap h_s h_i^{-1} B = B \cap h_j B$$

with some j, $2 \leq j \leq s - 1$. If $i = 1$, then since $|H| \geq 3$, there is an $h_j \in H$ with $2 \leq j \leq s - 1$. Now we have the contradiction

$$h_j x \in h_1 h_j B \cap h_s h_j B = h_j B \cap h_k B$$

with some k, $1 \leq j, k \leq s - 1$, $k \neq j$.

Assume that A is the weakly periodic subset

$$C\langle a \rangle \cup b\{e, a, a^2, \ldots, a^{r-1}\},$$

where C is a subset b, an element of G, the union is disjoint and the products are direct. We are done if C is the empty set or the translated cyclic part is missing. We may assume that this is not the case. Comparing the factorizations $G = AB$

and $G = (Aa)B$ we have that $B = Ba^r$. Consequently either B is periodic or $a^r = e$ in which case A is periodic.

Assume that A is the distorted cyclic subset

$$\{e, a, a^2, \ldots, a^{i-1}, a^i d, a^{i+1}, \ldots, a^{r-1}\}.$$

If $r = 2$ or $d = e$, then A is a cyclic subset and in this case either A or B is periodic. Thus we may assume that $r \geq 3$ and $d \neq e$. From the factorizations $G = AB$ and $G = (Aa)B$ it follows that the sets

$$B, aB, a^2 B, \ldots, a^{i-1} B, a^i d B, a^{i+1} B, \ldots, a^{r-1} B \tag{4}$$

and

$$aB, a^2 B, \ldots, a^i B, a^{i+1} dB, a^{i+2} B, \ldots, a^r B \tag{5}$$

form partitions of G. First we settle the special case when $i = r-1$. If $x \in a^{r-1} dB \cap a^r dB$, then $a^{-2} d^{-1} x \in a^{r-3} B \cap a^{r-2} B$ which violates (4). Hence $a^{r-1} dB \cap a^r dB = \emptyset$. If $x \in a^i B \cap a^r dB$, then $a^{-1} x \in a^{i-1} dB \cap a^{r-1} dB$ which violates (4). Hence $a^i B \cap a^r dB = \emptyset$ for each i, $1 \leq i \leq r - 2$. Consequently $a^r dB = B$. This implies that either B is periodic or $a^r d = e$. We are done if B is periodic so we may assume that $a^r d = e$. By (5), $a^i B \cap a^{r-1} B = \emptyset$ for each i, $0 \leq i \leq r - 2$. Consequently $a^{r-1} B = a^{r-1} dB$. Thus either B is periodic or $d = e$. As $d \neq e$ we are left with the possibility that B is periodic.

Next we prove the special case when $r = 3$. Now $A = \{e, ad, a^2\}$ or $A = \{e, a, a^2 d\}$. The second case has already been settled. In the first case A can be written in the form $A = \{e, a^2, a^4 d'\}$ which again reduces the problem to the previously settled case. Thus $r \geq 4$ and $1 \leq i \leq r - 2$ may be assumed for the remaining cases. We distinguish two subcases depending on whether $i = 1$ or $i \neq 1$. Consider first the $i \neq 1$ case. Comparing (4) and (5) gives that

$$B \cup a^i dB \cup a^{i+1} B = a^i B \cup a^{i+1} dB \cup a^r B.$$

We claim that in this equation $a^i B = a^i dB$ and so d is a period of B. In order to verify this claim note that by (4) $aB \cap a^{i+1} B = \emptyset$ and so $B \cap a^i B = \emptyset$. As $r \geq 4$ again by (4), there is a j, $0 \leq j \leq r - 2$ for which $a^j B \cap a^{j+1} B = \emptyset$ and so $a^i B \cap a^{i+1} B = \emptyset$. This proves the claim that $a^i B = a^i dB$. Finally consider the $i = 1$ case. From the factorization $G = AB$ we get that $G = A^{-1} B$ is also a factorization of G, where

$$A^{-1} = \{e, a^{-1} d^{-1}, a^{-1}, \ldots, a^{-(r-1)}\}.$$

A multiplication by a^{r-1} yields the factorization $G = (a^{r-1} A^{-1}) B$. Here $(a^{r-1} A^{-1})$ is a distorted cyclic subset with $i = r - 2 \geq 2$ since $r \geq 4$. This reduces the problem to the $i \neq 1$ case.

Assume that A is a distorted periodic subset such that D is the associated periodic subset with period a, that is, $A = (D \setminus \{d\}) \cup \{c\}$, where $d \in D$ and $c \in G$.

If $c \in D \setminus \{d\}$, then A is weakly periodic. Since this case has already been covered, we assume that $c \notin D \setminus \{d\}$. From the factorizations $G = AB$ and $G = (Aa)B$ it follows that $dB \cup acB = adB \cup cB$. If $dB \cap adB \neq \emptyset$, then $B \cap aB \neq \emptyset$ which is a contradiction. If $acB \cap cB \neq \emptyset$, then again $aB \cap B \neq \emptyset$. Thus $dB = cB$. Now either B is periodic or $d = c$ in which case $A = D$, that is, A is periodic.

This completes the proof.

Lemma 2.1.2. *Let G be a finite abelian group whose 2-component is cyclic and let H be a subgroup of G such that $H \neq \{e\}$ and $H \neq G$. Then there are nonsubgroup subsets $A_1, \ldots, A_s$ of G such that $G = HA_1 \cdots A_s$ is a factorization of G each, $|A_i|$ is a prime and each A_i is either cyclic or simulated.*

Proof. Set $L_0 = H$ and $L_s = G$ and consider the chain of subgroups

$$L_0 \subset L_1 \subset \cdots \subset L_s,$$

where $|L_i : L_{i-1}| = p(i)$ is a prime for each i, $1 \leq i \leq s$. It is sufficient to construct a nonsubgroup complete set of representatives A_i in L_i modulo L_{i-1} for each i, $1 \leq i \leq s$. As $|L_i/L_{i-1}| = p(i)$, the order of each element in $L_i \setminus L_{i-1}$ is divisible by $p(i)$. If there is an $a_i \in L_i \setminus L_{i-1}$ for which $a_i^{p(i)} \neq e$, then the subset

$$A_i = \{e, a_i, a_i^2, \ldots, a_i^{p(i)-1}\}$$

is a complete set of representatives in L_i modulo L_{i-1} and clearly A_i is not a subgroup of G. Thus we may assume that each element in $L_i \setminus L_{i-1}$ is of order $p(i)$. Since $L_i = \langle L_i \setminus L_{i-1} \rangle$, L_i is of type $(p(i), \ldots, p(i))$. Note that L_i is not cyclic since $|L_i| = |L_{i-1}| \cdot |H|$ and $i \geq 1$. As the 2-component of G is cyclic $p(i) = 2$ is not possible. Then choosing an $a_i \in L_i \setminus L_{i-1}$ and a $b_i \in L_{i-1} \setminus \{e\}$, the subset

$$A_i = \{e, a_i, a_i^2, \ldots, a_i^{p(i)-2}, a_i^{p(i)-1}b_i\}$$

is a complete set of representatives in L_i modulo L_{i-1} and A_i is not a subgroup of G.

This completes the proof.

Lemma 2.1.3. *Let $G = AB$ be a normed factorization of the finite abelian group G, where the 2-component of G is cyclic. Assume that A is not periodic and possesses the P property. Then A can be replaced by a product of nonsubgroup subsets of prime cardinality such that each factor is either cyclic or simulated.*

Proof. As A has the P property, either A or B is periodic. By the hypotheses of the lemma, B is periodic. Hence there is a normed subset Y and a subgroup H of G for which $B = YH$ is a factorization. We assume that H is the maximal possible with this property. If $|Y| = 1$, then $B = H$. By Lemma 2.1.2, A can be replaced by a product of nonsubgroup factors of prime cardinality that are either cyclic or simulated.

Assume that $|Y| \geq 2$ and use an induction on $|Y|$. From the factorization $G = AYH$ we get the factorization

$$G/H = (AH)/H \cdot (YH)/H$$

of the factor group G/H. By the P property of $(AH)/H$, either $(AH)/H$ or $(YH)/H$ is periodic. But $(YH)/H$ cannot be periodic by the maximality of H. Hence $(AH)/H$ is periodic. There are normalized subsets K and X of G such that $(KH)/H$ is a subgroup of G/H and

$$(AH)/H = (XH)/H \cdot (KH)/H$$

is a factorization. Now $L = KH$ is a subgroup of G. We choose L to be the largest with this property. We also have the factorization

$$G/L = (XL)/L \cdot (YL)/L$$

of the factor group G/L. Here $(XL)/L$ possesses the P property and it is not periodic because of the maximality of L. By Lemma 2.1.2, there are nonsubgroup subsets $A_1, \ldots, A_s$ of G such that $L = HA_1 \cdots A_s$ is a factorization of L each $|A_i|$ is a prime and each A_i is either cyclic or simulated. By the inductive assumption there are subsets $D_1, \ldots, D_r$ of G such that

$$G/L = (D_1 L)/L \cdots (D_r L)/L \cdot (YL)/L$$

is a factorization of G/L. Here each $(D_i L)/L$ consists of a prime number of cosets of L and each $(D_i L)/L$ is either a cyclic or a simulated subset of G/L. Replace the simulated ones by their associated subgroups. As $(D_i L)/L$ consists of a prime number of cosets of L, after the replacements each $(D_i L)/L$ is a cyclic subset of G/L. There are cyclic subsets $C_1, \ldots, C_r$ of G such that

$$G/L = (C_1 L)/L \cdots (C_r L)/L \cdot (YL)/L$$

is a factorization of G/L. It follows that $C_1 \cdots C_r Y$ is a complete set of representatives of G modulo L. Therefore

$$\begin{aligned} G &= C_1 \cdots C_r Y L \\ &= C_1 \cdots C_r Y H A_1 \cdots A_s \\ &= A_1 \cdots A_s C_1 \cdots C_r B \end{aligned}$$

are factorizations of G. Let $C_i = \{e, c_i, c_i^2, \ldots, c_i^{p(i)-1}\}$. If C_i is not a subgroup of G, then set $S_i = C_i$. If C_i is a subgroup of G, then set

$$S_i = \{e, c_i, c_i^2, \ldots, c_i^{p(i)-2}, c_i^{p(i)-1}l\}, \tag{6}$$

where $l \in L \setminus \{e\}$. Note that

$$G = S_1 \cdots S_r Y L = S_1 \cdots S_r Y H A_1 \cdots A_s = A_1 \cdots A_s S_1 \cdots S_r B$$

are factorizations of G. Clearly (6) is a nonsubgroup simulated subset of G if $p_i \geq 3$. If $p_i = 2$ and S_i is a subgroup of G, then $|c_i l| = 2$. Since the 2-component of G is cyclic G has only one element of order 2. Hence from the factorization $G = S_1 \cdots S_r Y L$ it follows that $|L|$ cannot be even. Therefore (6) is a nonsubgroup simulated subset of G.

This completes the proof.

The next exercise gives an example to show that Lemma 2.1.3 fails when the condition that G has cyclic 2-component is omitted.

Exercise 2.1.1. Let G have type $(2^3, 2)$ with generators x, y of orders $8, 2$ respectively. Let

$$A = \{e, x, x^4, x^5 y\}, \quad B = \{e, x^2\}\{e, y\}.$$

Verify that $G = AB$ is a factorization of G, A has property P, A is not periodic and A cannot be replaced by a product of nonperiodic cyclic or simulated subsets of prime order. (Hint: Write B in the form $B = B_1 H$, where $B_1 = \{e, x^2\}$, $H = \langle y \rangle$. Suppose there is a factorization $G = A_1 A_2 B_1 H$, where A_1, A_2 are normalized and $|A_1| = |A_2| = 2$. Consider the factor group G/H. Show that $(A_1 H)/H$ or $(A_2 H)/H$ is a subgroup of G/H. Then A_1 or A_2 is a subgroup of G.)

Now we come to the proof of the main result of the section.

Theorem 2.1.1. *Let G be a finite abelian group whose 2-component is cyclic. If $G = A_1 \cdots A_n$ is a factorization of G, where each A_i possesses the P property, then at least one of the factors must be periodic.*

Proof. Assume the contrary, that none of the factors is periodic. Set $A = A_1$ and $B = A_2 \cdots A_n$. Now Lemma 2.1.3 is applicable to the factorization $G = AB$ and gives that A_1 can be replaced by a product of nonsubgroup factors of prime cardinality each of which is either cyclic or simulated. Similarly, A_2 can be replaced by a product of nonsubgroup factors of prime cardinality each of which is either cyclic or simulated. Continuing in this way finally we get a factorization $G = B_1 \cdots B_m$, where each $|B_i|$ is a prime and each B_i is either cyclic or simulated and none of the factors is a subgroup of G. By Hajós's theorem it is not possible.

This completes the proof.

We close the section with the following open problem.

Problem 2.1.1. Does Theorem 2.1.1 hold with the condition on the 2-component omitted?

2.2 Direct product of elementary groups

Let $G = B A_1 \cdots A_n$ be a factorization of the finite abelian group G. Suppose that each $|A_i|$ is a prime, $|B|$ is a product of two primes, and each A_i is either cyclic or simulated. Does it follow that one of the factors $B, A_1, \ldots, A_n$ is periodic? The main result of this section gives an answer in the affirmative in the special case when G is a direct product of elementary p-groups. To consider factorizations where each factor has prime order except one whose order is a product of two primes was first proposed by A. D. Sands in 1963. Sections 6.2, 7.1, and 7.4 are also related to this problem of A. D. Sands. In Section 6.2 the group considered is cyclic. In Section 7.1 the groups have order p^4. In Section 7.4 the factor B has order 4.

Let $a_1 \in A_1, \ldots, a_n \in A_n$ and multiply the factorization $G = A_1 \cdots A_n$ by $(a_1 \cdots a_n)^{-1}$ to get the factorization

$$G = (a_1 \cdots a_n)^{-1} G = (a_1^{-1} A_1) \cdots (a_n^{-1} A_n).$$

This new factorization is normalized. Further if $a_i^{-1} A_i$ is periodic, then so is A_i. Thus when we prove the main result of the chapter we may focus our attention on normalized factorizations.

For a subgroup H of G we will use the notation $\nu(H) = a(1) + \cdots + a(t)$, where $p_1^{a(1)} \cdots p_t^{a(t)}$ is the canonical prime factorization of $|H|$.

Theorem 2.2.1. *Let G be a finite abelian group that is a direct product of elementary p-groups. Let $G = BA_1 \cdots A_n$ be a factorization of G, where each $|A_i|$ is a prime, $|B|$ is a product of two primes, and each A_i is either cyclic or simulated. Then at least one of the factors is periodic.*

Proof. We divide the proof into nine steps.

(1) We may assume that A_i is not a subgroup of G for each i, $1 \leq i \leq n$ since otherwise there is nothing to prove. In addition we may assume that B is normalized. If A_i is cyclic, then by Lemma 1.2.3, it can be replaced by a nonsubgroup subset of the form

$$A_i = \{e, h_i, h_i^2, \ldots, h_i^{p(i)-2}, h_i^{p(i)-1} d_i\}, \quad 1 \leq i \leq n,$$

where $H_i = \{e, h_i, h_i^2, \ldots, h_i^{p(i)-1}\}$ is the corresponding subgroup and $|d_i|$ is relatively prime to $p(i)$. Thus we may assume that if $p(i) \geq 3$, then A_i is simulated and if $p(i) = 2$, then A_i is cyclic and $|d_i|$ is odd. If A_i is a simulated subset, then it is in the above form and by Lemma 1.2.3 we may assume that $|d_i|$ is a prime.

We will need later the fact that if A_i is periodic, then $d_i = e$. In order to prove the claim assume that A_i is periodic. Since A_i is normalized and $|A_i| = p(i)$ is a prime, it follows that A_i is a subgroup of G. In the $p(i) \geq 3$ case $h_i \in A_i \cap H_i$ and so $A_i = H_i$. Therefore $h_i^{p(i)-1} d_i = h_i^{p(i)-1}$, that is, $d_i = e$. In the $p(i) = 2$ case $e = \left(h_i^{p(i)-1} d_i\right)^2 = h_i^2 d_i^2 = d_i^2$. Since now $|d_i|$ is odd, it follows that $d_i = e$.

(2) If the theorem holds, then B is periodic. We claim that the period is one of $d_1, \ldots, d_n$. Indeed, $B = A_0 H$, where H is a subgroup, A_0 is a normalized subset of G such that $|A_0|$, $|H|$ are primes and the product $A_0 H$ is direct. From the factorization

$$G = BA_1 \cdots A_n = A_0 H A_1 \cdots A_n$$

we get the factorization

$$G/H = (A_0H)/H \cdot (A_1H)/H \cdots (A_nH)/H$$

of the factor group G/H, where

$$(A_iH)/H = \{a_iH : a_i \in A_i\}.$$

By Rédei's theorem it follows that one of the factors is a subgroup of G/H, say $(A_iH)/H$. Now $A_iH = K$ is a subgroup of G. Considering the factor group G/K we can see that there is a factor, say A_j, such that $A_jA_iH = L$ is a subgroup of G. If $i \neq 0$, then from $A_iH = K$, by Lemma 1.2.3 we get that H and hence B is periodic with period d_i. If $i = 0$, then from $A_j(A_iH) = L$, by Lemma 1.2.3, it follows that $A_iH = B$ is periodic with period d_j.

(3) In the factorization $G = BA_1 \cdots A_n$ replace each A_i by H_i to get the factorization $G = BH_1 \cdots H_n$. By Lemma 1.2.3, this can be done. From this we can read off that the product $H_1 \cdots H_n$ is direct. Let $H = H_1 \cdots H_n$ and $N = \langle H, d_1, \ldots, d_n \rangle$. Clearly, $n \leq \nu(N) \leq n+2$. If $n = \nu(N)$, then $d_1, \ldots, d_n \in H$ and so $H = A_1 \cdots A_n$ is a factorization of H. By Rédei's theorem one of the factors is a subgroup of H and so is of G. This is a contradiction. We may assume that $n+1 \leq \nu(N) \leq n+2$. When $n+1 = \nu(N)$ we may assume that $d_1 \notin H$ and $d_2, \ldots, d_n \in \langle H, d_1 \rangle$ since this is only a matter of indexing the factors. Similarly, when $\nu(N) = n+2$ we may assume that $d_1, d_2 \notin H$ and $d_2, \ldots, d_n \in \langle H, d_1, d_2 \rangle$. In the remaining part of the proof we distinguish two cases.

Case (a): $n+1 = \nu(N)$ and $d_2, \ldots, d_n \in \langle H, d_1 \rangle$.

Case (b): $\nu(N) = n+2$ and $d_3, \ldots, d_n \in \langle H, d_1, d_2 \rangle$.

As G is a product of elementary p-groups, there is a subgroup M of G such that G is the direct product of M and H. Let $M = \langle x, y \rangle$, where $|x| = p$ and $|y| = q$. In Case (a) we may choose x to be d_1. In Case (b) we may choose x, y to be d_1, d_2 respectively.

From the factorization $G = BH$ it follows that B is a complete set of representatives modulo H in G. The elements of B can be listed by the following table.

$$
\begin{array}{cccc}
x^0y^0w_{0,0} & x^0y^1w_{0,1} & \cdots & x^0y^{q-1}w_{0,q-1} \\
x^1y^0w_{1,0} & x^1y^1w_{1,1} & \cdots & x^1y^{q-1}w_{1,q-1} \\
\vdots & \vdots & \ddots & \vdots \\
x^{p-1}y^0w_{p-1,0} & x^{p-1}y^1w_{p-1,1} & \cdots & x^{p-1}y^{q-1}w_{p-1,q-1}
\end{array}
\tag{1}
$$

Here $w_{i,j} \in H$ and $w_{0,0} = e$ as B is normalized.

(4) In the $n = 1$ case, by Lemma 1.2.3, B is periodic with period d_1. So we assume that $n \geq 2$ and proceed by induction on n. In the factorization $G = BA_1 \cdots A_n$ we will replace one or more A_i by the corresponding H_i. In order to describe the procedure let

$$\alpha(1), \ldots, \alpha(r), \beta(1), \ldots, \beta(s)$$

be a rearrangement of the indices $1, 2, \ldots, n$. In the factorization

$$G = BA_1 \cdots A_n = BA_{\alpha(1)} \cdots A_{\alpha(r)} A_{\beta(1)} \cdots A_{\beta(s)}$$

replace $A_{\beta(1)}, \ldots, A_{\beta(s)}$ by $H_{\beta(1)}, \ldots, H_{\beta(s)}$ respectively to get the factorization

$$G = BA_{\alpha(1)} \cdots A_{\alpha(r)} H_{\beta(1)} \cdots H_{\beta(s)}.$$

Let $K = H_{\beta(1)} \cdots H_{\beta(s)}$. The factorization $G = BA_{\alpha(1)} \cdots A_{\alpha(r)} K$ leads to the factorization

$$G/K = (BK)/K \cdot (A_{\alpha(1)}K)/K \cdots (A_{\alpha(r)}K)/K$$

of the factor group G/K, where

$$(AK)/K = \{aK : a \in A\}$$

for any subset A of G. By the inductive assumption one of the factors is periodic. If $(A_{\alpha(i)}K)/K$ is periodic, then by step (2) it follows that $d_{\alpha(i)} \in K$. If $(BK)/K$ is periodic, then by step (2) it is periodic with period $d_{\alpha(i)}K$ for some i, $1 \leq i \leq r$. Delete the K-parts of the elements of B and let B^* be the resulting subset of G. Delete the K-part of $d_{\alpha(i)}$ and let $d^*_{\alpha(i)}$ be the resulting element of G. We can see that B^* is periodic with period $d^*_{\alpha(i)}$. We may summarize the above argument saying that the factorization

$$G = BA_{\alpha(1)} \cdots A_{\alpha(r)} H_{\beta(1)} \cdots H_{\beta(s)}$$

implies that B^* is periodic with period $d^*_{\alpha(i)}$ or $d_{\alpha(i)} \in K$ for some i, $1 \leq i \leq r$.

(5) First we settle Case (a). We claim that in this case there is a permutation

$$\alpha(1), \ldots, \alpha(r), \beta(1), \ldots, \beta(s)$$

of the indices $1, 2, \ldots, n$ such that $\alpha(1) = 1$ and the elements of B are the following.

$$
\begin{array}{cccc}
x^0 y^0 u_0 w_0^0 v_0 & x^0 y^1 u_1 w_1^0 v_1 & \cdots & x^0 y^{q-1} u_{q-1} w_{q-1}^0 v_{q-1} \\
x^1 y^0 u_0 w_0^1 v_0 & x^1 y^1 u_1 w_1^1 v_1 & \cdots & x^1 y^{q-1} u_{q-1} w_{q-1}^1 v_{q-1} \\
\vdots & \vdots & \ddots & \vdots \\
x^{p-1} y^0 u_0 w_0^{p-1} v_0 & x^{p-1} y^1 u_1 w_1^{p-1} v_1 & \cdots & x^{p-1} y^{q-1} u_{q-1} w_{q-1}^{p-1} v_{q-1}
\end{array}
$$

Here

$$u_j \in H_{\alpha(1)} \cdots H_{\alpha(r)}, \quad v_j, w_j \in H_{\beta(1)} \cdots H_{\beta(s)}$$

and $u_0 = v_0 = w_0 = e$ as B is normed. Further

$$d_{\alpha(1)} = x, \qquad d_{\alpha(2)} = h_{\alpha(1)}, \ldots, d_{\alpha(r)} = h_{\alpha(r-1)}.$$

In other words the columns of the above table are the disjoint cosets

$$\langle xw_j \rangle (y^j u_j v_j), \quad 0 \leq j \leq q - 1.$$

The factorization $G = BA_1H_2 \cdots H_n$ gives that B^* is periodic with period $d_1^* = x^* = x$ or $d_1 = x \in K = H_2 \cdots H_n$. Since the latter one is impossible, B^* is periodic with period x. Therefore the H_1-parts of the elements

$$w_{0,j}, w_{1,j}, \ldots, w_{p-1,j}$$

in (1) are equal for each j, $0 \le j \le q - 1$. Let $u_0, u_1, \ldots, u_{q-1}$ be these common H_1-parts respectively. Now the elements of B are the following.

$$
\begin{array}{ccc}
x^0 y^0 u_0 w_{0,0} & \cdots & x^0 y^{q-1} u_{q-1} w_{0,q-1} \\
\vdots & \ddots & \vdots \\
x^{p-1} y^0 u_0 w_{p-1,0} & \cdots & x^{p-1} y^{q-1} u_{q-1} w_{p-1,q-1}
\end{array}
\tag{2}
$$

Here $u_j \in H_1$, $w_{i,j} \in H_2 \cdots H_n$ and $u_0 = w_{0,0} = e$ as B is normalized.

Multiply the factorization $G = BA_1 \cdots A_n$ by $(y^j u_j w_{0,j})^{-1}$ to get the factorization

$$G = [B(y^j u_j w_{0,j})^{-1}]A_1 \cdots A_n$$

for each j, $0 \le j \le q - 1$. Restricting the factorization

$$G = [B(y^j u_j w_{0,j})^{-1}]A_1 \cdots A_n$$

to the subgroup $L = \langle x, H \rangle$ we get the factorization

$$L = \big([B(y^j u_j w_{0,j})^{-1}] \cap L\big)A_1 \cdots A_n.$$

This is a normalized factorization and each factor has a prime number of elements. By Rédei's theorem at least one of the factors is a subgroup of L. Since none of $A_1, \ldots, A_n$ is a subgroup, $[B(y^j u_j w_{0,j})^{-1}] \cap L$ is a subgroup of L. Obviously,

$$[B(y^j u_j w_{0,j})^{-1}] \cap L = \langle x w_{1,j} w_{0,j}^{-1} \rangle.$$

After introducing the $w_j = w_{1,j} w_{0,j}^{-1}$ and $v_j = w_{0,j}$ notations, B is in the required form with

$$r = 1, \quad s = n - 1, \quad \alpha(1) = 1, \quad \beta(1) = 2, \ldots, \beta(s) = n.$$

(6) If $r = n$, then $v_j = w_j = e$ for each j, $0 \le j \le q - 1$ and so B is periodic with period x_1. We will show that if $r < n$, then B is periodic or r can be increased.

Now $s \ge 1$. From the factorization

$$G = BA_{\alpha(1)} \cdots A_{\alpha(r-1)} H_{\alpha(r)} A_{\beta(1)} \cdots A_{\beta(s)}$$

it follows that B^* is periodic with period $d_{\alpha(i)}^*$ or $d_{\beta(j)}^*$ or $d_{\alpha(i)} \in H_{\alpha(r)}$ or $d_{\beta(j)} \in H_{\alpha(r)}$ for some i, j, $1 \le i \le r - 1$, $1 \le j \le s$.

Assume first that B^* is periodic with period d^*, where d^* is $d^*_{\alpha(i)}$ or $d^*_{\beta(j)}$. Delete the $H_{\alpha(r)}$-part of u_j and let u_j^* be the resulting element. Notice that B^* is a union of the cosets

$$\langle xw_j\rangle(y^j u_j^* v_j), \quad 0 \le j \le q-1.$$

The $\langle y\rangle$-part of d^* is e. From $B^*d^* = B^*$ it follows that

$$\langle xw_j\rangle(y^j u_j^* v_j)d^* = \langle xw_j\rangle(y^j u_j^* v_j), \quad 0 \le j \le q-1.$$

This gives that B is periodic with period d^*.

Since $d_{\alpha(i)} \in \{x, h_{\alpha(1)}, \ldots, h_{\alpha(r-2)}\}$, the $d_{\alpha(i)} \in H_{\alpha(r)}$ possibility is excluded. Thus by relabelling we may assume that $d_{\beta(s)} \in H_{\alpha(r)}$. By Lemma 1.2.3, we may assume that $d_{\beta(s)} = h_{\alpha(r)}$. Setting $\alpha(r+1) = \beta(s)$ increases r and decreases s.

From the factorization

$$G = BA_{\alpha(1)} \cdots A_{\alpha(r)} A_{\alpha(r+1)} H_{\beta(1)} \cdots H_{\beta(s-1)}$$

it follows that B^* is periodic with period $d^*_{\alpha(i)}$ or $d_{\alpha(i)} \in K = H_{\beta(1)} \cdots H_{\beta(s-1)}$ for some i, $1 \le i \le r+1$. Since $d_{\alpha(i)} \in \{x, h_{\alpha(1)}, \ldots, h_{\alpha(r)}\}$, it follows that $d_{\alpha(i)} \notin K$ and so B^* is periodic with period x_1. This gives that the $(H_{\alpha(1)} \cdots H_{\alpha(r+1)})$-parts of the elements

$$w_{0,j}, w_{1,j}, \ldots, w_{p-1,j}$$

in (2) are equal for each j, $0 \le j \le q-1$. Thus the elements of B are in form (2), with

$$u_j \in H_{\alpha(1)} \cdots H_{\alpha(r+1)}, \quad w_{i,j} \in H_{\beta(1)} \cdots H_{\beta(s-1)}.$$

Finally, restricting the factorization $G = [B(x^j u_j w_{0,j})^{-1}]A_1 \cdots A_n$ to the subgroup $L = \langle x, H\rangle$ we can see that B is in the required form with an increased value of r.

(7) We may assume that Case (b) holds. We claim that there is a permutation

$$\alpha(1), \ldots, \alpha(r), \beta(1), \ldots, \beta(s), \gamma(1), \ldots, \gamma(t)$$

of the indices $1, 2, \ldots, n$ such that $\alpha(1) = 1$, $\beta(1) = 2$ and the elements of B can be written in the form

$$\begin{array}{ccc}
x^0 y^0 u_0 v_0 w_{0,0} & \cdots & x^0 y^{q-1} u_{q-1} v_0 w_{0,q-1} \\
\vdots & \ddots & \vdots \\
x^{p-1} y^0 u_0 v_{p-1} w_{p-1,0} & \cdots & x^{p-1} y^{q-1} u_{q-1} v_{p-1} w_{p-1,q-1},
\end{array} \tag{3}$$

where

$$u_j \in H_{\alpha(r)}, \quad v_i \in H_{\beta(s)}, \quad w_{i,j} \in H_{\gamma(1)} \cdots H_{\gamma(t)}.$$

Further

$$d_{\alpha(1)} = x, \quad d_{\alpha(2)} = h_{\alpha(1)}, \ldots, d_{\alpha(r)} = h_{\alpha(r-1)},$$

$$d_{\beta(1)} = y, \quad d_{\beta(2)} = h_{\beta(1)}, \ldots, d_{\beta(s)} = h_{\beta(s-1)},$$

and $u_0 = v_0 = w_{0,0} = e$ as B is normalized.

The factorization $G = BA_1 H_2 \cdots H_n$ gives that $d_1 = x \in H_2 \cdots H_n$ or B^* is periodic with period $d_1^* = x$. Only the second alternative can hold and so the H_1-parts of the elements

$$w_{0,j}, w_{1,j}, \ldots, w_{p-1,j}$$

in (1) are equal for each j, $0 \le j \le q - 1$. Let $u_0, u_1, \ldots, u_{q-1}$ be these common values respectively. Similarly the factorization $G = BH_1 A_2 H_3 \cdots H_n$ gives that the H_2-parts of the elements

$$w_{i,0}, w_{i,1}, \ldots, w_{i,q-1}$$

in (1) are equal for each i, $0 \le i \le p - 1$. Let $v_0, v_1, \ldots, v_{p-1}$ be these common values respectively. Now with the choices

$$r = 1, \quad s = 1, \quad t = n - 2,$$

$$\alpha(1) = 1, \quad \beta(1) = 2, \quad \gamma(1) = 3, \ldots, \gamma(n - 2) = n$$

the claim is established.

(8) Next we show that if $r + s = n$, then B is periodic. Now $t = 0$ and so $w_{i,j} = e$ for each i, j, $0 \le i \le p - 1$, $0 \le j \le q - 1$. If

$$e = u_0 = u_1 = \cdots = u_{q-1},$$

then B is periodic with period y. So we may assume that $u_j \ne e$ for some j, say $u_j = h_{\alpha(r)}^k$, $1 \le k \le p(\alpha(r)) - 1$. If

$$e = v_0 = v_1 = \cdots = v_{p-1},$$

then B is periodic with period x. So we may assume that $v_i \ne e$ for some i, say $v_i = h_{\beta(s)}^m$ $1 \le m \le p(\beta(s)) - 1$. By Lemma 1.2.3, in the factorization

$$G = BA_1 \cdots A_n = BA_{\alpha(1)} \cdots A_{\alpha(r)} A_{\beta(1)} \cdots A_{\beta(s)},$$

$A_{\alpha(1)}, A_{\alpha(r)}, A_{\beta(1)}, A_{\beta(s)}$ can be replaced by $A'_{\alpha(1)}, A'_{\alpha(r)}, A'_{\beta(1)}, A'_{\beta(s)}$ such that

$$h_{\alpha(1)}^{p(\alpha(1))-1} d_{\alpha(1)}^{-i} = h_{\alpha(1)}^{p(\alpha(1))-1} x^{-i} \in A'_{\alpha(1)},$$

$$h_{\beta(1)}^{p(\beta(1))-1} d_{\beta(1)}^{-j} = h_{\beta(1)}^{p(\beta(1))-1} y^{-j} \in A'_{\beta(1)},$$

$$h_{\alpha(r)}^{p(\alpha(r))-k} d_{\alpha(r)} = h_{\alpha(r)}^{p(\alpha(r))-k} h_{\alpha(r-1)} \in A'_{\alpha(r)},$$

$$h_{\beta(s)}^{p(\beta(s))-m} d_{\beta(s)} = h_{\beta(s)}^{p(\beta(s))-m} h_{\beta(s-1)} \in A'_{\beta(s)}.$$

Now the equation

$$\underbrace{\left(x^i y^j h^k_{\alpha(r)} h^m_{\beta(s)}\right)}_{\in B}$$

$$\underbrace{\left(h^{p(\alpha(1))-1}_{\alpha(1)} x^{-i}\right)}_{\in A'_{\alpha(1)}} \underbrace{\left(h^{p(\alpha(2))-1}_{\alpha(2)} h_{\alpha(1)}\right)}_{\in A_{\alpha(2)}} \cdots$$

$$\underbrace{\left(h^{p(\alpha(r-1))-1}_{\alpha(r-1)} h_{\alpha(r-2)}\right)}_{\in A_{\alpha(r-1)}} \underbrace{\left(h^{p(\alpha(r))-k}_{\alpha(r)} h_{\alpha(r-1)}\right)}_{\in A'_{\alpha(r)}}$$

$$\underbrace{\left(h^{p(\beta(1))-1}_{\beta(1)} y^{-j}\right)}_{\in A'_{\beta(1)}} \underbrace{\left(h^{p(\beta(2))-1}_{\beta(2)} h_{\beta(1)}\right)}_{\in A_{\beta(2)}} \cdots$$

$$\underbrace{\left(h^{p(\beta(s-1))-1}_{\beta(s-1)} h_{\beta(s-2)}\right)}_{\in A_{\beta(s-1)}} \underbrace{\left(h^{p(\beta(s))-m}_{\beta(s)} h_{\beta(s-1)}\right)}_{\in A'_{\beta(s)}} = e$$

contradicts the factorization

$$G = B A'_{\alpha(1)} A_{\alpha(2)} \cdots A_{\alpha(r-1)} A'_{\alpha(r)} A'_{\beta(1)} A_{\beta(2)} \cdots A_{\beta(s-1)} A'_{\beta(s)}.$$

This proves that B is periodic.

(9) Finally we claim that if $r + s < n$, then $r + s$ can be increased. Now $t \geq 1$.
The factorization

$$G = B A_{\alpha(1)} \cdots A_{\alpha(r)} A_{\beta(1)} \cdots A_{\beta(s)} H_{\gamma(1)} \cdots H_{\gamma(t)}$$

gives that B^* with period $d^*_{\alpha(i)}$ or $d^*_{\beta(j)}$ or $d_{\alpha(i)} \in H_{\gamma(1)} \cdots H_{\gamma(t)}$ or $d_{\beta(j)} \in H_{\gamma(1)} \cdots H_{\gamma(t)}$. Since

$$d_{\alpha(i)} \in \{x, h_{\alpha(1)}, \ldots, h_{\alpha(r-1)}\}, \qquad d_{\beta(j)} \in \{y, h_{\beta(1)}, \ldots, h_{\beta(s-1)}\}.$$

Only the B^* is periodic alternative can hold. The period can only be one of

$$x, h_{\alpha(1)}, \ldots, h_{\alpha(r-1)}, y, h_{\beta(1)}, \ldots, h_{\beta(s-1)}.$$

Note that B^* cannot be periodic with a period $h \in H_1 \cdots H_n$ since the $\langle x, y \rangle$-parts
of the elements of B^* are distinct. Therefore B^* is periodic with period x or y. If
B^* is periodic with period x, then

$$e = v_0 = v_1 = \cdots = v_{p-1}$$

and so the $H_{\beta(s)}$-parts are missing from B. If B^* is periodic with period y, then

$$e = u_0 = u_1 = \cdots = u_{q-1}$$

and so the $H_{\alpha(r)}$-parts are missing from B. For the sake of definiteness suppose that the $H_{\beta(s)}$-parts are missing from B. The factorization

$$G = BA_{\alpha(1)} \cdots A_{\alpha(r)} A_{\beta(1)} \cdots A_{\beta(s-1)} H_{\beta(s)} A_{\gamma(1)} \cdots A_{\gamma(t)}$$

gives that B^* is periodic or $d_{\alpha(i)} \in H_{\beta(s)}$ or $d_{\beta(j)} \in H_{\beta(s)}$ or $d_{\gamma(k)} \in H_{\beta(s)}$. If B^* is periodic, then B is periodic since B and B^* are equal. $d_{\alpha(i)} \in H_{\beta(s)}$ and $d_{\beta(j)} \in H_{\beta(s)}$ are contradictions. By relabelling we may assume that $d_{\gamma(t)} \in H_{\beta(s)}$. Using Lemma 1.2.3 we may assume that $d_{\gamma(t)} = h_{\beta(s)}$. We can decrease t and increase s by setting $\beta(s+1) = \gamma(t)$.

In the known way from the factorization

$$G = BH_{\alpha(1)} A_{\alpha(2)} \cdots A_{\alpha(r)} A_{\beta(1)} \cdots A_{\beta(s)} A_{\beta(s+1)} H_{\gamma(1)} \cdots H_{\gamma(t-1)}$$

it follows that B^* is periodic with period y. Consequently the $H_{\beta(s+1)}$-parts of the elements

$$w_{i,0}, w_{i,1}, \ldots, w_{i,p-1}$$

in (3) are equal for each i, $0 \le i \le p - 1$. Denoting these common $H_{\beta(s+1)}$-parts by $v_0, v_1, \ldots, v_{p-1}$ respectively we can see that B is in the required form with an increased value of s.

This completes the proof.

The next exercise is an application of Theorem 2.2.1.

Exercise 2.2.1. Let G be a group of type $(3, \ldots, 3)$. Show that if $G = A_1 \cdots A_n$ is a factorization of G such that $|B| = 9$, $|A_1| = \cdots = |A_n| = 3$, then at least one of the factors $B, A_1, \ldots, A_n$ must be periodic.

Construction (1) in the proof of Lemma 2.3.4 (in the next section) will show that the conditions of Theorem 2.2.1 cannot be removed in general.

2.3 Nonperiodic factorizations of p-groups

Let $(q_1, \ldots, q_n)$ be an n-tuple of integers. If from each factorization $G = A_1 \cdots A_n$ of G with $|A_1| = q_1, \ldots, |A_n| = q_n$ it follows that at least one of the factors is periodic, then we say the n-tuple $(q_1, \ldots, q_n)$ is a *periodicity forcing* factorization type for G. Rédei's theorem gives that if each q_i is a prime, then $(q_1, \ldots, q_n)$ is a periodicity forcing type for each finite abelian group. Motivated by this result we set forth the next problem. For a given finite abelian group determine, the periodicity forcing factorization types. First we tackle this problem for odd p-groups.

Observation 2.3.1. *Let G be a finite abelian group. Let H be a subgroup of G. Let $A, B \subset G$ such that $e \in A \cap B$, the product AB is direct, A, B are not periodic, the elements of B are pair-wise incongruent modulo H. Then the set AB is not periodic.*

Proof. Set $C = AB$ and let $g \in G$ such that $C = Cg$. Let $B = \{b_1, \ldots, b_s\}$. As the product is direct the sets $Ab_1, \ldots, Ab_s$ form a partition of C. In particular

$$C = Ab_1 \cup \cdots \cup Ab_s.$$

As $A \subset H$ and $b_1, \ldots, b_s$ are pair-wise incongruent modulo H, it follows that the sets $Ab_1, \ldots, Ab_s$ are in distinct cosets modulo H. Multiplying $Ab_1, \ldots, Ab_s$ by g permutes these sets. For each i, $1 \le i \le s$, there is a j, $1 \le j \le s$ such that $Ab_i g = Ab_j$. From this we must get $b_i g = b_j$ otherwise A is periodic with period $b_i g b_j^{-1}$. Hence $g = b_j b_i^{-1}$ and so

$$g \in \bigcap_{b \in B} Bb^{-1}.$$

From this we must get $g = e$ since otherwise by Lemma 1.2.1, B is periodic. This completes the proof.

The next lemma tells us that a nonperiodic factorization from a subgroup can be extended to a nonperiodic factorization of the group. The result holds for each finite abelian group. The 2-component of the group makes the situation unruly. To avoid this here we prove it only for p-groups with $p \ge 3$.

Lemma 2.3.1. *Let H be a subgroup of the finite abelian group G, where $|G : H| = p$ is an odd prime. If $(q_1, \ldots, q_n)$ is not a periodicity forcing factorization type for H, then $(q_1, \ldots, q_n, p)$ and $(q_1, \ldots, q_{i-1}, q_i p, q_{i+1}, \ldots, q_n)$ are not periodicity forcing factorization types for G, for each i, $1 \le i \le n$.*

Proof. Now the factor group G/H is cyclic and so there is an element $b \in G$ such that Hb is a generator element of G/H. Let $B = \{e, b, b^2, \ldots, b^{p-1}\}$ and note that $G = HB$ is a normalized factorization of G. As $(q_1, \ldots, q_n)$ is not a periodicity forcing factorization type for H, there is a normalized factorization $H = A_1 \cdots A_n$ with $|A_i| = q_i$, where none of the factors is periodic. Set $C = \{e, b, b^2, \ldots, b^{p-2}, b^{p-1}d\}$, where $d \in H \setminus \{e\}$ if B is a subgroup and $d = e$ when B is not a subgroup. In other words the subset C is simulated with $b^p = e$, $d \ne e$ or C is cyclic with $b^p \ne e$, $d = e$. Clearly, C is not a subgroup and the factorization $G = HC = A_1 \cdots A_n C$ shows that $(q_1, \ldots, q_n, p)$ is not a periodicity forcing factorization type for G. In the remaining part of the proof we would like to establish that $A_n C$ is not periodic and so the factorization $G = A_1 \cdots A_{n-1}(A_n C)$ gives that $(q_1, \ldots, q_{n-1}, q_n p)$ is not a periodicity forcing factorization type for G. This proves the lemma in the $i = n$ case. By relabelling we can cover all cases. Let us consider

$$A_n C = A_n \cup A_n b \cup \cdots \cup A_n b^{p-2} \cup A_n b^{p-1} d.$$

As $A_n \subset H$, the members of this partition are in distinct cosets of G modulo H. Now Observation 2.3.1 is applicable and gives that $A_n C$ is not periodic.

This completes the proof.

Observation 2.3.2. *Let G be a finite abelian group. Let H be a subgroup of G and let $k_1, \ldots, k_n$ be pair-wise incongruent elements of G modulo H. Let $K_1, \ldots, K_n$ be subgroups of H such that $K_1 = \cdots = K_{n-1} = L$ and $K_n = M$ and $L \cap M = \{e\}$. Then the set $A = k_1 K_1 \cup \cdots \cup k_n K_n$ is not periodic.*

Proof. Assume that A is periodic with period g. As the sets $k_1 K_1, \ldots, k_n K_n$ are in distinct cosets modulo H and as multiplying by g permutes the cosets modulo H, it follows that multiplying by g permutes the sets $k_1 K_1, \ldots, k_n K_n$. Let f be this permutation. Consider the cycle of f that contains $k_n K_n$. If this cycle has length 1, then $k_n K_n g = k_n K_n$ and we get $Mg = M$. Hence $g \in M \subset H$. This implies that f is the identity permutation and $k_1 K_1 g = k_1 K_1$ which gives $Lg = L$. Thus $g \in L \cap M = \{e\}$. We may assume that the length of the cycle containing $k_n K_n$ is not 1. There is an i, $1 \leq i \leq n-1$ such that $k_i K_i g = k_n K_n$. This gives $k_i g L = k_n M$, that is, a coset modulo L is equal to a coset modulo M. Since $L \cap M = \{e\}$, this is impossible. Therefore A is not periodic.
 This completes the proof.

The next lemmas are about factoring groups of order p^3, p^4, p^5.

Lemma 2.3.2. *Let $p \geq 5$ be a prime and G be a noncyclic abelian group of order p^3. There is a factorization $G = A_1 A_2$ such that $|A_1| = p^2$, $|A_2| = p$ and none of the factors is periodic.*

Proof. There are three nonisomorphic abelian groups of order p^3 with invariants

$$(p, p, p), \quad (p^2, p), \quad (p^3).$$

The third one corresponds to a cyclic group. For the noncyclic ones we will construct factorizations without periodic factors. We choose A_2 such that it differs from a subgroup in two elements. Then we try to find A_1 in the form

$$A_1 = k_0 K_0 \cup k_1 K_1 \cup \cdots \cup k_{p-1} K_{p-1},$$

where $k_i \in G$, $K_i \subset G$. If $K_i A_2$ is the same subgroup H of G for each i, $1 \leq i \leq p-1$ and $k_0, k_1, \ldots, k_{p-1}$ is a complete set of representatives modulo H, then $A_1 A_2$ is a factorization of G.

 The periodicity of the factors can be handled in two ways. If the factor has only a prime number of elements and contains e, then checking periodicity is reduced to checking if the factor is a subgroup. In other cases we can use Observation 2.3.2.

 (1) Let G be a group of type (p, p, p) with basis elements x, y, z. We choose

$$A_2 = \{e, yz, (yz)^2, \ldots, (yz)^{p-3}, y^{p-2} z^{p-1}, y^{p-1} z^{p-2}\}$$

and k_i, K_i as defined in Table 1.

i	k_i	K_i
0	e	$\langle z \rangle$
1	x	$\langle y \rangle$
$\vdots$	$\vdots$	$\vdots$
$p-1$	x^{p-1}	$\langle y \rangle$

Table 1

i	k_i	K_i
0	e	$\langle y \rangle$
1	x	$\langle x^p \rangle$
$\vdots$	$\vdots$	$\vdots$
$p-1$	x^{p-1}	$\langle x^p \rangle$

Table 2

A routine computation shows that $K_i A_2 = \langle y, z \rangle$ for each i, $0 \le i \le p-1$. For the sake of a compact notation we will denote the cyclic subset

$$\{e, a, a^2, \ldots, a^{p-1}\}$$

by $[a, p]$ and we will denote the simulated subset

$$\{e, a, a^2, \ldots, a^{p-2}, a^{p-1}b\}$$

by $[\langle a \rangle, b]$.

(2) Let G be of type (p^2, p) with basis elements x, y. Consider the factorization
tion

$$G = [x, p]\langle x^p \rangle \langle y \rangle$$

and use x, x^p, y in a similar way as we used x, y, z in the earlier construction. Set

$$A_2 = \{e, x^p y, (x^p y)^2, \ldots, (x^p y)^{p-3}, (x^p)^{p-2} y^{p-1}, (x^p)^{p-1} y^{p-2}\}$$

and choose k_i, K_i as defined in Table 2. A routine computation shows that $K_i A_2 = \langle x^p, y \rangle$ for each i, $0 \le i \le p-1$.

This completes the proof.

Lemma 2.3.3. *Let p be a prime $p \ge 3$. Let G be a noncyclic abelian group of order p^4. There is a factorization $G = A_1 A_2$ such that $|A_1| = |A_2| = p^2$ and none of the factors is periodic.*

Proof. There are five nonisomorphic abelian groups of order p^4 with invariants

$$(p, p, p, p), \ (p^2, p, p),$$

$$(p^2, p^2), \ (p^3, p), \ (p^4).$$

The last one is cyclic and the others are noncyclic. Let G be a noncyclic group of order p^4. There is a normalized factorization $G = A_1 A_2$ such that $|A_1| = |A_2| = p^2$. We try to find A_1 and A_2 in the next forms.

$$A_1 = k_0 K_0 \cup k_1 K_1 \cup \cdots \cup k_{p-1} K_{p-1},$$
$$A_2 = l_0 L_0 \cup l_1 L_1 \cup \cdots \cup l_{p-1} L_{p-1}.$$

i	K_i	L_i
0	$\langle zu \rangle$	$\langle z^2 u \rangle$
1	$\langle z \rangle$	$\langle u \rangle$
$\vdots$	$\vdots$	$\vdots$
$p-1$	$\langle z \rangle$	$\langle u \rangle$

i	K_i	L_i
0	$\langle yz \rangle$	$\langle y^2 z \rangle$
1	$\langle y \rangle$	$\langle z \rangle$
$\vdots$	$\vdots$	$\vdots$
$p-1$	$\langle y \rangle$	$\langle z \rangle$

Table 3   Table 4

Here k_i, l_j are elements and K_i, L_j are subsets of G. The product $A_1 A_2$ is the union of the sets

$$k_i l_j K_i L_j, \ 0 \le i, j \le p-1.$$

If each $K_i L_j$ is equal to a subgroup H and the elements $k_i l_j$ form a complete set of representatives of G modulo H, then $A_1 A_2$ is a factorization of G.

(1) Let G be of type (p, p, p, p) with basis elements x, y, z, u. Set $k_i = x^i$, $l_i = y^i$, $0 \le i \le p-1$, and K_i, L_i as indicated in Table 3.

One can verify that $K_i L_j = \langle z, u \rangle$ for each i, j, $0 \le i, j \le p-1$.

(2) Let G be of type (p^2, p, p) with basis elements x, y, z. Consider the factorization

$$G = [x, p]\langle x^p \rangle \langle y \rangle \langle z \rangle$$

and use the elements x, x^p, y, z as we used the basis elements x, y, z, u in the previous construction. Set $k_i = x^i$, $l_i = (x^p)^i$, $0 \le i \le p-1$, and K_i, L_i by Table 4. It is a routine computation to verify that $K_i L_j = \langle y, z \rangle$ for each i, j, $0 \le i, j \le p-1$.

(3) Let G be of type (p^2, p^2) with basis elements x, y. Factor G as

$$G = [x, p]\langle x^p \rangle [y, p]\langle y^p \rangle$$

and use the elements x, x^p, y, y^p as we used the basis elements x, y, z, u in construction (1). Set $k_i = x^i$, $l_i = y^i$, $0 \le i \le p-1$, and K_i, L_i as shown in Table 5.

The reader can verify easily that $K_i L_j = \langle x^p, y^p \rangle$ for each $i, j, 0 \le i, j \le p-1$.

(4) Let G be of type (p^3, p) with basis elements x, y. Consider the factorization

$$G = [x, p][x^p, p]\langle x^{p^2} \rangle \langle y \rangle$$

and use the elements x, x^p, x^{p^2}, y as we used the basis elements x, y, z, u in construction (1). Set $k_i = x^i$, $l_i = (x^p)^i$, $0 \le i \le p-1$, and K_i, L_i as given in Table 6. It is easy to check that $K_i L_j = \langle x^{p^2}, y \rangle$ for each $i, j, 0 \le i, j \le p-1$.

This completes the proof.

i	K_i	L_i
0	$\langle x^p y^p \rangle$	$\langle x^p y^{2p} \rangle$
1	$\langle x^p \rangle$	$\langle y^p \rangle$
$\vdots$	$\vdots$	$\vdots$
$p-1$	$\langle x^p \rangle$	$\langle y^p \rangle$

Table 5

i	K_i	L_i
0	$\langle x^{p^2} y \rangle$	$\langle x^{2p^2} y \rangle$
1	$\langle x^{p^2} \rangle$	$\langle y \rangle$
$\vdots$	$\vdots$	$\vdots$
$p-1$	$\langle x^{p^2} \rangle$	$\langle y \rangle$

Table 6

Lemma 2.3.4. *Let G be an abelian group of order p^5 not of type (p^5), (p^4,p), where p is a prime $p \geq 3$. There is a factorization $G = BA_1 A_2$ such that $|B| = p^3$, $|A_1| = |A_2| = p$ and none of the factors is periodic.*

Proof. There are seven nonisomorphic abelian groups of order p^5 with invariants

$$(p^5),\ (p^4,p),\ (p^3,p^2),\ (p^3,p,p),$$

$$(p^2,p^2,p),\ (p^2,p,p,p),\ (p,p,p,p,p).$$

The first one corresponds to a cyclic group. For the last five cases we will exhibit a suitable factorization. There is a common underlying pattern in all constructions. We choose A_1, A_2 to be simulated or cyclic. Then we try to find B in the form

$$B = k_0 K_0 L_0 \cup k_1 K_1 L_1 \cup \cdots \cup k_{p-1} K_{p-1} L_{p-1},$$

where k_i is an element and K_i, L_i are subsets of G. If each $K_i L_i A_1 A_2$ is the same subgroup H of G and the elements k_i form a complete set of representatives of G modulo H, then $G = BA_1 A_2$ is a factorization G. The subsets K_i, L_i will be either subgroups or simulated subsets. When we verify $K_i L_i A_1 A_2 = H$ from time to time we use the following observation. Let K be a subgroup and let $A = \{h_1, \ldots, h_{t-1}, h_t d\}$ be a simulated subset of G, where $H = \{h_1, \ldots, h_t\}$ is the corresponding subgroup and $d \in K$. Then $KA = KH$. Indeed, now $Kd = K$ and so

$$
\begin{aligned}
KA &= K h_1 \cup \cdots \cup K h_{t-1} \cup K h_t d \\
&= K h_1 \cup \cdots \cup K h_t \\
&= KH.
\end{aligned}
$$

(1) Let G have invariants (p,p,p,p,p) and let x, y, z, u, v be a basis for G. Set $k_i = x^i$, $0 \leq i \leq p-1$ and K_i, L_i by Table 7.

Next we can check that $K_i L_i A_1 A_2 = \langle y, z, u, v \rangle$. As $z \neq e$ and $y \neq e$, it follows that A_1 and A_2 are not periodic.

(2) Let G have invariants (p^2, p, p, p) and let x, y, z, u be a basis for G, where $|x| = p^2$, $|y| = |z| = |u| = p$. We factor G into the form

$$G = [x,p]\langle x^p \rangle \langle y \rangle \langle z \rangle \langle u \rangle$$

$$A_1 = [\langle u \rangle, y], \ A_2 = [\langle v \rangle, z],$$

i	K_i	L_i
0	$\langle y \rangle$	$[\langle z \rangle, u]$
1	$[\langle y \rangle, v]$	$\langle z \rangle$
2	$\langle y \rangle$	$\langle z \rangle$
$\vdots$	$\vdots$	$\vdots$
$p-1$	$\langle y \rangle$	$\langle z \rangle$

Table 7

$$A_1 = [\langle z \rangle, x^p], \ A_2 = [\langle u \rangle, y],$$

i	K_i	L_i
0	$\langle x^p \rangle$	$[\langle y \rangle, z]$
1	$[\langle x^p \rangle, u]$	$\langle y \rangle$
2	$\langle x^p \rangle$	$\langle y \rangle$
$\vdots$	$\vdots$	$\vdots$
$p-1$	$\langle x^p \rangle$	$\langle y \rangle$

Table 8

$$A_1 = [x, p], \ A_2 = [y, p],$$

i	K_i	L_i
0	$\langle x^p \rangle$	$[\langle y^p \rangle, x]$
1	$[\langle x^p \rangle, y]$	$\langle y^p \rangle$
2	$\langle x^p \rangle$	$\langle y^p \rangle$
$\vdots$	$\vdots$	$\vdots$
$p-1$	$\langle x^p \rangle$	$\langle y^p \rangle$

Table 9

$$A_1 = [x^p, p], \ A_2 = [\langle y \rangle, x^p]$$

i	K_i	L_i
0	$\langle x^{p^2} \rangle$	$[\langle y \rangle, x^p]$
1	$[\langle x^{p^2} \rangle, z]$	$\langle y \rangle$
2	$\langle x^{p^2} \rangle$	$\langle y \rangle$
$\vdots$	$\vdots$	$\vdots$
$p-1$	$\langle x^{p^2} \rangle$	$\langle y \rangle$

Table 10

$$A_1 = [x^p, p], \ A_2 = [y, p],$$

i	K_i	L_i
0	$\langle x^{p^2} \rangle$	$[\langle y^p \rangle, x^p]$
1	$[\langle x^{p^2} \rangle, y]$	$\langle y^p \rangle$
2	$\langle x^{p^2} \rangle$	$\langle y^p \rangle$
$\vdots$	$\vdots$	$\vdots$
$p-1$	$\langle x^{p^2} \rangle$	$\langle y^p \rangle$

Table 11

and we use x, x^p, y, z, u like we used the basis elements in construction (1). Set $k_i = x^i$, $0 \leq i \leq p - 1$ and Table 8 shows the choices for K_i, L_i. It is a routine exercise to check that $K_i L_i A_1 A_2 = \langle x^p, y, z, u \rangle$.

(3) Let G have invariants (p^2, p^2, p) and let x, y, z be a basis for G, where $|x| = |y| = p^2$ and $|z| = p$. As a starting point we factor G into the form

$$G = [x, p]\langle x^p \rangle [y, p]\langle y^p \rangle \langle z \rangle$$

and we use x, x^p, y, y^p, z in a similar manner as we used the basis elements in construction (1). Set $k_i = z^i$, $0 \leq i \leq p - 1$ and K_i, L_i by Table 9. Then $K_i L_i A_1 A_2 = \langle x, y \rangle$ and B is not periodic.

(4) Let G have invariants (p^3, p, p) and let x, y, z be a basis for G, where $|x| = p^3$ and $|y| = |z| = p$. Let us start with the factorization

$$G = [x, p][x^p, p]\langle x^{p^2} \rangle \langle y \rangle \langle z \rangle$$

and use x, x^p, x^{p^2}, y, z in similar fashion as we used the basis elements in construction (1). Set $k_i = x^i$, $0 \leq i \leq p - 1$ and K_i, L_i as in Table 10. Then $K_i L_i A_1 A_2 = \langle x^p, y, z \rangle$ and B is not periodic.

(5) Let G have invariants (p^3, p^2) and let x, y be a basis for G, where $|x| = p^3$ and $|y| = p^2$. Let us start with the factorization

$$G = [x, p][x^p, p]\langle x^{p^2} \rangle [y, p]\langle y^p \rangle$$

and use x, x^p, x^{p^2}, y, y^p in the same manner as we used the basis elements in construction (1). Set $k_i = x^i$, $0 \leq i \leq p - 1$, Table 11 contains the choices for K_i, L_i. Then $K_i L_i A_1 A_2 = \langle x^p, y \rangle$ and B is not periodic.

This completes the proof.

In constructions (1), (2), (4) the $p \geq 3$ condition makes sure that the sets A_1, A_2 are not subgroups. In constructions (3), (5) the $p \geq 3$ condition is not needed.

We draw some conclusions from these constructions. The first theorem is a straightforward consequence of Lemma 2.3.1 and Lemma 2.3.2. It states that if $p \geq 5$, then the only nontrivial periodicity forcing factorization type for noncyclic p-groups is the one guaranteed by Rédei's theorem. In the $p = 3$ case the factorization type $(3, \ldots, 3)$ is periodicity forcing by Rédei's theorem. The only possibility for another periodicity forcing factorization type is $(3^2, 3, \ldots, 3)$. When G is an elementary 3-group this is in fact a periodicity forcing factorization type as can be seen by Theorem 2.2.1. However for other 3-groups the problem remains open.

Theorem 2.3.1. *Let G be a noncyclic abelian group of order p^m, where $p \geq 5$ is a prime. Let $q_1 \geq \cdots \geq q_n \geq p$ be p-powers such that $q_1 \cdots q_n = p^m$. If $(q_1, \ldots, q_n)$ is not (p^m), $(p, \ldots, p)$, then there is a factorization $G = A_1 \cdots A_n$ with $|A_i| = q_i$ and none of the factors is periodic.*

Proof. As $(q_1, \ldots, q_n) \neq (p^m)$, we get that $n \geq 2$. This and $(q_1, \ldots, q_n) \neq (p, \ldots, p)$ gives that $q_1 \geq p^2$, $q_2 \geq p$ and $m \geq 3$. The group G is not cyclic and $|G| \geq p^3$. So there is a chain of subgroups

$$H_3 \subset H_4 \subset \cdots \subset H_m = G$$

such that $|H_i| = p^i$. By Lemma 2.3.2, (p^2, p) is not a periodicity forcing factorization type for H_3. Thus Lemma 2.3.1 gives that (p^2, p, p), (p^3, p), (p^2, p^2) are not periodicity forcing factorization types for H_4. Continuing in this way finally we get that $(q_1, \ldots, q_n)$ is not a periodicity forcing factorization type for $H_m = G$.

This completes the proof.

Theorem 2.3.2. *Let G be an abelian group of order 3^m not of type (3^m), $(3^{m-1}, 3)$ and let $q_1 \geq \cdots \geq q_n \geq 3$ be 3-powers such that $q_1 \cdots q_n = 3^m$. If $(q_1, \ldots, q_n)$ is not*

$$(3, \ldots, 3),\ (3^2, 3 \ldots, 3),\ (3^{m-1}, 3),\ (3^m),$$

then there is a factorization $G = A_1 \cdots A_n$ with $|A_i| = q_i$ and none of the factors is periodic.

Proof. As $(q_1, \ldots, q_n) \neq (3^m)$, we get that $n \geq 2$. This and $(q_1, \ldots, q_n) \neq (3, \ldots, 3)$ imply that $q_1 \geq 3^2$, $q_2 \geq 3$ and $m \geq 3$. Further $(q_1, \ldots, q_n) \neq (3^{m-1}, 3)$ gives that $m \geq 4$ and either $q_1 \geq 3^2$, $q_2 \geq 3^2$ or $q_1 \geq 3^2$, $q_2 = \cdots = q_n = 3$, $n \geq 3$.

First we deal with the case when $q_1 \geq 3^2$, $q_2 \geq 3^2$ and $m \geq 4$. Now $|G| \geq 3^4$ and G is not of type (3^m), $(3^{m-1}, 3)$. Thus there is a chain of subgroups

$$H_4 \subset H_5 \subset \cdots \subset H_m = G$$

such that $|H_i| = 3^i$. By Lemma 2.3.3, $(3^2, 3^2)$ is not a periodicity forcing factorization type for H_4. Using Lemma 2.3.1 we can conclude that $(q_1, \ldots, q_n)$ is not a periodicity forcing factorization type for $H_m = G$.

Turn to the case when $q_1 \geq 3^2$, $q_2 = \cdots = q_n = 3$, $n \geq 3$. As $(q_1, \ldots, q_n) \neq (3^2, 3, \ldots, 3)$, it follows that $q_1 \geq 3^3$ and $m \geq 5$. Now G is not cyclic and $|G| \geq 3^5$. Thus there is a chain of subgroups

$$H_5 \subset H_6 \subset \cdots \subset H_m = G$$

such that $|H_i| = 3^i$. By Lemma 2.3.4, $(3^3, 3, 3)$ is not a periodicity forcing factorization type for H_5. Using Lemma 2.3.1 we can conclude that $(q_1, \ldots, q_n)$ is not a periodicity forcing factorization type for $H_m = G$.

This completes the proof.

Chapter 3

The machinery

3.1 Vanishing sums of roots of unity

In this section and in the next two we will develop the basic machinery of the factorization theory of finite abelian groups. We start with vanishing sums of roots of unity.

Lemma 3.1.1. *Let ρ be a primitive nth root of unity and let $a(1), \ldots, a(k)$ be integers. If n is a prime power, say $n = p^e$, and*

$$\rho^{a(1)} + \cdots + \rho^{a(k)} = 0, \tag{1}$$

then k must be a multiple of p.

Proof. We may assume that $0 \le a(i) \le p^e - 1$ for each i, $1 \le i \le k$. Construct the polynomial

$$P(x) = x^{a(1)} + \cdots + x^{a(k)}$$

associated with the vanishing sum (1). Keep in mind that $a(1), \ldots, a(k)$ are not necessarily distinct numbers. Clearly $\deg P(x) \le p^e - 1$ and $P(x)$ is not the zero polynomial since $P(1) = k$. Hence $\deg P(x) \ge 0$. (We adopt the convention that the degree of the zero polynomial is $-\infty$.) The (p^e)th cyclotomic polynomial

$$F(x) = 1 + x^{p^{e-1}} + x^{2p^{e-1}} + \cdots + x^{(p-1)p^{e-1}}$$

is irreducible over the field of rational numbers. Therefore as ρ is a common root of $P(x)$ and $F(x)$, it follows that $F(x)$ divides $P(x)$ over the rationals. In other words there is a polynomial $Q(x)$ with rational coefficients such that $P(x) = F(x)Q(x)$. Here

$$\begin{aligned}
0 &\le \deg Q(x) = \deg P(x) - \deg F(x) \\
&\le (p^e - 1) - (p-1)p^{e-1} = p^{e-1} - 1.
\end{aligned}$$

Thus the nonzero terms of $Q(x)$ occur among the nonzero terms of $P(x)$. If the coefficient of x^a is nonzero in $Q(x)$, then

$$x^a, x^{a+p^{e-1}}, x^{a+2p^{e-1}}, \ldots, x^{a+(p-1)p^{e-1}}$$

have equal nonzero coefficients in $P(x)$. From this it follows that the numbers $a(1), \ldots, a(k)$ are clumped into clusters of p elements.

This completes the proof.

From the previous proof we can read off the following fact. As ρ is a primitive (p^e)th root of unity, $\sigma = \rho^{p^{e-1}}$ is a primitive pth root of unity. The sum $1 + \sigma + \sigma^2 + \cdots + \sigma^{p-1}$ vanishes. The sum (1) is in fact of form

$$\alpha(1 + \sigma + \sigma^2 + \cdots + \sigma^{p-1}),$$

where α is a nonempty sum of powers of ρ.

Exercise 3.1.1. Let n be an integer $n \geq 2$. Show that if σ is a primitive nth root of unity, then

$$1 + \sigma + \sigma^2 + \cdots + \sigma^{n-1} = 0.$$

(Hint: Let $\alpha = 1 + \sigma + \sigma^2 + \cdots + \sigma^{n-1}$ and show that $\alpha = \sigma\alpha$.)

Besides vanishing sums of roots of unity it is expedient to consider equal sums of roots of unity.

Lemma 3.1.2. *Let ρ be a primitive nth root of unity and let $a(1), \ldots, a(k), b(1), \ldots, b(l)$ be integers. If n is a prime power, say $n = p^e$, and if*

$$\rho^{a(1)} + \cdots + \rho^{a(k)} = \rho^{b(1)} + \cdots + \rho^{b(l)}, \tag{2}$$

where no $\rho^{a(i)}$ can be cancelled against a $\rho^{b(j)}$, then

$$\rho^{a(1)} + \cdots + \rho^{a(k)} = \rho^{b(1)} + \cdots + \rho^{b(l)} = 0.$$

Proof. We may assume that $0 \leq a(i) \leq p^e - 1$ for each i, $1 \leq i \leq k$ and similarly $0 \leq b(i) \leq p^e - 1$ for each i, $1 \leq i \leq l$. Consider the polynomial

$$P(x) = x^{a(1)} + \cdots + x^{a(k)} - x^{b(1)} - \cdots - x^{b(l)}$$

associated with (2). The argument used in the proof of Lemma 3.1.1 gives that the list of integers $a(1), \ldots, a(k)$ is a union of lists of the forms

$$a, a + p^{e-1}, a + 2p^{e-1}, \ldots, a + (p-1)p^{e-1}.$$

Similarly, the list of integers $b(1), \ldots, b(l)$ is a union of lists of the forms

$$b, b + p^{e-1}, b + 2p^{e-1}, \ldots, b + (p-1)p^{e-1}.$$

This completes the proof.

In the next lemma n is a product of two prime powers.

Lemma 3.1.3. *Let ρ be a primitive nth root of unity and let $a(1), \ldots, a(k)$ be integers. If n is a product of two prime powers, say $n = p^e q^f$, and if*

$$\rho^{a(1)} + \cdots + \rho^{a(k)} = 0, \tag{3}$$

then

$$\rho^{a(1)} + \cdots + \rho^{a(k)} =$$

$$\lambda(1 + \delta + \delta^2 + \cdots + \delta^{p-1}) + \mu(1 + \varepsilon + \varepsilon^2 + \cdots + \varepsilon^{q-1}),$$

where λ, μ are sums of powers of ρ and δ, ε are primitive pth and qth roots of unity respectively.

The $\lambda = 0$ and $\mu = 0$ cases, that is, the case of empty sums is included. In addition k is a linear combination of p and q with nonnegative integer coefficients.

Proof. Define the numbers $b(i)$ and $c(i)$ by

$$a(i) \equiv b(i) \pmod{p^e}, \qquad 0 \le b(i) \le p^e - 1,$$
$$a(i) \equiv c(i) \pmod{q^f}, \qquad 0 \le c(i) \le q^f - 1,$$

for each i, $1 \le i \le k$. Clearly $a(i)$ determines $b(i)$ and $c(i)$ uniquely. Conversely, by the Chinese remainder theorem the pair $(b(i), c(i))$ determines $a(i)$ uniquely. Let $b'(1), \ldots, b'(s)$ be all distinct numbers among $b(1), \ldots, b(k)$. Write ρ in the form $\rho = \sigma\tau$, where σ is a primitive (p^e)th root of unity and τ is a primitive (q^f)th root of unity. Now

$$
\begin{aligned}
0 &= \sum_{i=1}^{k} \rho^{a(i)} \\
&= \sum_{i=1}^{k} \sigma^{b(i)} \tau^{c(i)} \\
&= \sum_{i=1}^{s} \alpha_i \sigma^{b'(i)},
\end{aligned}
$$

that is,

$$0 = \sum_{i=1}^{s} \alpha_i \sigma^{b'(i)}, \tag{4}$$

where α_i is a nonempty sum of powers of τ. Construct the polynomial

$$P(x) = \sum_{i=1}^{s} \alpha_i x^{b'(i)}$$

associated with (4).

   If $\alpha_i = 0$ for some i, $1 \leq i \leq s$, then by Lemma 3.1.1, α_i contains a partial sum

$$\beta(1 + \varepsilon + \varepsilon^2 + \cdots + \varepsilon^{q-1}),$$

where β is a sum of powers of τ and ε is a primitive qth root of unity. Hence (4) contains a nonempty vanishing sum of roots of unity. After canceling this partial sum we get a new vanishing sum of roots of unity containing fewer terms and so this reduces the proof to a simpler case.

   Assume that $\alpha_i \neq 0$ for each i, $1 \leq i \leq s$. Now $P(x)$ is not the zero polynomial. The (p^e)th cyclotomic polynomial is irreducible over the (q^f)th cyclotomic field since p is relatively prime to q. The argument we saw in the proof of Lemma 3.1.1 gives that if the coefficient of $x^{b'}$ in $P(x)$ is nonzero, then the coefficients of

$$x^{b'}, x^{b'+p^{e-1}}, x^{b'+2p^{e-1}}, \ldots, x^{b'+(p-1)p^{e-1}}$$

are nonzero and all are equal. Let these coefficients be $\alpha_0, \alpha_1, \ldots, \alpha_{p-1}$. First assume that each term in $\alpha_i = \alpha_j$ cancels for each i and j. Now (4) contains a partial sum

$$\gamma \sigma^{b'}(1 + \delta + \delta^2 + \cdots + \delta^{p-1}),$$

where γ is a common term of $\alpha_0, \ldots, \alpha_{p-1}$ and δ is a primitive pth root of unity. Canceling this vanishing partial sum reduces the proof to a simpler case. Secondly assume that for some i and j some terms remain after all possible cancellations. Let $\alpha_i' = \alpha_j'$ the resulting equation, where at least one side is not empty. By Lemma 3.1.2, $\alpha_i' = \alpha_j' = 0$ and so as in the case of $\alpha_i = 0$ we can reduce the proof to a simpler case.

   This completes the proof.

   Let ρ be a primitive nth root of unity, where $n = p^e q^f r^g$. Does

$$\rho^{a(1)} + \cdots + \rho^{a(k)} = 0$$

imply that this vanishing sum is of form

$$\lambda(1 + \sigma + \cdots + \sigma^{p-1}) + \mu(1 + \tau + \cdots + \tau^{q-1}) + \nu(1 + \varepsilon + \cdots + \varepsilon^{r-1}),$$

where λ, μ, ν are sums of powers of ρ and $\sigma, \tau, \varepsilon$ are primitive pth, qth, rth roots of unity? (Of course the $\lambda = 0$, $\mu = 0$, $\nu = 0$ cases are included.) The next exercise helps to verify that the answer for this question is "no".

Exercise 3.1.2. Let $\rho = \sigma \tau \varepsilon$, where $\sigma, \tau, \varepsilon$ are primitive 2nd, 3rd, 5th roots of unity. Show that

$$(1 + \varepsilon + \varepsilon^2 + \varepsilon^3 + \varepsilon^4) + \sigma \varepsilon^4(1 + \tau + \tau^2) - \varepsilon^4(1 + \sigma)$$

is a vanishing linear combination of powers of ρ with coefficients 0 and 1. Then show that this sum cannot be represented in the form

$$\lambda(1 + \varepsilon + \varepsilon^2 + \varepsilon^3 + \varepsilon^4) + \mu(1 + \tau + \tau^2) + \nu(1 + \sigma),$$

where λ, μ, ν are sums of powers of ρ. (λ, μ, ν can be empty sums that are by definition zero.)

The next lemma establishes a lower bound for the number of terms of a vanishing sum of roots of unity.

Lemma 3.1.4. *Let ρ be a primitive nth root of unity and let $a(1), \ldots, a(k)$ be integers. If the least prime divisor of n is p and if*

$$\rho^{a(1)} + \cdots + \rho^{a(k)} = 0, \tag{5}$$

then $k \geq p$.

Proof. If n is a prime power, then the result follows from Lemma 3.1.1. So we proceed by an induction on the number of the distinct prime divisors of n. Let $n = p^e r$, where r is relatively prime to p. Hence r has fewer distinct prime divisors than n. Let $\rho = \sigma\tau$, where σ and τ are primitive (p^e)th and rth roots of unity respectively. Define the numbers $b(i)$ and $c(i)$ by

$$a(i) \equiv b(i) \pmod{p^e}, \qquad 0 \leq b(i) \leq p^e - 1,$$
$$a(i) \equiv c(i) \pmod{r}, \qquad 0 \leq c(i) \leq r - 1,$$

for each i, $1 \leq i \leq k$. Let $b'(1), \ldots, b'(s)$ be all the distinct numbers among $b(1), \ldots, b(k)$. Now

$$\begin{aligned}
0 &= \sum_{i=1}^{k} \rho^{a(i)} \\
&= \sum_{i=1}^{k} \sigma^{b(i)} \tau^{c(i)} \\
&= \sum_{i=1}^{s} \alpha_i \sigma^{b'(i)},
\end{aligned}$$

where α_i is a nonempty sum of powers of τ. Construct the corresponding polynomial

$$P(x) = \sum_{i=1}^{s} \alpha_i x^{b'(i)}.$$

If $\alpha_i = 0$ for some i, $1 \leq i \leq s$, then by the inductive assumption α_i consists of at least p terms and so we are done. We assume that $\alpha_i \neq 0$ for each i, $1 \leq i \leq s$. Consequently $P(x)$ is not the zero polynomial. The argument we used in the proof of Lemma 3.1.1 now gives that if the coefficient of $x^{b'}$ is not zero in $P(x)$, then the coefficients of

$$x^{b'}, x^{b' + p^{e-1}}, x^{b' + 2p^{e-1}}, \ldots, x^{b' + (p-1)p^{e-1}}$$

are nonzero and all are equal. We found again p terms in (5).

   This completes the proof.

Exercise 3.1.3. Let ρ be a primitive nth root of unity and let $a(1), \ldots, a(p)$ be integers, where p is the least prime divisor of n. Prove that if

$$\rho^{a(1)} + \cdots + \rho^{a(p)} = 0,$$

and $a(1) = 0$, then $\rho^{a(1)}, \ldots, \rho^{a(p)}$ and

$$1, e^{2\pi i/p}, e^{2\cdot 2\pi i/p}, \ldots, e^{(p-1)2\pi i/p}$$

are rearrangements of each other.

   The next lemma extends the previous result.

Lemma 3.1.5. *Let ρ be a primitive nth root of unity and let $a(1), \ldots, a(k)$, $b(1), \ldots$, $b(l)$ be integers. If p is the least prime divisor of n and there are no terms in the equation*

$$\rho^{a(1)} + \cdots + \rho^{a(k)} = \rho^{b(1)} + \cdots + \rho^{b(l)} \tag{6}$$

that can be cancelled against a term on the other side, then the inequality $\max(k, l) \geq p$ *holds.*

   Putting it differently, if $k, l < p$, then from (6) it follows that $k = l$, that is,

$$\rho^{a(1)} + \cdots + \rho^{a(k)} = \rho^{b(1)} + \cdots + \rho^{b(k)},$$

and there is a permutation π of $1, 2, \ldots, k$ such that $a(i) \equiv b(\pi(i)) \pmod{n}$.

Proof. If n is a prime power, then the result follows from Lemma 3.1.2. So we proceed by an induction on the number of the distinct prime divisors of n. Let $n = p^e r$, where p and r are relatively prime. Let q be the least prime divisor of r. Clearly $p < q$ and r has fewer distinct prime divisors than n. Define the numbers $c(i), d(i)$

$$a(i) \equiv c(i) \pmod{p^e}, \qquad 0 \leq c(i) \leq p^e - 1,$$
$$a(i) \equiv d(i) \pmod{r}, \qquad 0 \leq d(i) \leq r - 1,$$

for each i, $1 \leq i \leq k$. Similarly define $g(j), h(j)$ by

$$b(j) \equiv g(j) \pmod{p^e}, \qquad 0 \leq g(j) \leq p^e - 1,$$
$$b(j) \equiv h(j) \pmod{r}, \qquad 0 \leq h(j) \leq r - 1,$$

for each j, $1 \leq j \leq l$. Let $\rho = \sigma\tau$, where σ and τ are primitive (p^e)th and rth roots of unity. Further let $c'(1), \ldots, c'(s)$ denote all the distinct numbers among $c(1), \ldots, c(k)$ and similarly let $g'(1), \ldots, g'(t)$ be the distinct numbers among $g(1), \ldots, g(l)$. Using this notation (6) can be written in the forms

$$\sum_{i=1}^{k} \sigma^{c(i)} \tau^{d(i)} = \sum_{j=1}^{l} \sigma^{g(j)} \tau^{h(j)},$$

$$\sum_{i=1}^{s} \alpha_i \sigma^{c'(i)} = \sum_{j=1}^{t} \beta_j \sigma^{g'(j)}, \tag{7}$$

where α_i and β_j are nonempty sums of powers of τ. Let $f(1), \ldots, f(w)$ be the elements of

$$\{c'(1), \ldots, c'(s)\} \cap \{g'(1), \ldots, g'(t)\}.$$

We may assume that

$$\{c'(1), \ldots, c'(u)\} = \{c'(1), \ldots, c'(s)\} \setminus \{f(1), \ldots, f(w)\},$$

$$\{g'(1), \ldots, g'(v)\} = \{g'(1), \ldots, g'(t)\} \setminus \{f(1), \ldots, f(w)\}$$

since this is only a matter of indexing the terms. Finally we write (7) in the form

$$0 = \sum_{i=1}^{u} \alpha_i \sigma^{c'(i)} - \sum_{j=1}^{v} \beta_j \sigma^{g'(j)} + \sum_{m=1}^{w} \gamma_m \sigma^{f(m)}, \tag{8}$$

where

$$\gamma_m = \alpha - \beta, \quad \alpha \in \{\alpha_{u+1}, \ldots, \alpha_s\}, \quad \beta \in \{\beta_{v+1}, \ldots, \beta_t\}.$$

Consider the polynomial

$$P(x) = \sum_{i=1}^{u} \alpha_i x^{c'(i)} - \sum_{j=1}^{v} \beta_j x^{g'(j)} + \sum_{m=1}^{w} \gamma_m x^{f(m)}$$

associated with (8).

If $\alpha_i = 0$ for some i, $1 \leq i \leq u$, then by Lemma 3.1.4, α_i contains at least q terms and so $\max(k, l) \geq q > p$. The case when $\beta_j = 0$ for some j, $1 \leq j \leq v$ can be settled in a similar way.

If $\gamma_m = 0$ for some m, $1 \leq m \leq w$, then we have an equation $\alpha = \beta$. We show that in this equation no terms can be cancelled against terms on the other side. To show this assume the contrary, that $\tau^{d(i)}$ is a term of α and $\tau^{h(j)}$ is a term of β and $\tau^{d(i)} = \tau^{h(j)}$. From this it follows that $\sigma^{c(i)} \tau^{d(i)} = \sigma^{g(j)} \tau^{h(j)}$ since $c(i) = g(j) = f(m)$. This gives the contradiction that $\rho^{a(i)}$ can be cancelled against $\rho^{b(j)}$ in (6). From $\alpha = \beta$ by the induction assumption we get $\max(k, l) \geq q > p$.

Thus we may assume in the remaining cases that $P(x)$ is not the zero polynomial. The divisibility by the (p^e)th cyclotomic polynomial in the known way gives that if the coefficient of x^{λ} is nonzero in $P(x)$, then the coefficients of

$$x^{\lambda}, x^{\lambda+p^{e-1}}, x^{\lambda+2p^{e-1}}, \ldots, x^{\lambda+(p-1)p^{e-1}}$$

are all equal. In this way we found at least p terms on one side of (6).

This completes the proof.

Exercise 3.1.4. Let ρ be a primitive nth root of unity and let $a(1), \ldots, a(k)$ and $b(1), \ldots, b(l)$ be integers. Assume that p is the least prime divisor of n. Show that if $k < p$, $l < p$ and

$$\rho^{a(1)} + \cdots + \rho^{a(k)} = \rho^{b(1)} + \cdots + \rho^{b(l)},$$

then $k = l$.

Exercise 3.1.5. Let $n = pq$, where p and q are distinct primes and let ρ be a primitive nth root of unity. Verify that

$$\rho^q + \rho^{2q} + \cdots + \rho^{(p-1)q} = \rho^p + \rho^{2p} + \cdots + \rho^{(q-1)p}.$$

Show that there are no terms on the left side that can be cancelled against a term on the right side. Finally choose p and q that $p + 2 = q$. Now $\max(p - 1, q - 1) = q - 1 = p + 1$ which shows that the inequality $\max(k, l) \geq p$ in Lemma 3.1.5 cannot be replaced by a sharper one in general.

Exercise 3.1.6. Let ρ, σ, and τ be distinct nth roots of unity. Show that from $\rho + \sigma = \tau$ it follows that 6 divides n.

Lemma 3.1.6. *Let ρ be a primitive nth root of unity and let $a(1), \ldots, a(k)$ be integers. If p is the least prime divisor of n and*

$$\rho^{a(1)} + \cdots + \rho^{a(k)} = 0,$$

where $1 \leq k \leq 2p - 1$, then k divides n.

Proof. If n is a power of p, then by Lemma 3.1.1, $p \mid k$. But as $1 \leq k \leq 2p - 1$, we have $k = p$ and so $k \mid n$. We proceed by induction on the number of the distinct prime divisors of n. Let $n = p^e r$, where r is prime to p. Let q be the least prime factor of r. Obviously, $q > p$ and r has fewer distinct prime divisors than n. Write ρ in the form $\rho = \sigma\tau$, where σ and τ are primitive (p^e)th and rth roots of unity. Define the numbers $b(i)$ and $c(i)$ by

$$
\begin{aligned}
a(i) &\equiv b(i) \pmod{p^e}, & 0 \leq b(i) \leq p^e - 1, \\
a(i) &\equiv c(i) \pmod{r}, & 0 \leq c(i) \leq r - 1,
\end{aligned}
$$

for each i, $1 \leq i \leq k$. Denote the distinct numbers occurring among $b(1), \ldots, b(k)$ by $b'(1), \ldots, b'(s)$. Using this notation

$$0 = \sum_{i=1}^{k} \rho^{a(i)} = \sum_{i=1}^{k} \sigma^{b(i)}\tau^{c(i)} = \sum_{i=1}^{s} \alpha_i \sigma^{b'(i)},$$

where α_i is a sum of $t(i)$ powers of τ. Consider the associated polynomial

$$P(x) = \sum_{i=1}^{s} \alpha_i x^{b'(i)}.$$

If $P(x)$ is the zero polynomial, then $\alpha_1 = \cdots = \alpha_s = 0$ and so by Lemma 4, $t(i) \geq q$. Hence from

$$sq \leq t(1) + \cdots + t(s) = m \leq 2p - 1 < 2q - 1$$

it follows that $s = 1$ and then $t(1) = m \leq 2p - 1 < 2q - 1$. From $\alpha_1 = 0$ by the induction assumption it follows that $k \mid r$ and hence $k \mid n$.

Assume that $P(x)$ is not the zero polynomial. The divisibility of $P(x)$ by the (p^e)th cyclotomic polynomial gives that if the coefficient of $x^{b'}$ is nonzero in $P(x)$, then the coefficients of

$$x^{b'}, x^{b'+p^{e-1}}, x^{b'+2p^{e-1}}, \ldots, x^{b'+(p-1)p^{e-1}}$$

are equal and so $p \mid s$. From $p \leq s \leq k \leq 2p - 1$ it follows that $s = p$ and so $\alpha_1 = \cdots = \alpha_p$. Let $t(i)$ be one of the smallest among $t(1), \ldots, t(p)$ and let $t(j)$ be one of the largest among $t(1), \ldots, t(p)$. By $t(1) + \cdots + t(p) = k \leq 2p - 1$, it follows that $t(i) \geq 1$ and $t(j) \leq p < q$. Therefore the terms in the equation $\alpha_i = \alpha_l$ can be cancelled against each other by Lemma 3.1.5 for each l, $1 \leq l \leq p$, $l \neq i$. Thus $t(1) = \cdots = t(p) = 1$, that is $k = p$. Hence $k \mid n$.

This completes the proof.

Exercise 3.1.7. Let ρ be a primitive nth root of unity and let $a(1), \ldots, a(k)$ be integers and let p be the least prime divisor of n. Show that if $a(1) = 0$ and

$$\rho^{a(1)} + \cdots + \rho^{a(k)} = 0,$$

where $1 \leq k \leq 2p - 1$, then $k = p$ and $\rho^{a(1)}, \ldots, \rho^{a(p)}$ is a rearrangement of

$$1, e^{2\pi i/p}, e^{2 \cdot 2\pi i/p}, \ldots, e^{(p-1)2\pi i/p}.$$

In the proof of the next result we will draw on the complex plane instead of relying on cyclotomic polynomials. We will refer to this result as the four roots of unity lemma.

Lemma 3.1.7. *Let ρ_1, ρ_2, ρ_3, ρ_4 be complex numbers such that $|\rho_1| = |\rho_2| = |\rho_3| = |\rho_4| = 1$.*

(1) If $\rho_1 - \rho_2 = \rho_3 - \rho_4$. Then either $\rho_1 = \rho_2$, $\rho_3 = \rho_4$ or $\rho_1 = \rho_3$, $\rho_2 = \rho_4$.

(2) If $\rho_1 + \rho_2 + \rho_3 + \rho_4 = 0$, then there is a permutation $\sigma_1, \sigma_2, \sigma_3, \sigma_4$ of $\rho_1, \rho_2, \rho_3, \rho_4$ for which $\sigma_1 + \sigma_2 = 0$, $\sigma_3 + \sigma_4 = 0$.

(3) If $\rho_1 + \rho_2 = \rho_3 + \rho_4$ and the orders of ρ_1, ρ_2, ρ_3, ρ_4 are odd, then ρ_1, ρ_2 is a permutation of ρ_3, ρ_4.

The reader may note that part (3) is a special case of Lemma 3.1.5.

Proof. (1) If $\rho_1 - \rho_2 = 0$, then $\rho_3 - \rho_4 = 0$ and we get $\rho_1 = \rho_2$, $\rho_3 = \rho_4$. We assume that $\rho_1 - \rho_2 \neq 0$, $\rho_3 - \rho_4 \neq 0$. Using the complex numbers ρ_1, ρ_2, ρ_3, ρ_4 we draw a closed polygon with sides P_1P_2, P_2P_3, P_3P_4, P_4P_1 on the complex plane such that

$$\overrightarrow{P_1P_2} = \rho_1, \ \overrightarrow{P_2P_3} = -\rho_2, \ \overrightarrow{P_3P_4} = -\rho_3, \ \overrightarrow{P_4P_1} = \rho_4.$$

From $\rho_1 - \rho_2 \neq 0$ it follows that $P_1 \neq P_3$. If P_1, P_2, P_3 are not collinear, then the closed polygon forms a parallelogram all of whose sides are of length 1. Thus $\rho_1 = \rho_3$, $\rho_2 = \rho_4$ must hold. If P_1, P_2, P_3 are collinear, then $P_2 = P_4$ and the closed polygon forms a degenerate parallelogram with all sides length 1. Again it follows that $\rho_1 = \rho_3$, $\rho_2 = \rho_4$.

(2) Draw a closed polygon with sides P_1P_2, P_2P_3, P_3P_4, P_4P_1 such that

$$\overrightarrow{P_1P_2} = \rho_1, \ \overrightarrow{P_2P_3} = \rho_2, \ \overrightarrow{P_3P_4} = \rho_3, \ \overrightarrow{P_4P_1} = \rho_4.$$

If $P_1 = P_3$, then $\rho_1 + \rho_2 = 0$ and consequently $\rho_3 + \rho_4 = 0$. The choice $\sigma_1 = \rho_1$, $\sigma_2 = \rho_2$, $\sigma_3 = \rho_3$, $\sigma_4 = \rho_4$ satisfies the conclusion of the lemma. Suppose $P_1 \neq P_3$. If P_1, P_2, P_3 are not collinear, then P_1, P_2, P_3, P_4 are the vertices of a parallelogram whose sides all have length 1. Now $\overrightarrow{P_1P_2} = -\overrightarrow{P_3P_4}$, $\overrightarrow{P_2P_3} = -\overrightarrow{P_4P_1}$, that is, $\rho_1 = -\rho_3$, $\rho_2 = -\rho_4$. We can choose $\sigma_1 = \rho_1$, $\sigma_2 = \rho_3$, $\sigma_3 = \rho_2$, $\sigma_4 = \rho_4$. If P_1, P_2, P_3 are collinear, then the parallelogram is degenerate. But we can argue as earlier.

(3) Draw a closed polygon with sides P_1P_2, P_2P_3, P_3P_4, P_4P_1 such that

$$\overrightarrow{P_1P_2} = \rho_1, \ \overrightarrow{P_2P_3} = \rho_2, \ \overrightarrow{P_3P_4} = -\rho_3, \ \overrightarrow{P_4P_1} = -\rho_4.$$

If $P_1 = P_3$, then $\rho_1 + \rho_2 = 0$. This can happen only if the orders of ρ_1, ρ_2 are even. This is not the case. So $P_1 \neq P_3$. If P_1, P_2, P_3 are not collinear, then the polygon is a parallelogram whose sides all have length 1. It follows that $\rho_1 = \rho_3$, $\rho_2 = \rho_4$. If P_1, P_2, P_3 are collinear, then $P_2 = P_4$ and the parallelogram is degenerate. It follows that $\rho_1 = \rho_4$, $\rho_2 = \rho_3$.

This completes the proof.

3.2 Annihilators

A function χ from the abelian group G to the complex numbers of absolute value 1 is called a *character* of G if $\chi(gh) = \chi(g)\chi(h)$ for each $g, h \in G$. The character for which $\chi(g) = 1$ for each $g \in G$ is called the *principal* character of G. For any subset A of G define $\chi(A)$ to be

$$\chi(A) = \sum_{a \in A} \chi(a).$$

If $A = \emptyset$, then let $\chi(A)$ be 0. Sometimes we have to consider sequences of elements of G. We call a sequence of elements of G a *multiset* since it might contain elements with multiplicity. The definition of $\chi(A)$ extends to multisets.

Exercise 3.2.1. Show that if A_1, A_2 are subsets and χ is a character of G, then $\chi(A_1 A_2) = \chi(A_1)\chi(A_2)$. (Keep in mind that the product $A_1 A_2$ in general may contain elements with multiplicity so it is in general not a set but rather a multiset.) Verify that in general if $A_1, \ldots, A_n$ are subsets and χ is a character of G, then $\chi(A_1 \cdots A_n) = \chi(A_1) \cdots \chi(A_n)$.

The following observation is the key for applying characters to study factorizations.

Observation 3.2.1. *If A and B are subsets of the finite abelian group G and if $\chi(A) = \chi(B)$ for each character of G, then $A = B$.*

Proof. Let $g_1, \ldots, g_n$ be all the elements of G and let $\chi_1, \ldots, \chi_n$ be all the characters of G. The relation $\chi(A) - \chi(B) = 0$ corresponds to the system of equations

$$(x_1 - y_1)\chi_1(g_1) + \quad \cdots \quad + (x_n - y_n)\chi_1(g_n) = 0,$$
$$\vdots \qquad \ddots \qquad \vdots$$
$$(x_1 - y_1)\chi_n(g_1) + \quad \cdots \quad + (x_n - y_n)\chi_n(g_n) = 0.$$

Here

$$x_i = \begin{cases} 1, & \text{if } g_i \in A, \\ 0, & \text{if } g_i \notin A, \end{cases} \qquad y_i = \begin{cases} 1, & \text{if } g_i \in B, \\ 0, & \text{if } g_i \notin B. \end{cases}$$

By the orthogonality relations of the characters the matrix $(\chi_i(g_j))$ is orthogonal and hence the system of equations has only the trivial solution $x_1 - y_1 = \cdots = x_n - y_n = 0$.

This completes the proof.

We can test now by means of characters whether the product of two subsets A and B of a given finite abelian group provides a factorization of G. Namely, the product AB is direct and gives G if and only if $\chi(G) = \chi(A)\chi(B)$ for each character χ of G. Indeed, from $\chi(A)\chi(B) = \chi(AB)$ it follows that the product AB is equal to G. When χ is the principal character of G the relation $\chi(G) = \chi(A)\chi(B)$ simply reduces to $|G| = |A||B|$ and so the directness of the product is ensured.

We can reformulate this observation. If χ is a nonprincipal character of G, then $\chi(G) = 0$ and consequently $0 = \chi(A)\chi(B)$ for each nonprincipal character χ of G. Conversely, if $0 = \chi(A)\chi(B)$ for each nonprincipal character χ of G and in addition $|G| = |A||B|$ holds, then $\chi(G) = \chi(A)\chi(B)$ for each character χ of G. Therefore $G = AB$ is a factorization of G.

Character test for replacement. Let A and A' be subsets of a finite abelian group G such that

    (i) $|A| = |A'|$,

    (ii) $\chi(A) = 0$ implies $\chi(A') = 0$.

Then A can be replaced by A' in each factorization. In other words $G = A'B$ is a factorization whenever $G = AB$ is.

We give a name for the set of characters χ of G for which $\chi(A) = 0$. We call this set of characters the *annihilator* set of A or simply the annihilator of A and we denote it by $\mathrm{Ann}(A)$. We will need the following variant of Observation 3.2.1.

Observation 3.2.2. *Let A be a subset of G. Suppose that $\chi(A) = 0$ for each non-principal character χ of G. Then $A = G$.*

Proof. Let $|A| = r$, $|G| = n$. The multiset nA contains the elements of A with multiplicity n. Similarly, the multiset rG contains the elements of G with multiplicity r. Note that $\chi(nA) = \chi(rG)$ holds for each character χ of G. Indeed, for a nonprincipal character χ of G it holds because $\chi(nA) = n\chi(A) = 0$, $\chi(rG) = r\chi(G) = 0$. For the principal character of G the equation $\chi(nA) = \chi(rG)$ is equivalent to $n|A| = r|G|$. Using the ideas in the proof of Observation 3.2.1 we get that the multisets nA and rG are equal. It follows that $A = G$.

This completes the proof.

Exercise 3.2.2. Let $G = AB$ be a factorization of the finite abelian group G such that $|A| \geq 2$ and $|B| \geq 2$. Prove that $\mathrm{Ann}(A) \neq \emptyset$. (Hint: Assume the contrary and show that $G = A$ which is a contradiction.)

Exercise 3.2.3. Let $G = \langle x \rangle \times \langle y \rangle$, with $|x| = 9$ and $|y| = 5$. Find the annihilator of

$$A_1 = \langle x^3 y \rangle, \qquad\qquad A_2 = \{e, xy, (xy)^2\},$$
$$A_3 = \{e, xy, (xy)^2, (xy)^3\}, \qquad A_4 = \{e, y, y^2, y^3, xy^4\},$$
$$A_5 = \{e, y, y^2, xy^3\}.$$

Exercise 3.2.4. Let H be a subgroup and let χ be a character of the finite abelian group G. Show that $\chi(H) = |H|$ if $\chi(h) = 1$ for each element $h \in H$ and $\chi(H) = 0$ otherwise.

Lemma 3.2.1. *Let A be a normalized subset of the finite abelian group G such that $|A| = p$ is the least prime divisor of $|G|$. Then $\chi(a_{|p'}) = 1$ for each $\chi \in \mathrm{Ann}(A)$ and each $a \in A$.*

Proof. Let χ be a character of G for which $\chi \in \mathrm{Ann}(A)$, that is, for which

$$
\begin{aligned}
0 &= \chi(A) \\
  &= \sum_{a \in A} \chi(a) \\
  &= \sum_{a \in A} \chi(a_{|p})\chi(a_{|p'}).
\end{aligned}
$$

There exists a minimal nonnegative integer n such that each $\chi(a_{|p})$ is a power of a fixed primitive (p^n)th root of unity, say ρ. There is a minimal positive integer r such that each $\chi(a_{|p'})$ is a power of a fixed primitive rth root of unity, say σ.

If $r = 1$, then $\chi(a_{|p'}) = 1$ for each $a \in A$ and so we are done. In the remaining part we assume that $r \neq 1$. Let q be the least prime divisor of r. Clearly $p < q$. If $n = 0$, then from

$$0 = \sum_{a \in A} \chi(a_{|p'})$$

by Lemma 3.1.4, we get the contradiction $p = |A| \geq q$. Thus $n \geq 1$. Let

$$\chi(a_{|p}) = \rho^{t(a)} \quad \text{for each} \quad a \in A,$$

where $0 \leq t(a) \leq p^n - 1$. Denote by $r(1), \ldots, r(s)$ all the distinct numbers among $t(a)$, $a \in A$. Now

$$0 = \sum_{i=1}^{s} \alpha_i \rho^{r(i)},$$

where α_i is a nonempty sum of powers of σ. Consider the polynomial

$$A(x) = \sum_{i=1}^{s} \alpha_i x^{r(i)}.$$

If $\alpha_i = 0$ for some i, $1 \leq i \leq s$, then by Lemma 3.1.4 we get that α_i consists of at least q terms. This is a contradiction since α_i can contain at most p terms. Thus we may assume that $\alpha_i \neq 0$ for each i, $1 \leq i \leq s$. Therefore $A(x)$ is not the zero polynomial. The (p^n)th cyclotomic polynomial divides $A(x)$. From this in the known way it follows that as the coefficient of 1 is not zero in $A(x)$, the coefficients of

$$1, x^{p^{n-1}}, x^{2p^{n-1}}, \ldots, x^{(p-1)p^{n-1}}$$

are nonzero and equal. Consequently each coefficient is a single root of unity and hence they are only a rearrangement of $\chi(a_{|p'})$, $a \in A$. As these coefficients are equal and $\chi(e)$ is a term in $\chi(A)$, we get the sought result.

This completes the proof.

We introduce the following notation. Choose an $a \in A \setminus \{e\}$ and set $A_a = \{e, a, a^2, \ldots, a^{p-1}\}$. Let $A_p = \{a_{|p} : a \in A\}$.

Corollary 3.2.1. *Let A be a normalized subset of the finite abelian group G such that $|A| = p$ is the least prime divisor of $|G|$. Then $\mathrm{Ann}(A) \subset \mathrm{Ann}(A_p)$ and $\mathrm{Ann}(A) \subset \mathrm{Ann}(A_a)$.*

Proof. Choose a $\chi \in \mathrm{Ann}(A)$. Now

$$
\begin{aligned}
0 &= \chi(A) = \sum_{a \in A} \chi(a) = \sum_{a \in A} \chi(a_{|p})\chi(a_{|p'}) = \sum_{a \in A} \chi(a_{|p}) \\
&= \chi(A_p).
\end{aligned}
$$

This proves the first statement.

To prove the second assertion choose a $\chi \in \mathrm{Ann}(A_p)$. Now

$$0 = \chi(A_p) = \sum_{a \in A} \chi(a_{|p})$$

implies that $\chi(a_{|p})$ is a primitive pth root of unity for each $a \in A \setminus \{e\}$. Thus

$$\begin{aligned}
\chi(A_a) &= \sum_{i=0}^{p-1} \chi(a^i) \\
&= \sum_{i=0}^{p-1} \chi(a_{|p}^i)\chi(a_{|p'}^i) \\
&= \sum_{i=0}^{p-1} [\chi(a_{|p})]^i \\
&= 0.
\end{aligned}$$

This completes the proof.

Exercise 3.2.5. Let A be a normalized subset of the finite abelian group G and assume that $|A| < 2p - 1$, where p is the least prime divisor of $|G|$. Prove that $\chi(a_{|p'}) = 1$ for each $\chi \in \mathrm{Ann}(A)$ and each $a \in A$. (Hint: Mimic the proof of the previous lemma but apply Lemma 3.1.6 in place of Lemma 3.1.4.)

Lemma 3.2.2. *Let A be a normalized subset of the finite abelian group G. Assume that $|A| = p$ and A contains only (p, q)-elements, where p and q are distinct primes. Then $\chi(a_{|q}) = 1$ for each $\chi \in \mathrm{Ann}(A)$ and each $a \in A$.*

Proof. Let χ be a character of G for which $\chi \in \mathrm{Ann}(A)$, that is, for which

$$\begin{aligned}
0 &= \chi(A) \\
&= \sum_{a \in A} \chi(a) \\
&= \sum_{a \in A} \chi(a_{|p})\chi(a_{|q}).
\end{aligned}$$

There exists a minimal integer n such that each $\chi(a_{|p})$ is a power of a fixed primitive (p^n)th root of unity, say ρ. Similarly there exists a minimal integer m such that each $\chi(a_{|q})$ is a power of a fixed primitive (q^m)th root of unity, say σ.

If $m = 0$, then $\chi(a_{|q}) = 1$ for each $a \in A$ and so we are done. Hence we assume that $m \geq 1$. If $n = 0$, then from

$$0 = \sum_{a \in A} \chi(a_{|q})$$

by Lemma 3.1.1, we get the contradiction that q divides $p = |A|$. Thus $n \geq 1$. Let

$$\chi(a_{|p}) = \rho^{t(a)} \quad \text{for each} \quad a \in A,$$

where $0 \le t(a) \le p^n - 1$. Let us denote all the distinct numbers occurring among the exponents $t(a)$, $a \in A$, by $r(1), \ldots, r(s)$.

$$0 = \sum_{i=1}^{s} \alpha_i \rho^{r(i)},$$

where α_i is a nonempty sum of powers of σ. Consider the polynomial

$$A(x) = \sum_{i=1}^{s} \alpha_i x^{r(i)}.$$

If $\alpha_1 = \cdots = \alpha_s = 0$, then by Lemma 3.1.2, each of the number of terms in α_i is a multiple of q for each i, $1 \le i \le s$. This gives the contradiction that q divides $p = |A|$. Therefore $A(x)$ is not the zero polynomial. The (p^n)th cyclotomic polynomial divides $A(x)$. Now in the known way we can see that the coefficients of

$$1, x^{p^{n-1}}, x^{2p^{n-1}}, \ldots, x^{(p-1)p^{n-1}}$$

are nonzero and equal. Thus each coefficient must be a single power of σ. The coefficients form a rearrangement of $\chi(a_{|q})$, $a \in A$. Since they are equal and since $\chi(e)$ is a term of $\chi(A)$, we get the sought result.

This completes the proof.

Lemma 3.2.2 has a corollary in a similar manner as Lemma 3.2.1 has Corollary 3.2.1. We state this consequence of Lemma 3.2.2 without proof.

Corollary 3.2.2. *Let A be a normalized subset of the finite abelian group G. Assume that $|A| = p$ and A contains only (p, q)-elements, where p and q are distinct primes. Then $\mathrm{Ann}(A) \subset \mathrm{Ann}(A_p)$ and $\mathrm{Ann}(A) \subset \mathrm{Ann}(A_a)$.*

The next lemma is a common generalization of the previous two.

Lemma 3.2.3. *Let A be a normalized subset of the finite abelian group G such that $|A| = p$ is a prime. Suppose that $a_{|q} = e$ for each, except at most one, prime divisor q of $|G|$ for which $q < p$. Then $\chi(a_{|p'}) = 1$ for each $\chi \in \mathrm{Ann}(A)$ and each $a \in A$.*

Proof. Assume that there is a prime divisor q of $|G|$ such that $q < p$ and $a_{|q} \ne e$ for some $a \in A$. First we show that $\chi(a_{|q}) = 1$ for each $\chi \in \mathrm{Ann}(A)$ and $a \in A$. To do so let χ be a character of G for which $\chi \in \mathrm{Ann}(A)$, that is, for which

$$\begin{aligned} 0 &= \chi(A) \\ &= \sum_{a \in A} \chi(a) \\ &= \sum_{a \in A} \chi(a_{|q}) \chi(a_{|q'}). \end{aligned}$$

There is a minimal nonnegative integer m such that all $\chi(a_{|q})$ is a power of a fixed primitive (q^m)th root of unity, say ρ. Similarly there is a minimal positive integer r such that all $\chi(a_{|q'})$ is a power of a fixed primitive rth root of unity, say σ. Let t be the least prime divisor of r. Obviously, $t \geq p > q$. Let

$$\chi(a_{|q}) = \rho^{t(a)} \quad \text{for each} \quad a \in A.$$

Let $r(1), \ldots, r(s)$ be all the distinct numbers among the exponents $t(a)$, $a \in A$. Now

$$0 = \chi(A) = \sum_{i=1}^{s} \alpha_i \rho^{r(i)},$$

where α_i is a nonempty sum of powers of σ. Consider the associated polynomial

$$A(x) = \sum_{i=1}^{s} \alpha_i x^{r(i)}.$$

If $\alpha_i = 0$ for some i, $1 \leq i \leq s$, then from this by Lemma 3.1.4 it follows that α_i has at least t terms. As $t \geq p$, we have that α_i consists of precisely p terms and hence $s = 1$ and each $t(a) = 0$. Therefore $\chi(a_{|q}) = 1$ for each $a \in A$. This is the sought result. Thus in the remaining part we may assume that $\alpha_i \neq 0$ for each i, $1 \leq i \leq s$. Consequently, $A(x)$ is not the zero polynomial. The (q^m)th cyclotomic polynomial divides $A(x)$. From this in the known way we can deduce that if the coefficient of x^r in $A(x)$ is nonzero, then the coefficients of

$$x^r, x^{r+q^{m-1}}, x^{r+2q^{m-1}}, \ldots, x^{r+(q-1)q^{m-1}}$$

are equal and nonzero. Clearly each coefficient must have fewer than p terms as all coefficients in $A(x)$ together have at most p terms. Let α and α^* be two such coefficients. Cancel the terms against each other on the two sides of the equation $\alpha = \alpha^*$. By Lemma 3.1.5, it follows that each term is cancelable, that is, α and α^* have the same number of terms. This gives the contradiction that q divides p

So far we have proved that $\chi(a_{|q}) = 1$ for each $a \in A$. Now we turn to prove that $\chi(a_{|p'}) = 1$ for each $a \in A$. To do so let χ be a character of G with $\chi \in \mathrm{Ann}(A)$, that is, for which

$$\begin{aligned}
0 \;&=\; \chi(A) \\
&=\; \sum_{a \in A} \chi(a) \\
&=\; \sum_{a \in A} \chi(a_{|p}) \chi(a_{|p'}).
\end{aligned}$$

There is a minimal nonnegative integer n such that each $\chi(a_{|p})$ is a power of a fixed primitive (p^n)th root of unity, say ρ. Similarly there is a minimal nonnegative integer r with the property that each $\chi(a_{|p'})$ is a power of a fixed primitive rth root of unity, say σ. Let t be the least prime divisor of r. Obviously $t > p$. Let

$$\chi(a_{|p}) = \rho^{t(a)} \quad \text{for each} \quad a \in A,$$

where $0 \le t(a) \le p^n - 1$. Further denote $r(1), \ldots, r(s)$ all the distinct numbers among the exponents $t(a)$, $a \in A$. Now

$$0 = \chi(A) = \sum_{i=1}^{s} \alpha_i \rho^{r(i)},$$

where α_i is a nonempty sum of powers of σ. Consider the associated polynomial

$$A(x) = \sum_{i=1}^{s} \alpha_i x^{r(i)}.$$

If $\alpha_i = 0$ for some i, $1 \le i \le s$, then by Lemma 3.1.5, it follows that α_i consists of at least t terms. This is a contradiction since α_i can contain at most p terms and $p < t$. Thus $\alpha_i \neq 0$ for each i, $1 \le i \le s$ and hence $A(x)$ is not the zero polynomial. The (p^n)th cyclotomic polynomial divides $A(x)$ so by the known argument we get that, if the coefficient of x^r in $A(x)$ is nonzero, then the coefficients of

$$x^r, x^{r+p^{n-1}}, x^{r+2p^{n-1}}, \ldots, x^{r+(p-1)p^{n-1}}$$

are equal and nonzero. Hence the number of the nonzero terms in $A(x)$ is a multiple of p. It follows that $s = p$ and so each α_i consists of one term and all are equal. In particular all are equal to $\chi(e)$.

This completes the proof.

Let A be a normalized subset of the finite abelian group G. If $|A| = p$ is the least prime divisor of $|G|$, then there is no such prime divisor q of $|G|$ for which $q < p$ and $a_{|q} \neq e$ for some $a \in A$ and so Lemma 3.2.3 implies Lemma 3.2.1. If $|A| = p$ is the second smallest prime divisor of $|G|$, then there is at most one prime divisor q of $|G|$ with $q < p$ and $a_{|q} \neq e$ for some $a \in A$ and so Lemma 3.2.3 is applicable. Thus Lemma 3.2.3 generalizes Lemma 3.2.1. If $|A| = p$ is a prime and A contains only (p, q)-elements, then there is at most one prime divisor q of $|G|$ with the property that $q < p$ and $a_{|q} \neq e$ for some $a \in A$. Therefore Lemma 3.2.3 implies Lemma 3.2.2 as well.

Lemma 3.2.4. *Let A be a normalized subset of the finite abelian group G. Let p be a prime divisor of $|G|$ such that $|A| = k < p$. Then* $\mathrm{Ann}(A) \subset \mathrm{Ann}(A')$, *where* $A' = \{a_{|p'} : a \in A\}$.

Proof. Let χ be a character of G such that $\chi \in \mathrm{Ann}(A)$. Then

$$0 = \chi(A) = \sum_{a \in A} \chi(a) = \sum_{a \in A} \chi(a_{|p})\chi(a_{|p'}).$$

There is a minimal nonnegative integer n with the property that each $\chi(a_{|p})$ is a power of a fixed primitive (p^n)th root of unity, say ρ. Similarly there is a minimal positive integer r such that each $\chi(a_{|p'})$ is a power of a fixed primitive rth root of unity, say σ. If $n = 0$, then

$$0 = \sum_{a \in A} \chi(a_{|p'}) = \chi(A')$$

is the sought result. Thus we may assume that $n \geq 1$. Let

$$\chi(a_{|p}) = \rho^{t(a)} \quad \text{for each} \quad a \in A,$$

where $0 \leq t(a) \leq p^n - 1$. Let $r(1), \ldots, r(s)$ be all the distinct numbers among the exponents $t(a)$, $a \in A$. Now

$$0 = \chi(A) = \sum_{i=1}^{s} \alpha_i \rho^{r(i)},$$

where α_i is a nonempty sum of powers of σ. Consider the associated polynomial

$$A(x) = \sum_{i=1}^{s} \alpha_i x^{r(i)}.$$

Suppose that $A(x)$ is not the zero polynomial. The (p^n)th cyclotomic polynomial divides $A(x)$. The known argument gives that the number of the nonzero terms in $A(x)$ is a multiple of p. This leads to the contradiction $k \geq s \geq p$. Thus $A(x)$ is the zero polynomial and so

$$0 = A(1) = \sum_{a \in A} \chi(a_{|p'}) = \chi(A').$$

This completes the proof.

Lemma 3.2.5. *If A is a subset of the finite abelian group G and $A^k = \{a^k : a \in A\}$, then $\mathrm{Ann}(A) \subset \mathrm{Ann}(A^k)$ whenever k is relatively prime to $|G|$.*

Proof. Let χ be a character of G such that $\chi \in \mathrm{Ann}(A)$. Now

$$0 = \chi(A) = \sum_{a \in A} \chi(a).$$

There is a minimal positive integer n with the property that each $\chi(a)$ is a power of a fixed primitive nth root of unity, say ρ. In the case $n = 1$ we get the contradiction $0 = \chi(A) = |A|$. Thus $n \geq 2$. Let

$$\chi(a) = \rho^{t(a)} \quad \text{for each} \quad a \in A.$$

This means that

$$0 = \chi(A) = \sum_{a \in A} \rho^{t(a)}.$$

Consider the associated polynomial

$$A(x) = \sum_{a \in A} x^{t(a)}.$$

Note that ρ is a common root of $A(x)$ and the nth cyclotomic polynomial $F(x)$ which is irreducible over the field of rational numbers. Consequently $F(x)$ divides $A(x)$ over the rationals. As k is relatively prime to $|G|$ and as n is a divisor of $|G|$, it follows that ρ^k is also a primitive nth root of unity and hence ρ^k is also a root of $F(x)$. Thus

$$0 = \sum_{a \in A} \rho^{kt(a)} = \chi(A^k)$$

is the sought result.

This completes the proof.

The periodicity of a subset can be tested by means of characters.

Lemma 3.2.6. *Let A be a subset and let H be a subgroup of a finite abelian group G. If there is a subset B of G such that the product HB is direct and is equal to A, then $\mathrm{Ann}(H) \subset \mathrm{Ann}(A)$. Conversely if $\mathrm{Ann}(H) \subset \mathrm{Ann}(A)$, then there is a subset B of G such that the product BH is direct and is equal to A.*

Proof. Assume that there is a subset B of G such that the product HB is direct and is equal to A. From the equation $A = HB$ we get that $\chi(A) = \chi(h)\chi(B)$ for each character χ of G. Clearly $\chi(H) = 0$ implies $\chi(A) = 0$, that is, $\mathrm{Ann}(H) \subset \mathrm{Ann}(A)$.

Assume next that $\mathrm{Ann}(H) \subset \mathrm{Ann}(A)$. First show that $hA = A$ for each $h \in H$. It is enough to verify that $\chi(hA) = \chi(A)$ for each character χ of G. This is immediate when $\chi(h) = 1$. When $\chi(h) \neq 1$, then $\chi(H) = 0$ and so $\chi(A) = 0$. Therefore $\chi(hA) = \chi(h)\chi(A) = \chi(A)$.

Elements of $H \setminus \{e\}$ are periods of A. The subset A is a union of cosets modulo H. Hence there is a subset B of G for which the product BH is direct and gives A.

This completes the proof.

Let G be a finite abelian group. Assume that the subset A of G is expressible in the form $A = BH \cup CK$ such that H, K are subgroups, B, C are subsets of G and in addition the products are direct and the union is disjoint. Then $\chi(A) = \chi(B)\chi(H) + \chi(C)\chi(K)$ for each character χ of G. Hence $\mathrm{Ann}(H) \cap \mathrm{Ann}(K) \subset \mathrm{Ann}(A)$. The converse of this is also true and this is the content of the next lemma. We will refer to this result as the partition lemma.

Lemma 3.2.7. *Let H, K be subgroups and let A be a subset of the finite abelian group G satisfying*

$$\operatorname{Ann}(H) \cap \operatorname{Ann}(K) \subset \operatorname{Ann}(A).$$

Then there are subsets B and C of G such that

$$A = BH \cup CK,$$

where the product are direct and the union is disjoint.

Proof. The subset $A' = a^{-1}A$ contains the identity element e for each $a \in A$. First we prove that A' contains H or K. To prove that, assume the contrary that there exist $h \in H$ and $k \in K$ with $h \notin A'$ and $k \notin A'$. Note that

$$A' \cup A'h^{-1}k^{-1} = A'h^{-1} \cup A'k^{-1} \tag{1}$$

is an equation of multisets as

$$\chi(A') + \chi(A'h^{-1}k^{-1}) = \chi(A'h^{-1}) + \chi(A'k^{-1}) \tag{2}$$

holds for each character χ of G. This is clear when $\chi(h^{-1}) = 1$ or $\chi(k^{-1}) = 1$. If $\chi(h^{-1}) \neq 1$ and $\chi(k^{-1}) \neq 1$, then $\chi(H) = 0$ and $\chi(K) = 0$. Hence $\chi(A) = \chi(A') = 0$ and thus (2) holds.

As $e \in A'$, the left side of (1) contains e. Since $h \notin A'$ and $k \notin A'$, the right side cannot contain e. Thus A contains a coset modulo H or a coset modulo K. For the concreteness assume that $aH \subset A$ for some $a \in A$. Repeat the whole argument with $A \setminus aH$ in place of A. Continuing in this way finally we have that A is a disjoint union of cosets modulo H and cosets modulo K.

This completes the proof.

The next exercise illustrates that the representations of a subset A in the form described in Lemma 3.2.7 can be essentially very different.

Exercise 3.2.6. Let $G = \langle x \rangle \times \langle y \rangle$, where $|x| = p$ and $|y| = q$ are primes. Define $H = \langle x \rangle$, $K = \langle y \rangle$, $A = G$. Verify that $A = BH \cup CK$ with

$$B = \emptyset, \ \ C = \{e, xy, (xy)^2, \ldots, (xy)^{p-1}\}.$$

Similarly verify that $A = B'H \cup C'K$ with

$$B' = \{e, xy, (xy)^2, \ldots, (xy)^{q-1}\}, \ \ C' = \emptyset.$$

(Hint: $G = CK$ follows as $C \equiv H \pmod{K}$ and similarly $G = B'H$ follows as $B' \equiv K \pmod{H}$.)

One might be tempted to think that the partition lemma holds in the next more general form. Let A be a subset and let $H_1, \ldots, H_n$ be subgroups of the finite abelian group G such that

$$\operatorname{Ann}(H_1) \cap \cdots \cap \operatorname{Ann}(H_n) \subset \operatorname{Ann}(A).$$

Then A can be represented in the form

$$A = B_1 H_1 \cup \cdots \cup B_n H_n$$

with suitable subsets $B_1, \ldots, B_n$ of G, where the products are direct and the unions are disjoint.

The following exercises shatter this belief. (The counterexamples for Keller's conjecture discribed in Section 1.5 can also be used for this purpose.)

Exercise 3.2.7. Let $G = \langle x \rangle \times \langle y \rangle$, $|x| = |y| = 2$,

$$H_1 = \langle x \rangle, \ H_2 = \langle y \rangle, \ H_3 = \langle xy \rangle, \ A = \{e\}.$$

Check that

$$\mathrm{Ann}(H_1) \cap \mathrm{Ann}(H_2) \cap \mathrm{Ann}(H_3) \subset \mathrm{Ann}(A).$$

On the other hand A cannot be a union of cosets modulo H_1, H_2, H_3.

The next exercise will show that there are counterexamples even if G is cyclic.

Exercise 3.2.8. Let $G = \langle x \rangle \times \langle y \rangle \times \langle z \rangle$, $|x| = 2$, $|y| = 3$, $|z| = 5$ and

$$H_1 = \langle z \rangle, \quad H_2 = \langle y \rangle, \quad H_3 = \langle x \rangle$$

and

$$A = \left(H_1 \cup xz^4 H_2 \right) \setminus z^4 H_3.$$

Confirm that

$$\mathrm{Ann}(H_1) \cap \mathrm{Ann}(H_2) \cap \mathrm{Ann}(H_3) \subset \mathrm{Ann}(A)$$

and that A cannot contain any coset modulo any subgroup of G.

Exercise 3.2.9. Let G be a group of type $(2, 2, 2)$ with basis elements x, y, z. Set

$$H_1 = \langle x \rangle, \quad H_2 = \langle y \rangle, \quad H_3 = \langle z \rangle,$$

$$A = \{e, xyz\}.$$

Show that

$$\mathrm{Ann}(H_1) \cap \mathrm{Ann}(H_2) \cap \mathrm{Ann}(H_3) \subset \mathrm{Ann}(A).$$

On the other hand A cannot be a union of cosets modulo H_1, H_2, H_3.

Lemma 3.2.7 can be generalized replacing the subset A of G by a nonnegative element of the group ring $Z(G)$. We say that an element A of $Z(G)$ is nonnegative if each coefficient of the elements of G in the representation of A is nonnegative. The condition of nonnegativity of A cannot be removed. This is the content of the next exercises. If A is a subset of G, then $\overline{A}$ stands for the element

$$\sum_{a \in A} a$$

of the group ring $Z(G)$.

Exercise 3.2.10. Let H, K be subgroups of G and let A be a nonnegative element of $Z(G)$. If $\mathrm{Ann}(H) \cap \mathrm{Ann}(K) \subset \mathrm{Ann}(A)$, then there are nonnegative elements B and C of $Z(G)$ such that $A = B\overline{H} + C\overline{K}$. Prove this statement.

Exercise 3.2.11. Let $G = \langle x \rangle \times \langle y \rangle$, with $|x| = |y| = 2$. Verify that

$$0 = (e - xy)(e - x)(e - y).$$

Thus with the choices $A = e - xy$, $H = \langle x \rangle$, $K = \langle y \rangle$,

$$\mathrm{Ann}(H) \cap \mathrm{Ann}(K) \subset \mathrm{Ann}(A)$$

holds and A is not in the form $A = B\overline{H} + C\overline{K}$ where B and C are nonnegative elements of $Z(G)$.

However, if B and C may contain negative coefficients, then the representation of A in the above form is possible. Note that $e - xy = e + x - x(e + y)$. So $B = e$, $C = -x$ do the job.

3.3   Zero divisors

Let $g_1, \ldots, g_n$ be all the elements of the finite abelian group G. Let $x_1, \ldots, x_n$ be integers and consider the formal sum $x_1 g_1 + \cdots + x_n g_n$. Addition and multiplication is defined among formal sums. The set of all formal sums with these operations furnishes a ring and we denote this ring by $Z(G)$. If χ is a character of G, then by

$$\chi(x_1 g_1 + \cdots + x_n g_n) = x_1 \chi(g_1) + \cdots + x_n \chi(g_n),$$

χ can be extended to be a character of $Z(G)$. The subset A of G naturally corresponds to the element

$$\overline{A} = \sum_{a \in A} a$$

of $Z(G)$. The fact that the product $A_1 \cdots A_n$ is a factorization of G can be expressed such that the product $\overline{A}_1 \cdots \overline{A}_n$ is equal to $\overline{G}$. In fact the group ring is a very adequate tool to deal with factorizations and multiple factorizations.

To formulate our results we need some notation. If A is an element of $Z(G)$ and

$$A = x_1 g_1 + \cdots + x_n g_n,$$

then all the elements g_i whose coefficients x_i are nonzero, that is, the support of A is denoted by $\{A\}$. The generatum (or span) of $\{A\}$ is denoted by $\langle A \rangle$. In this chapter $|A|$ denotes

$$|x_1| + \cdots + |x_n|.$$

With a little abuse of the notation we also use $\langle A \rangle$ to denote the subgroup generated (or spanned) by the subset A of G and $\langle A_1, \ldots, A_n \rangle$ denotes the subgroup generated by the supports of $A_i \in Z(G)$ or subsets $A_i \in G$. We use the notation $\nu(n)$ to denote the number of not necessarily distinct prime factors of n, that is $\nu(n) = \alpha(1) + \cdots + \alpha(k)$ if $n = p_1^{\alpha(1)} \cdots p_k^{\alpha(k)}$, where $p_1, \ldots, p_k$ are distinct primes. The number of the not necessarily distinct prime factors of $|\langle A_1, \ldots, A_n \rangle|$ is denoted by $\gamma(A_1, \ldots, A_n)$.

In the first zero divisor lemma the following type of elements A_i of $Z(G)$ occur,

$$A_i = e - a_i. \tag{1}$$

We assign the number $m_i = \nu(|A_i|)$ to A_i. In this case $m_i = 1$ as $|A_i| = 2$ is a prime. Let $C_1, \ldots, C_n$ be elements of $Z(G)$. We say that the product $C_1 \cdots C_n$ is primitive if $C_1 \cdots C_n = 0$ and no factor C_i can be cancelled without destroying the equation. In other words no partial product of $C_1 \cdots C_n$ is equal to zero.

Lemma 3.3.1. *Let $B, A_1, \ldots, A_n \in Z(G)$, where A_i's are of type (1). Assume that the product*

$$BA_1 \cdots A_n = 0 \tag{2}$$

is primitive. Then

$$\gamma(B, A_1, \ldots, A_n) - \gamma(B) < m_1 + \cdots + m_n. \tag{3}$$

Proof. We proceed by induction on n. Let $n = 1$. As the product $BA_1 = 0$ is primitive it follows that $B \neq 0$ and $A \neq 0$. In other words A and B are zero divisors in $Z(G)$.

The equation $BA_1 = 0$ is equivalent to $B = Ba_1$. The support of B is not empty so there are elements b and c in the support of B such that $c = a_1 b$. Hence $a_1 = cb^{-1}$. Thus $a_1 \in \langle B \rangle$. This means that $\gamma(B, A_1) = \gamma(B)$, that is $\gamma(B, A_1) - \gamma(B) = 0$. Now $m_1 = 1$ and we get $\gamma(B, A_1) - \gamma(B) < m_1$, as required.

Let $n \geq 2$. By the inductive assumption on n from

$$(BA_1 \cdots A_s)A_{s+1} \cdots A_n = 0$$

we have that

$$\gamma(BA_1 \cdots A_s, A_{s+1}, \ldots, A_n) - \gamma(BA_1 \cdots A_s) <$$

$$m_{s+1} + \cdots + m_n \tag{4}$$

for each s, $1 \leq s \leq n - 1$.

If $H, K, L \subset G$ and $\langle H \rangle \subset \langle K \rangle$, then the index $|\langle K, L \rangle : \langle H, L \rangle|$ divides the index $|\langle K \rangle : \langle H \rangle|$ and so $\gamma(K, L) - \gamma(H, L) \leq \gamma(K) - \gamma(H)$. Applying this observation in the special case

$$H = \{BA_1 \cdots A_s\},$$

$$K = \{B\} \cup \{A_1\} \cup \cdots \cup \{A_s\},$$
$$L = \{A_{s+1}\} \cup \cdots \cup \{A_n\}$$

we have that

$$\gamma(B, A_1, \ldots, A_n) - \gamma(BA_1 \cdots A_s, A_{s+1}, \ldots, A_n) \leq$$
$$\gamma(B, A_1, \ldots, A_s) - \gamma(BA_1 \cdots A_s).$$

Adding this to (4) we obtain

$$\gamma(B, A_1, \ldots, A_n) - \gamma(B, A_1, \ldots, A_s) < m_{s+1} + \cdots + m_n. \tag{5}$$

Of course this holds not only for the subset $\{1, 2, \ldots, s\}$ of the indices but for each proper subset of $\{1, 2, \ldots, n\}$. For example

$$\gamma(B, A_1, \ldots, A_n) - \gamma(B, A_n) < m_1 + \cdots + m_{n-1}. \tag{6}$$

Consider again the product $BA_1 \cdots A_n = 0$. If $\gamma(A_i) = m_i$ for some i, $1 \leq i \leq n$, say $\gamma(A_1) = m_1$, then $\gamma(B, A_1) - \gamma(B) \leq m_1$. Adding this to the $s = 1$ special case of (5) we have the desired result. We may assume that $\gamma(A_i) \geq m_i + 1$ for each i, $1 \leq i \leq n$ and for a fixed n we proceed by induction on the quantity

$$\gamma(A_1) + \cdots + \gamma(A_n). \tag{7}$$

If $|a_n|$ is a prime, then $\gamma(A_n) = 1$, that is $\gamma(A_n) = m_n$ and we are done. Thus we assume that $|a_n| = pm$, where p is a prime and $m \geq 2$. Set $A'_n = e - a_n^p$. We claim that $BA_1 \cdots A_{n-1} A'_n = 0$. To prove the claim we show that

$$\chi(BA_1 \cdots A_{n-1} A'_n) = 0 \tag{8}$$

for each character χ of G. The equation (8) holds when

$$\chi(BA_1 \cdots A_{n-1}) = 0.$$

When $\chi(BA_1 \cdots A_{n-1}) \neq 0$, then apply χ to (2) to get

$$\chi(BA_1 \cdots A_{n-1})\chi(A_n) = 0.$$

It follows that $\chi(a_n) = 1$, then $\chi(a_n^p) = 1$ and so equation (8) holds. It can happen that some factors are cancelable in the equation $BA_1 \cdots A_{n-1} A'_n = 0$. After the possible cancellations and reindexing the factors if this is necessary, we have the primitive product $BA_1 \cdots A_t A'_n = 0$. Here $t = 0$ is possible but A'_n surely remains. Now $\gamma(A'_n) + 1 = \gamma(A_n)$. If $t = n - 1$ we use the inductive assumption on (7). If $t \leq n - 2$ we use the inductive assumption on n to get

$$\gamma(B, A_1, \ldots, A_t, A'_n) - \gamma(B) < m_1 + \cdots + m_t + m'_n.$$

Here $m'_n = 1$ hence

$$\gamma(B, A_1, \ldots, A_t, A'_n) - \gamma(B) \le m_1 + \cdots + m_t$$

and

$$\gamma(B, A_1, \ldots, A_t) - \gamma(B) \le m_1 + \cdots + m_t.$$

In the $t \ge 1$ case, adding this to the $s = t$ special case of (5), we get (3). In the $t = 0$ case $\gamma(B, A'_n) - \gamma(B) < m'_n$ holds, that is $\gamma(B, A'_n) - \gamma(B) = 0$. It follows that $\gamma(B, A_n) - \gamma(B) \le 1$. As $m_n = 1$ this is equivalent to $\gamma(B, A_n) - \gamma(B) \le m_n$. Adding this to (6) we have (3).

This completes the proof.

To a subset A of G we assign the subgroup

$$H = \bigcap_{\chi(A)=0} \mathrm{Ker}\chi.$$

We will deal with normalized subsets A of a finite abelian group G satisfying the following conditions.

(i) $|A| \ge 2$.

(ii) There is a subgroup K of G for which $K = AH$ is a factorization of K.

One can verify that the cyclic subset

$$A = \{e, a, a^2, \ldots, a^{r-1}\}$$

with $r \ge 2$ satisfies the conditions (i), (ii). Here $K = \langle a \rangle$, $H = \langle a^r \rangle$.

The simulated subset

$$A = (L \setminus \{l\}) \cup \{ld\},$$

where L is a subgroup of G, $|L| \ge 2$, $l \in L$, $d \in G$ satisfies conditions (i), (ii). Here $H = \langle d \rangle$, $K = L \times H$.

If G is a finite abelian group and 6 is relatively prime to $|G|$, then the distorted cyclic subset

$$A = \{e, a, a^2, \ldots, a^{i-1}, a^i d, a^{i+1}, \ldots, a^{r-1}\}$$

with $r \ge 2$, $d \in G$ satisfies condition (i), (ii). Here $H = \langle d \rangle \times \langle a^r \rangle$, $K = \langle d \rangle \times \langle a \rangle$. (This assertion will be formally restated and proved as Lemma 5.4.1.)

In the second zero divisor lemma two types of elements A_i of $Z(G)$ occur. Namely,

$$A_i = e - a_i, \tag{9}$$

$$A_i = \sum_{a \in A_i^*} a, \tag{10}$$

where A_i^* is a normalized subset of G satisfying conditions (i) and (ii) with subgroups H_i and K_i of G associated with A_i^*. We assign the number $m_i = \nu(|A_i|)$ to A_i.

Lemma 3.3.2. *Let $B, A_1, \ldots, A_n \in Z(G)$, where A_i's are of types (9) or (10). Assume that the product*

$$BA_1 \cdots A_n = 0 \tag{11}$$

is primitive. Then

$$\gamma(B, A_1, \ldots, A_n) - \gamma(B) < m_1 + \cdots + m_n. \tag{12}$$

Proof. We proceed by induction on n. Let $n = 1$. As the product $BA_1 = 0$ is primitive we have $B \neq 0$. If A_1 is of type (9), then by Lemma 3.3.1 we are done. We assume that A_1 is of type (10). We claim that $B = Bh_1$ for each $h_1 \in H_1 \setminus \{e\}$. To prove the claim we show that

$$\chi(B) = \chi(Bh_1) \tag{13}$$

for each character χ of G. The equation (13) holds when $\chi(B) = 0$. If $\chi(B) \neq 0$, then apply χ to $BA_1 = 0$ to get $\chi(B)\chi(A_1) = 0$. This implies $\chi(A_1) = 0$, then $\chi(h_1) = 1$ and again the equation (13) holds. From $B \neq 0$ and $B = Bh_1$ it follows that $h_1 \in \langle B \rangle$ and so $H_1 \subset \langle B \rangle$. The index $\big|\langle A_1, B \rangle : \langle H_1, B \rangle\big| = \big|\langle B, A_1 \rangle : \langle B \rangle\big|$ divides $\big|\langle A_1 \rangle : H_1\big|$ which in turn divides $|K_1 : H_1|$ as $A_1 \subset K_1$. Then the index $|K_1 : H_1|$ divides $|A_1^*|$ since $K_1 = A_1^* H_1$ is a factorization of K_1. Thus $\gamma(B, A_1) - \gamma(B) \leq m_1$. The equation $BA_1 = 0$ is equivalent to $B(A_1 - e) = -B$. This gives that there is an $a \in A_1^* \setminus \{e\}$ with $a \in \langle B \rangle$ and we get that $\gamma(B, A_1) - \gamma(B) < m_1$.

Let $n \geq 2$. By the inductive assumption on n from

$$(BA_1 \cdots A_s)A_{s+1} \cdots A_n = 0$$

in the same way as in the proof of Lemma 3.3.1 we get that

$$\gamma(B, A_1, \ldots, A_n) - \gamma(B, A_1, \ldots, A_s) < m_{s+1} + \cdots + m_n, \tag{14}$$

$$\gamma(B, A_1, \ldots, A_n) - \gamma(B, A_n) < m_1 + \cdots + m_{n-1}. \tag{15}$$

Consider $BA_1 \cdots A_n = 0$. Let k be the number of factors of type (10) among $A_1, \ldots, A_n$. If $k = 0$, then by Lemma 3.3.1 we are done. We may assume that A_n is of type (10) since this is only a matter of reordering the factors $A_1, \ldots, A_n$. For a fixed n we proceed by induction on k. We can verify that $BA_1 \cdots A_{n-1}(e - h_n) = 0$ for each $h_n \in H_n \setminus \{e\}$. After canceling and reindexing the factors if this is necessary we get the primitive product $BA_1 \cdots A_t(e - h_n) = 0$. Here $t = 0$ is possible but $e - h_n$ surely remains. If $t = n - 1$, then we use the inductive assumption on k. If $t \leq k - 2$, then we use the inductive assumption on n to get

$$\gamma(B, A_1, \ldots, A_t, h_n) - \gamma(B) < m_1 + \cdots + m_t + 1,$$

that is

$$\gamma(B, A_1, \ldots, A_t, h_n) - \gamma(B) \le m_1 + \cdots + m_t$$

and so

$$\gamma(B, A_1, \ldots, A_t) - \gamma(B) \le m_1 + \cdots + m_t.$$

If $t \ge 1$ for some $h_n \in H_n \setminus \{e\}$, then adding this to the $s = t$ special case of (14) we are done. If $t = 0$ for each $h_n \in H_n \setminus \{e\}$, then $B(e - h_n) = 0$, for each $h_n \in H_n \setminus \{e\}$. In the way we have seen in the first part of the proof we can deduce that $H_n \subset \langle B \rangle$. Further $\gamma(B, A_n) - \gamma(B) \le m_n$. Adding this to (15) we get (12).
This completes the proof.

As an application we give a new proof for Hajós's theorem. In fact we prove a generalization of Hajós's theorem.

Theorem 3.3.1. *Let G be a finite abelian group and*

$$G = A_1 \cdots A_n \tag{16}$$

a normalized factorization of G, where each A_i satisfies conditions (i), (ii). Then at least one of the factors $A_1, \ldots, A_n$ is a subgroup of G.

Proof. Let H_i and K_i be the subgroups associated with the subset A_i. If $H_i = \{e\}$ for some i, $1 \le i \le n$, then from the factorization $K_i = A_i H_i$ it follows that $K_i = A_i$ and so A_i is a subgroup of G. Thus we may assume that $|H_i| \ge 2$ for each i, $1 \le i \le n$.
If $n = 1$, then $G = A_1$ and so A_1 is a subgroup of G. We assume that $n \ge 2$ and proceed by induction on n. For a subset A of G we define $\overline{A}$ to be the element

$$\sum_{a \in A} a$$

of the group ring $Z(G)$. The factorization (16) corresponds to the equation

$$\overline{G} = \overline{A}_1 \cdots \overline{A}_n \tag{17}$$

in the group ring $Z(G)$. Choose an $h \in H_n \setminus \{e\}$. From (17) we obtain

$$0 = \overline{A}_1 \cdots \overline{A}_{n-1}(e - h).$$

After cancellations and reordering we have

$$0 = \overline{A}_1 \cdots \overline{A}_t(e - h). \tag{18}$$

Here $t \ge 1$ since $e - h \ne 0$ and clearly $e - h$ cannot be cancelled. We may apply Lemma 3.3.2 to (18) with $B = e$. So

$$\gamma(\overline{A}_1, \ldots, \overline{A}_t, h) < m_1 + \cdots + m_t + 1,$$

that is,

$$\gamma(\overline{A}_1, \ldots, \overline{A}_t) \leq m_1 + \cdots + m_t. \tag{19}$$

Let $L = \langle A_1, \ldots, A_t \rangle$ and $M = A_1 \cdots A_t$ and $N = A_{t+1} \cdots A_n$. Restricting the factorization $G = MN$ to L we have the factorization $L = G \cap L = M(N \cap L)$. This implies that $|M|$ divides $|L|$. From (19) we deduce that $M = L$, that is, $L = A_1 \cdots A_t$ is a factorization of the subgroup L. Note that $t \leq n - 1$ and the assumptions of the theorem hold with the subgroups $K_i' = K_i \cap L$ and $H_i' = H_i \cap L$. By the inductive assumption, one of the factors $A_1, \ldots, A_t$ is a subgroup of L and so is of G.

This completes the proof.

Let A be normalized subset of the finite abelian group G such that
(iii) $|A| = p$ is a prime divisor of $|G|$.
(iv) $\chi(a_{|p'}) = 1$ for each $\chi \in \mathrm{Ann}(A)$ and each $a \in A$.
One can verify that the cyclic subset $A = \{e, a, a^2, \ldots, a^{r-1}\}$ of the finite abelian group G, where r is a prime divisor of $|G|$ satisfies conditions (iii), (iv).

It was proved in Lemma 3.2.2 that if A is a normalized subset of the finite abelian group G, $|A| = p$ is a prime divisor of $|G|$ and A contains only (p, q)-elements, where q is a prime, then A satisfies conditions (iii), (iv).

In the third zero divisor lemma three types of elements A_i of $Z(G)$ occur:

$$A_i = e - a_i, \tag{20}$$

$$A_i = \sum_{a \in A_i^*} a, \tag{21}$$

where A_i^* is a normalized subset of G satisfying conditions (i), (ii).

$$A_i = \sum_{a \in A_i^*} a, \tag{22}$$

where A_i^* is a normalized subset of G satisfying conditions (iii), (iv). We assign the number $m_i = \nu(|A_i|)$ to A_i.

Lemma 3.3.3. *Let $B, A_1, \ldots, A_n \in Z(G)$, where each A_i is one of types (20), (21), (22). Assume that the product*

$$BA_1 \cdots A_n = 0 \tag{23}$$

is primitive. Then

$$\gamma(B, A_1, \ldots, A_n) - \gamma(B) < m_1 + \cdots + m_n. \tag{24}$$

Proof. We proceed by induction on n. Let $n = 1$. As the product $BA_1 = 0$ is primitive we have $B \neq 0$. If A_1 is of type (20) or (21), then by Lemma 3.3.2 we are done. We assume that A_1 is of type (21). Suppose

$$A_1 = e + a_{1,1} + \cdots + a_{1,r(1)-1}$$

and set

$$A_1^{(i)} = e + a_{1,i} + a_{1,i}^2 + \cdots + a_{1,i}^{r(1)-1}.$$

We know from Corollary 3.2.2 that $\mathrm{Ann}(A_1) \subset \mathrm{Ann}(A_1^{(i)})$ holds. Using this from $BA_1 = 0$ it follows that $BA_1^{(i)} = 0$. Lemma 3.3.2 gives that $a_{1,i} \in \langle B \rangle$. Since this holds for each i, $1 \leq i \leq r(1) - 1$ $A_1^* \subset \langle B \rangle$. Consequently $\gamma(B, A_1) - \gamma(B) = 0$ and so $\gamma(B, A_1) - \gamma(B) < m_1$.

Let $n \geq 2$. By the inductive assumption on n from

$$(BA_1 \cdots A_s)A_{s+1} \cdots A_n = 0$$

in the same way as in the proof of Lemma 3.3.1 we get that

$$\gamma(B, A_1, \ldots, A_n) - \gamma(B, A_1, \ldots, A_s) < m_{s+1} + \cdots + m_n, \tag{25}$$

$$\gamma(B, A_1, \ldots, A_n) - \gamma(B, A_n) < m_1 + \cdots + m_{n-1}. \tag{26}$$

Consider $BA_1 \cdots A_n = 0$. Let k be the number of factors of type (22) among $A_1, \ldots, A_n$. If $k = 0$, then by Lemma 3.3.2 we are done. We may assume that $k \geq 1$ and A_n is of type (22). For a fixed n we proceed by induction on k. Let

$$A_n = e + a_{n,1} + \cdots + a_{n,r(n)-1}$$

and set

$$A_n^{(i)} = e + a_{n,i} + a_{n,i}^1 + \cdots + a_{n,i}^{r(n)-1}.$$

Since $\mathrm{Ann}(A_1) \subset \mathrm{Ann}(A_1^{(i)})$ from $BA_1 \cdots A_{n-1}A_n = 0$ it follows that $BA_1 \cdots A_{n-1}A_n^{(i)} = 0$. After canceling and possibly reindexing the factors we get the primitive product $BA_1 \cdots A_t A_n^{(i)} = 0$. Here $t = 0$ is possible but $A_n^{(i)}$ surely remains. If $t = n - 1$, then we use the inductive assumption on k. If $t \leq k - 2$, then we use the inductive assumption on n to get

$$\gamma(B, A_1, \ldots, A_t, A_n^{(i)}) - \gamma(B) < m_1 + \cdots + m_t + 1,$$

that is,

$$\gamma(B, A_1, \ldots, A_t, A_n^{(i)}) - \gamma(B) \leq m_1 + \cdots + m_t$$

and so

$$\gamma(B, A_1, \ldots, A_t) - \gamma(B) \leq m_1 + \cdots + m_t.$$

If $t \geq 1$ for some i, $1 \leq i \leq r(n) - 1$, then adding this to the $s = t$ special case of (25) we are done. If $t = 0$ for each i, $1 \leq i \leq r(n) - 1$, then $BA_n^{(i)} = 0$, for each i,

$1 \leq i \leq r(n) - 1$. By Lemma 3.3.2, $a_{n,i} \in \langle B \rangle$. Consequently, $A_n^* \subset \langle B \rangle$. Therefore $\gamma(B, A_n) - \gamma(B) \leq m_n$. Adding this to (26) we get (24).

This completes the proof.

We prove Rédei's theorem using the third zero divisor lemma.

Theorem 3.3.2. *Let $G = A_1 \cdots A_n$ be a normalized factorization of the finite abelian group G, where each $|A_i|$ is a prime. Then at least one of the factors $A_1, \ldots, A_n$ must be a subgroup of G.*

Proof. We proceed by induction on n. The $n = 1$ case is trivial so we assume that $n \geq 2$. The following observation is used several times in the course of the proof. If the factor A_1 can be replaced by a subgroup H of G in the factorization $G = A_1 \cdots A_n$, then from the resulting factorization $G = H A_2 \cdots A_n$ we get the factorization

$$G/H = (A_2 H)/H \cdots (A_n H)/H.$$

From this by the inductive assumption it follows that one factor $(A_i H)/H$ is a subgroup of G/H or in other words $H A_i$ is a subgroup of G. Continuing in this way we get that there is a permutation $B_2, \ldots, B_n$ of the factors $A_2, \ldots, A_n$ such that the partial products

$$H, \ HB_2, \ HB_2B_3, \ldots, HB_2 \cdots B_n \tag{27}$$

are subgroups of G.

First deal with the case when G is a finite elementary abelian p-group, that is, when G is of type $(p, \ldots, p)$ for some prime p. The case $n = 2$ is settled in Lemma 1.4.5 and also in Appendix C. So we assume that $n \geq 3$. In the factorization $G = A_1 \cdots A_n$ the factor A_1 can be replaced by a subgroup H. There is a permutation $B_2, \ldots, B_n$ of the factors $A_2, \ldots A_n$ such that the partial products (27) are subgroups of G. For the simpler notation set $B_1 = A_1$. Consider the subgroup $K = HB_2 \cdots B_{n-1}$ from the chain (27). In the factorization $G = KB_n$ the factor B_n is replaceable by a subgroup M. From the resulting $G = KM$ factorization we can see that $K \cap M = \{e\}$. The factorization $G = B_1 \cdots B_{n-1} M$ also gives rise to a chain of subgroups of partial products. There is an index j such that MB_j is a subgroup of G. If $j \neq 1$, then $K \cap MB_j = B_j$ is a subgroup. Hence we assume that $MB_1 = N$ is a subgroup of G. If $B_n \not\subset N$, then in the factorizations $G = KB_n$ and $G = B_1 \cdots B_n$ the factor B_n can be replaced by a subgroup L of G for which $K \cap L = L \cap N = \{e\}$. There is an index j such that LB_j is a subgroup of G. If $j \neq 1$, then $K \cap LB_j = B_j$ is a subgroup of G. Thus we assume that LB_1 is a subgroup of G. But now $N \cap LB_1 = B_1$ is a subgroup. Consequently we may assume that $B_n \subset N$. In this case the direct product $B_1 B_n$ gives N. This reduces the proof to the $n = 2$ case.

Suppose that G is not an elementary p-group. Consider a factor A_i with $|A_i| = p$. If each $a \in A_i$ has order p, then call A_i a type 1 factor, otherwise call A_i a type 2 factor. By Lemma 1.4.3, each type 2 factor can be replaced by a nonsubgroup cyclic subset. As G is not an elementary p-group there must be a factor of type 2, say A_n. In the factorization $G = A_1 \cdots A_n$ the factor A_n can be replaced by the cyclic subset $C = \{e, a, a^2, \cdots, a^{p-1}\}$. The resulting factorization $G = A_1 \cdots A_{n-1}C$ corresponds to the equation $\overline{G} = \overline{A}_1 \ldots \overline{A}_{n-1}\overline{C}$. Multiplying this equation by $(e - a)$ we get $0 = \overline{A}_1 \cdots \overline{A}_{n-1}(e - a^p)$. After all possible cancellations and reindexing the factors if this is necessary we have $0 = \overline{A}_1 \cdots \overline{A}_m(e - a^p)$. Here $m \geq 1$ since $e - a^p \neq 0$. Further $(e - a^p)$ cannot be cancelled since $A_1 \cdots A_m \neq 0$. Lemma 3.3.3 is applicable with $B = e$ to give the conclusion $\gamma(A_1, \ldots, A_m, a^p) < m + 1$ from which it follows that

$$\gamma(A_1, \ldots, A_m) \leq m. \tag{28}$$

Set $T = A_1 \cdots A_m$ and $S = A_{m+1} \cdots A_n$. Restricting the factorization $G = TS$ to $\langle T \rangle$ we get the factorization $\langle T \rangle = G \cap \langle T \rangle = T(S \cap \langle T \rangle)$. Thus $|\langle T \rangle|$ is a multiple of $|T|$. On the other hand by (28), $|\langle T \rangle|$ is a product of at most m primes and $|T|$ is a product of m primes. Therefore $T = \langle T \rangle$. As $m \leq n - 1$ by the induction assumption it follows that one of the factors $A_1, \ldots, A_m$ is a subgroup of G.

  This completes the proof.

Chapter 4

Four characterization results

4.1 Factorizations by nonsubgroup factors

Examples show that Rédei's theorem does not hold for normalized factorizations with general factors. For instance let G be the cyclic group of order 8 with generator x and let

$$A_1 = \{e, x^2\}, \quad A_2 = \{e, x, x^4, x^5\}.$$

Now $G = A_1 A_2$ is a normalized factorization of G and the factors A_1 and A_2 are not subgroups of G.

Exercise 4.1.1. Let $G = \langle x \rangle \times \langle y \rangle \times \langle z \rangle$, where $|x| = p$, $|y| = q$, $|z| = r$ are distinct primes. Verify that $G = AB$ is a factorization of G, where

$$A = \{e, y, \ldots, y^{q-2}, xy^{q-1}\},$$
$$B = \langle x \rangle \{e, z, \ldots, z^{r-2}, yz^{r-1}\}.$$

Show that neither A nor B is a subgroup of G.

However, for finite groups of type $(2, \ldots, 2)$ a sharper result than Rédei's theorem holds. By Exercise 1.2.8, in each factorization of a finite group of type $(2, \ldots, 2)$ by subsets of cardinality 4 at least one of the factors is a subgroup of the group.

We set forth the following problem. A group G of type $(t_1, \ldots, t_s)$ is given. For which integers $q_1, \ldots, q_n$ does it follow that in each normalized factorization $G = A_1 \cdots A_n$, with $|A_1| = q_1, \ldots, |A_n| = q_n$ at least one of the factors A_i is a subgroup of G. In short what are the subgroup forcing factorization types for a given finite abelian group? (The $n = 1$ case is trivial so we assume that $n \geq 2$.) The answer is the next theorem.

Theorem 4.1.1. *Let $G = A_1 \cdots A_n$ be a normalized factorization of the finite abelian group G, where $|A_1| = q_1, \ldots, |A_n| = q_n$, $n \geq 2$. In the following cases at least one* *of the factors must be a subgroup of G.*

(1) G is of type $(2,\ldots,2)$ and either $q_i = 2$ for some i, $1 \leq i \leq n$ and the other q's are arbitrary or $q_1 = \cdots = q_n = 4$.

(2) G is of type $(2,\ldots,2,2p)$, where p is a prime and $q_1,\ldots,q_n$ is a rearrangement of $2,\ldots,2,2p$ or $2,\ldots,2,p$. (The $p = 2$ case is included.)

(3) G is not of the types $(2,\ldots,2)$ and $(2,\ldots,2,2p)$, where p is a prime and each q_i is a prime.

In all the other cases there are normalized factorizations without subgroup factors.

Theorem 4.1.1 is a typical structure theorem in the sense we detailed in the beginning of Section 1.3. Namely, if H_1 is a subgroup of G and if the normalized subset A_1 is a complete set of representatives modulo H_1, then $G = A_1 H_1$ is a normalized factorization of G. Similarly, if H_2 is a subgroup of H_1 and if the normalized subset A_2 is a complete set of representatives modulo H_2, then $H_1 = A_2 H_2$ is a normalized factorization of H_1 and so $G = A_1 A_2 H_2$ is a normalized factorization of G. In this way we can construct a large stock of factorizations of G. Note that in the resulting factorizations at least one of the factors is always a subgroup of G. On the other hand by Theorem 4.1.1 in cases (1), (2), and (3) G, does not have any other type of normalized factorizations. In short, there is a straightforward way to construct normalized factorizations of G and by Theorem 4.1.1 in cases (1), (2), and (3) there are no further normalized factorizations of G.

The proof of Theorem 4.1.1 naturally splits into two parts. First to prove that in the cases (1), (2), and (3) one of the factors must be a subgroup. Then to exhibit examples with nonsubgroup factors in the remaining cases.

Lemma 4.1.1. *Let G be a group of type $(2,\ldots,2,2p)$, where p is a prime. If $G = A_1 \cdots A_n$ is a normalized factorization of G such that $|A_1| = \cdots = |A_{n-1}| = 2$ and $|A_n| = 2p$, then at least one of the factors A_i is a subgroup of G.*

Proof. Now G is a direct product of subgroups of types $(2,\ldots,2)$ and $(2p)$. Let x be a generator of the $(2p)$ type direct factor. Let $A_i = \{e, a_i\}$ for each i, $1 \leq i \leq n-1$ and consider a character χ of G for which $\chi(x)$ is a primitive $(2p)$th root of unity. Since χ is not the principal character of G, $\chi(A_i) = 0$ for some i, $1 \leq i \leq n$. If $1 \leq i \leq n-1$, then $0 = \chi(A_i) = 1 + \chi(a_i)$, that is, $\chi(a_i) = -1$ and so a_i is a second order element of G. Hence A_i is a subgroup of G. Thus suppose that $\chi(A_n) = 0$ for each character χ of G for which $\chi(x^2)$ is a primitive pth root of unity. Now by Lemma 3.2.6, A_n can be factored in the form $A_n = \langle x^2 \rangle \{e, a\}$. As $a = bx^2$, where b is a second order element, A_n is a subgroup of G.

This completes the proof.

To construct factorizations without subgroup factors we need some preparation. Let G be a finite abelian group and $q_1,\ldots,q_n$ be integers such that $q_1 \cdots q_n = |G|$, $q_1 \geq 2,\ldots,q_n \geq 2$ and q_n is composite, say $q_n = uv$, $u \geq 2$ and $v \geq 2$. In order to construct a normalized factorization $G = A_1 \cdots A_n$ with $|A_1| = q_1,\ldots,$

$|A_n| = q_n$ consider a chain of subgroups

$$\{e\} = K_0 \subset K_1 \subset \cdots \subset K_n \subset K_{n+1} = G \tag{1}$$

such that

$$|K_1 : K_0| = u,$$

$$|K_2 : K_1| = q_1, \ldots, |K_n : K_{n-1}| = q_{n-1},$$

$$|K_{n+1} : K_n| = v.$$

Choose B_i to be a complete set of representatives modulo K_i in K_{i+1} for each i, $1 \leq i \leq n$ such that $e \in B_i$. Clearly $K_{i+1} = K_i B_i$ is a normalized factorization of K_{i+1} for each i, $1 \leq i \leq n$. Hence

$$G = K_1 B_1 B_2 \cdots B_n = B_1 B_2 \cdots B_{n-1}(K_1 B_n)$$

is a normalized factorization of G and

$$|B_1| = q_1, \ldots, |B_{n-1}| = q_{n-1}, |K_1 B_n| = uv = q_n.$$

There is a great degree of freedom when choosing B_i and we try to exploit this to avoid the occurrence of subgroups among the factors.

Lemma 4.1.2. *Let T be a finite abelian group and let H, L be subgroups of T with $L \subset H$ and $L \neq H$. If $|T : H| \geq 3$ or if $|T : H| = 2$ and H/L is not of type $(2, \ldots, 2)$, then there is a complete set of representatives B of T modulo H such that $e \in B$ and the product LB is direct and is not a subgroup of T.*

Proof. First consider the case $|T : H| \geq 3$. Let B be a complete set of representatives modulo H in T with $e \in B$. If LB is not a subgroup of T, then we are done. Thus suppose that $LB = M$ is a subgroup of T. There are elements $b \in B \setminus \{e\}$ and $h \in H \setminus L$. Let $C = \{bh\} \cup (B \setminus \{b\})$. Clearly C is also a complete set of representatives modulo H in T. If LC is not a subgroup of T we are done. Thus suppose that $LC = N$ is a subgroup of T. Then $M \cap N = LB \cap LC = L(B \cap C)$ is a subgroup of M and so $|M \cap N|$ divides $|M|$. As $|M \cap N| = |L|(|B| - 1)$ and $|M| = |L||B|$, we have $|B| - 1$ divides $|B|$. Thus $|B| = 2$. On the other hand $|T : H| = |B| \neq 2$ which is a contradiction.

Turn to the case when H/L is not of type $(2, \ldots, 2)$. As before we are done unless $|B| = 2$. If $b^2 \notin L$, then LB is not a subgroup of T and we are done. Thus we may assume that $b^2 \in L$. As H/L is not of type $(2, \ldots, 2)$, h can be chosen such that $h^2 \notin L$. If $LC = L\{e, bh\}$ is a subgroup of T, then $(bh)^2 \in L$. Since $(bh)^2 = b^2 h^2$ and $b^2 \in L$, we have that $h^2 \in L$ which is a contradiction.

This completes the proof.

We are now in position to prove Theorem 4.1.1.

Proof of Theorem 4.1.1. Consider a normalized factorization $G = A_1 \cdots A_n$ of a finite abelian group G.

If G is of type $(2, \ldots, 2)$ and if $|A_i| = 2$ for some i, $1 \le i \le n$, then A_i is a subgroup of G since each element of $G \setminus \{e\}$ is of second order.

If G is of type $(2, \ldots, 2)$ and $|A_1| = \cdots = |A_n| = 4$, then by Exercise 1.2.8, one of the factors A_i is a subgroup of G.

If G is of type $(2, \ldots, 2, 2p)$, where p is a prime and $|A_1| = \cdots = |A_{n-1}| = 2$ and $|A_n| = 2p$, then by Lemma 4.1.1, one of the factors A_i is a subgroup of G.

Finally, if each $|A_i|$ is a prime, then by Rédei's theorem one of the factors A_i is a subgroup of G no matter what is the type of G.

In order to construct nonsubgroup normalized factors $B_1, \ldots, B_{n-1}, K_1 B_n$ we consider a chain of subgroups (1). Then we will apply Lemma 4.1.2 with the choice

$$L = K_0 = \{e\}, \quad H = K_i, \quad T = K_{i+1}, \quad 1 \le i \le n - 1$$

and

$$L = K_1, \quad H = K_n, \quad T = K_{n+1} = G.$$

Let G be a finite group of type $(2, \ldots, 2)$. Since the case when $q_i = 2$ for some i, $1 \le i \le n$ is settled suppose that $q_1 \ge 4, \ldots, q_n \ge 4$. Since the case $q_1 = \cdots = q_n = 4$ is also settled we may assume that $q_i \ge 8$ for some i, $1 \le i \le n$, say $q_n \ge 8$. Let $u = 2$ and $v = q_n/2$. Now $|K_{i+1} : K_i| \ge 4$ for each i, $1 \le i \le n$ and so Lemma 4.1.2 is applicable.

Let G be of type $(2, \ldots, 2, 2p)$, where p is an odd prime. The case when $q_1, \ldots, q_n$ is a rearrangement of $2, \ldots, 2, 2p$ is settled so we assume that this is not the case. We may assume that $q_n = 2^t$ or $q_n = 2^t p$, where $t \ge 2$ since this is only a matter of indexing the numbers $q_1, \ldots, q_n$.

First consider the case $q_n = 2^t$. Let $u = 2$ and $v = q_n/2$. There is a q_i that is divisible by p. Assume that p divides q_1. Now $K_2, \ldots, K_n$ and K_{n+1}/K_1 are not of type $(2, \ldots, 2)$ and so Lemma 4.1.2 is applicable.

Secondly consider the case $q_n = 2^t p$, $t \ge 2$. Let $u = p$ and $v = 2^t$. Now $K_2 \ldots, K_n$ are not of type $(2, \ldots, 2)$ and $|K_{n+1} : K_n| = 2^t \ge 4$. Hence Lemma 4.1.2 is applicable.

Let G be of type $(2, \ldots, 2, 4)$. The case when $q_1, \ldots, q_n$ is a rearrangement of $2, \ldots, 2, 4$ is settled so we assume that $q_i \ge 8$ for some i, $1 \le i \le n$ or $q_i = 4$ and $q_j = 4$ for some i and j, $1 \le i < j \le n$.

First assume that $q_1 = 4$ and $q_n = 4$. Let $u = 2$ and $v = 2$. Let K_1 be of type (2) and be a direct factor of $K_{n+1} = G$. As G is of type $(2, \ldots, 2, 4)$ and $|K_2| = 8$, we can choose $K_2, \ldots, K_n$ such that they are of type $(2, 4)$ or $(2, \ldots, 2, 4)$ Hence $K_2, \ldots, K_n$ and K_{n+1}/K_1 are not of type $(2, \ldots, 2)$ and so Lemma 4.1.2 is applicable.

Secondly assume that $q_n \ge 8$. Let $u = 2$ and $v = q_n/2$. Let K_1 of type (2) such that K_1 is the unique subgroup of the (4) type direct factor of $K_{n+1} = G$. As G is of type $(2, \ldots, 2, 4)$ we can choose $K_2, \ldots, K_n$ such that they are of types (4) or $(2, \ldots, 2, 4)$. Now $K_2, \ldots, K_n$ are not of type $(2, \ldots, 2)$ and $|K_{n+1} : K_1| = q_n/2 \ge 4$. Therefore Lemma 4.1.2 is applicable.

Let G be a finite abelian group not of types $(2, \ldots, 2)$ and $(2, \ldots, 2, 2p)$, where p is a prime. The case when $q_1, \ldots, q_n$ are primes is settled by Rédei's theorem so we assume that some q_i is composite, say $q_n = uv$, $u \geq 2$ and $v \geq 2$.

First suppose that some q_i, $1 \leq i \leq n - 1$ is divisible by an odd prime. We may assume that this is q_1. Now $K_2, \ldots, K_n$ and K_{n+1}/K_1 are not of type $(2, \ldots, 2)$ and so Lemma 4.1.2 is applicable.

Secondly suppose that q_n is divisible by an odd prime. We may assume that u is an odd prime. Now $K_2, \ldots, K_n$ is not of type $(2, \ldots, 2)$ and as G is not of types $(2, \ldots, 2)$ and $(2, \ldots, 2, 2p)$, K_{n+1}/K_1 is not of type $(2, \ldots, 2)$. Therefore Lemma 4.1.2 is applicable.

Thirdly we may assume that each $q_1, \ldots, q_n$ is a power of two and so G is a 2-group. Let $u = 2$ and $v = q_n/2$. As G is not of types $(2, \ldots, 2)$ and $(2, \ldots, 2, 4)$, we can choose $K_1, \ldots, K_n$ such that K_1 is of type (2) and $K_2, \ldots, K_n$ and K_{n+1}/K_1 are not of type $(2, \ldots, 2)$. Therefore Lemma 4.1.2 is applicable.

This completes the proof.

There are finite abelian groups G such that in every normalized factorization of G one of the factors is always a subgroup of G. In the remaining part of the section we will characterize these groups.

Exercise 4.1.2. Let H be a subgroup of the finite abelian group G. Prove that if H has a normalized factorization into nonsubgroup factors so does G. In other words prove that if in each normalized factorization of G one of the factors is always a subgroup of G, then each subgroup H of G possesses this property.

Let U be the family of groups whose type is one of

$$(p^2), \quad (p, p), \quad (p, q), \quad (4, 2), \quad (2, 2, 2, 2), \tag{2}$$

where p and q are distinct primes or a subgroup of such a group. An alternative way to describe U is that it contains the groups of types (2) and it is closed under the operation of forming subgroups.

Exercise 4.1.3. Show that if G is a member of the U family and $G = A_1 \cdots A_n$ is a normalized factorization of G, then at least one of the factors must be a subgroup of G.

Exercise 4.1.4. Show that if G has a subgroup H such that $H \neq G$ and H is of one types (2), then there is a normalized factorization $G = AB$ such that neither A nor B is a subgroup of G.

4.2 Multiple simulated factorizations

Let $A_1, \ldots, A_n$ be subsets of the finite abelian group G. If each element g of G is expressible in precisely k ways in the form

$$g = a_1 \cdots a_n, \qquad a_1 \in A_1, \ldots, a_n \in A_n,$$

then we say that the product $A_1 \cdots A_n$ is a k-*factorization* of G. The 1-factorization is the familiar concept of factorization.

Let $A_1, \ldots, A_n$ be simulated subsets of G and let $H_1, \ldots, H_n$ be the subgroups of G associated with $A_1, \ldots, A_n$ respectively. By Theorem 1.2.1, if $A_1 \cdots A_n$ is a normalized factorization of G, then one of the factors is a subgroup of G, that is, there is an i, $1 \leq i \leq n$ such that $A_i = H_i$. This section deals with normalized k-factorizations of G by simulated subsets. We know that in the $k = 1$ case one of the factors is a subgroup. The next example shows that this is not necessarily the case when $k \geq 2$.

Exercise 4.2.1. Let G be of type $(2, 2, 2)$ with basis x, y, z. The subsets A_1, A_2, A_3 are simulated. The corresponding subgroups are H_1, H_2, H_3.

$$
\begin{aligned}
A_1 &= \{e, y, z, xyz\}, & H_1 &= \{e, y, z, yz\}, \\
A_2 &= \{e, x, z, xyz\}, & H_2 &= \{e, x, z, xz\}, \\
A_3 &= \{e, x, z, zy\}, & H_3 &= \{e, x, z, xz\}.
\end{aligned}
$$

Prove that the product $A_1 A_2 A_3$ is an 8-factorization of G and none of the factors is a subgroup of G.

We are interested in the following two problems.

(1) Characterize all the finite abelian groups for which in any k-factorization $A_1 \cdots A_n$ of G by simulated factors at least one of the factors is always a subgroup.

(2) Characterize all the k, n values for which there is a k-factorization $A_1 \cdots A_n$ of some group G by simulated factors such that none of the factors is a subgroup. We call such (k, n) pair an admissible pair.

The next theorems give the solution of these problems.

Theorem 4.2.1. *Let $A_1 \cdots A_n$ be a k-factorization of the finite abelian group G by simulated subsets. If G is cyclic or is of type $(2, 2)$, then at least one of the factors is a subgroup. In all the other cases there is a k-factorization $A_1 \cdots A_n$ of G by nonsubgroup simulated subsets.*

Theorem 4.2.2. *Let $A_1 \cdots A_n$ be a k-factorization of the finite abelian group G by simulated subsets. If*
$$n \leq 2,\ k \geq 1,$$
$$n = 3,\ k \text{ is odd},$$
$$n \geq 4,\ k = 1,$$
then at least one of the factors is a subgroup. In all the other cases there is a k-factorization $A_1 \cdots A_n$ of G by nonsubgroup simulated subsets.

Before turning to the proofs we make some preparations. The group ring $Z(G)$ provides an adequate tool to deal with k-factorizations. We identify the subset A of G with the element

$$
\overline{A} = \sum_{a \in A} a
$$

of $Z(G)$. The product $A_1 \cdots A_n$ is a k-factorization of G if and only if

$$k\overline{G} = \overline{A}_1 \cdots \overline{A}_n.$$

Characters of $Z(G)$ which are linear extensions of characters of G can be used to study multiple factorizations.

The equation $k\overline{G} = \overline{A}_1 \cdots \overline{A}_n$ holds if and only if $\chi(k\overline{G}) = \chi(\overline{A}_1 \cdots \overline{A}_n)$ for each character χ of G. For the principal character this reduces to $k|G| = |A_1| \cdots |A_n|$. For nonprincipal characters we have $0 = \chi(\overline{A}_1) \cdots \chi(\overline{A}_n)$. Therefore the product $A_1 \cdots A_n$ is a k-factorization of G if and only if $k|G| = |A_1| \cdots |A_n|$, and for each nonprincipal character χ of G there is an i, $1 \le i \le n$ such that $\chi(\overline{A}_i) = 0$.

Let A be a simulated subset of G and let H be the subgroup of G associated with A. There are elements $a \in A$ and $h \in H$ such that $a \notin H$ and $h \notin A$. Further there is an element $d \in G$ for which $a = hd$. The subgroup H and the element d do not determine A uniquely since there are different possible choices for h. For our purposes each of these choices plays the same role, so we will not specify A more closely than giving H and d.

Lemma 4.2.1. *Let A be a simulated subset of a finite abelian group G and let H be the subgroup of G associated with A. Let χ be a character of G. Then with the notation introduced above $\chi(A) = 0$ implies $\chi(d) = 1$ and $\chi(H) = 0$ and conversely $\chi(H) = 0$ and $\chi(d) = 1$ imply $\chi(A) = 0$.*

Proof. Note that

$$\begin{aligned}
\chi(\overline{H}) - \chi(\overline{A}) &= \chi(h) - \chi(hd) \\
&= \chi(h)\left[1 - \chi(d)\right].
\end{aligned}$$

From this it follows that if $\chi(\overline{H}) = 0$ and $\chi(d) = 1$, then $\chi(\overline{A}) = 0$. We show that if $\chi(\overline{A}) = 0$, then $\chi(\overline{H}) = 0$ and $\chi(d) = 1$. To show this suppose that $\chi(\overline{A}) = 0$ and $\chi(\overline{H}) \ne 0$. Since H is a subgroup of G it follows that $\chi(h) = 1$ for each $h \in H$. This leads to the contradiction

$$\begin{aligned}
3 &\le |H| \\
&= \chi(\overline{H}) \\
&= \chi(\overline{H}) - \chi(\overline{A}) \\
&= \chi(h)\left[1 - \chi(d)\right] \\
&\le 2.
\end{aligned}$$

Thus $\chi(\overline{H}) = 0$ and $\chi(d) = 1$.

This completes the proof.

Characters can be used to study k-factorizations by simulated subsets. If the product $A_1 \cdots A_n$ is a k-factorization of G, then $k|G| = |A_1| \cdots |A_n|$ and for each nonprincipal character χ of G there is an i, $1 \le i \le n$ for which $\chi(\overline{H}_i) = 0$ and

i	H_i	d_i
1	$\langle y, z \rangle$	x
2	$\langle x, z \rangle$	y
3	$\langle x, z \rangle$	xy

Table 1

i	H_i	d_i
1	$\langle xy \rangle$	y
2	$\langle xy \rangle$	x
3	$\langle y \rangle$	xy
4	$\langle y \rangle$	xy^2
$\vdots$	$\vdots$	$\vdots$
$\alpha + 1$	$\langle y \rangle$	$xy^{2^{\alpha-1}}$

Table 2

$\chi(d_i) = 1$. Conversely, if there are subgroups $H_1, \ldots, H_n$ and elements $d_1, \ldots, d_n$ of G such that $k|G| = |H_1| \cdots |H_n|$ and for each nonprincipal character χ of G there is an i, $1 \leq i \leq n$ for which $\chi(\overline{H}_i) = 0$ and $\chi(d_i) = 1$, then the product $A_1 \cdots A_n$ is a k-factorization of G.

Proof of Theorem 4.2.1. Let $A_1 \cdots A_n$ be a k-factorization of the finite abelian group G by simulated factors. Since H_i is a subgroup of G, $|H_i| = |A_i|$ is a divisor of $|G|$. Thus if G is of type $(2,2)$, then $|A_i| = 4$ and so $A_i = G$. If G is cyclic, then there is a character χ of G whose kernel is $\{e\}$. There is an i, $1 \leq i \leq n$ such that $\chi(\overline{H}_i) = 0$ and $\chi(d_i) = 1$. From $\chi(d_i) = 1$ it follows, $d_i = e$. Therefore $A_i = H_i$.

For the remaining finite abelian groups we exhibit a k-factorization by non-subgroup simulated factors. We start with three special cases.

Suppose G is of type $(2, 2, p^\alpha)$, where p is a prime and $\alpha \geq 1$. Let x, y, z be a basis of G and define A_1, A_2, A_3 is indicated in Table 1.

Clearly, A_i is not a subgroup of G since $d_i \notin H_i$. Let χ be a nonprincipal character of G. We show that there is an i, $1 \leq i \leq 3$ such that $\chi(\overline{H}_i) = 0$ and $\chi(d_i) = 1$. Note that one of $\chi(x), \chi(y), \chi(xy)$ is always 1. (See Exercise 4.2.2 after the proof.) If $\chi(x) = 1$, then $\chi(\overline{H}_1) \neq 0$ only if $\chi(y) = 1$ and $\chi(z) = 1$. But this is impossible since χ is not the principal character. The $\chi(y) = 1$ and $\chi(xy) = 1$ cases are similar.

Suppose that G is of type $(2, 2^\alpha)$, $\alpha \geq 2$ with basis x, y. Define the simulated subsets $A_1, \ldots, A_{\alpha+1}$ in the way Table 2 shows. Clearly $A_i \neq H_i$ since $d_i \notin H_i$. Let χ be a nonprincipal character of G. If $\chi(y) = 1$, then $\chi(\overline{H}_1) \neq 0$ only if $\chi(xy) = 1$. But now $\chi(x) = 1$ and χ is the principal character. The $\chi(x) = 1$ case is similar. Thus we may suppose that $\chi(x) \neq 1$ and $\chi(y) \neq 1$. One of $\chi(xy), \chi(xy^2), \ldots, \chi(xy^{2^{\alpha-1}})$ is 1. In any case there is an i, $1 \leq i \leq \alpha + 1$ such that $\chi(\overline{H}_i) = 0$ and $\chi(d_i) = 1$.

Suppose that G is of type (p^α, p^β), where p is an odd prime $\alpha \geq 1$ and $\beta \geq 1$. Let x, y be a basis of G. The definitions of $A_1, \ldots, A_{p+1}$ can be found in Table 3.

i	H_i	d_i
1	$\langle xy \rangle$	$y^{p^{\beta-1}}$
2	$\langle y \rangle$	$x^{p^{\alpha-1}}$
3	$\langle y \rangle$	$x^{p^{\alpha-1}} y^{p^{\beta-1}}$
4	$\langle y \rangle$	$x^{p^{\alpha-1}} y^{2p^{\beta-1}}$
$\vdots$	$\vdots$	$\vdots$
$p+1$	$\langle y \rangle$	$x^{p^{\alpha-1}} y^{(p-1)p^{\beta-1}}$

Table 3

Let χ be a nonprincipal character of G. If $\chi(y^{p^{\beta-1}}) = 1$, then $\chi(\overline{H}_1) \neq 0$ only if $\chi(x) = 1$. Hence $\chi(x^{p^{\alpha-1}}) = 1$. Now $\chi(\overline{H}_2) \neq 0$ only if $\chi(y) = 1$. But this is impossible since χ is not the principal character. The case when $\chi(y^{p^{\beta-1}}) = 1$ is similar. Thus we may assume that $\chi(x^{p^{\alpha-1}}) \neq 1$ and $\chi(y^{p^{\beta-1}}) \neq 1$. In this case one of $\chi(x^{p^{\alpha-1}} y^{ip^{\beta-1}})$ must be 1. Therefore $\chi(\overline{H}_{i+2}) = 0$ and $\chi(d_{i+2}) = \chi(x^{p^{\alpha-1}} y^{ip^{\beta-1}}) = 1$.

Finally, let G be of type $(q_1, \ldots, q_s)$. By the fundamental theorem of finite abelian groups we may assume that $q_1, \ldots, q_s$ are prime powers. If G is noncyclic and is not of type $(2, 2)$, then $G = K \times L$, where K is one of the types

$$\begin{aligned}
(2, 2, p^\alpha), \quad & \alpha \geq 1, \\
(2, 2^\beta), \quad & \beta \geq 2, \\
(p^\alpha, p^\beta), \quad & p \geq 3, \ \alpha \geq 1, \ \beta \geq 1.
\end{aligned}$$

By the previous constructions K has a k-factorization $A_1 \cdots A_n$ by nonsubgroup simulated factors. Define the simulated subset A_{n+1} by $H_{n+1} = L$ and $d_{n+1} \in K \setminus \{e\}$. Clearly $A_1 \cdots A_n A_{n+1}$ is a k-factorization of G.

This completes the proof.

Proof of Theorem 4.2.2. Let $A_1 \cdots A_n$ be a k-factorization of the finite abelian group G by simulated factors. For each nonprincipal character χ of G there is an i, $1 \leq i \leq n$ such that $\chi(\overline{H}_i) = 0$ and $\chi(d_i) = 1$. From this it follows that an equation $X_1 \cdots X_n = 0$ holds in $Z(G)$, where $X_i = \overline{H}_i$ or $X_i = e - d_i$.

If $n = 1$, then $e - d_1 = 0$ and so $d_1 = e$. Thus $A_1 = H_1$ no matter what is the value of k.

Suppose $n = 2$. From $(e - d_1)(e - d_2) = 0$ it follows that $e + d_1 d_2 = d_1 + d_2$ and so $e = d_1$ or $e = d_2$. Thus $A_1 = H_1$ or $A_2 = H_2$ independently of the value of k.

Let $n = 3$. Consider the equation

$$(e - d_1)(e - d_2)(e - d_3) = 0. \tag{1}$$

This equation may still hold after cancelling some factors. If such a cancellation is possible, say $e - d_1 = 0$ or $(e - d_1)(e - d_2) = 0$, then as before $d_i = e$ and $A_i = H_i$ for some i. Thus we may assume that no factor can be cancelled from (1) without destroying the equation. From (1) after multiplying out we have

$$e + d_1 d_2 + d_1 d_3 + d_2 d_3 = d_1 + d_2 + d_3 + d_1 d_2 d_3.$$

Some term on the right side is equal to e. Clearly $e = d_1 d_2 d_3$ since $e - d_i \neq 0$. Now

$$d_1 d_2 + d_1 d_3 + d_2 d_3 = d_1 + d_2 + d_3.$$

Some term on the left-hand side is equal to d_1. Once again $e - d_i \neq 0$ and so $d_2 d_3 = d_1$. Continuing in this way we have $d_1^2 = d_2^2 = e$ and $d_3 = d_1 d_2$.

Consider $\overline{H}_1 (e - d_2)(e - d_3) = 0$. After multiplying out,

$$\overline{H}_1(e + d_2 d_3) = \overline{H}_1(d_2 + d_3), \quad \text{or} \quad \overline{H}_1(e + d_1) = \overline{H}_1(d_2 + d_3).$$

G is a disjoint union of cosets modulo H_1. Hence $H_1 = H_1 d_2$ or $H_1 = H_1 d_3$. Thus $d_2 \in H_1$ or $d_3 \in H_1$. Similarly, from $(e - d_1)\overline{H}_2(e - d_3) = 0$ we have $d_1 \in H_2$ or $d_3 \in H_2$, and from $(e - d_1)(e - d_2)\overline{H}_3 = 0$ we have $d_1 \in H_3$ or $d_2 \in H_3$.

Suppose that two different H's contain the same d, say $d_3 \in H_1$ and $d_3 \in H_2$. Now $\langle d_3 \rangle \subset H_1 \cap H_2$ and so $H_1 H_2 H_3$ is a k-factorization of G with even k.

Suppose that no two H's contain the same d, say $d_3 \in H_1$, $d_1 \in H_2$, $d_2 \in H_3$. Now $\langle d_1 d_2 \rangle \subset H_1 \cap H_2 H_3$ shows that in the k-factorization $H_1 H_2 H_3$ of G, k is even.

We exhibit an example to show that each even k occurs. Let G be of type $(2, 2, 2, 2, 2t)$, $t \geq 1$ with basis x, y, u, v, w. Define the nonsubgroup simulated subsets A_1, A_2, A_3 by Table 4. Let χ be a nonprincipal character of G.

If $\chi(x) = 1$, then $\chi(\overline{H}_1) \neq 0$ only if $\chi(y) = 1$ and $\chi(u) = 1$. Now $\chi(\overline{H}_2) \neq 0$ only if $\chi(v) = 1$. Since χ is not the principal character $\chi(w) \neq 1$ and so $\chi(\overline{H}_3) = 0$ and $\chi(d_3) = \chi(xy) = 1$.

If $\chi(y) = 1$, then $\chi(\overline{H}_2) \neq 0$ only if $\chi(x) = 1$, and this reduces the problem to the previous case.

If $\chi(xy) = 1$, then $\chi(\overline{H}_3) \neq 0$ only if $\chi(x) = 1$, which reduces the problem again to the first case.

In the resulting k-factorization $A_1 A_2 A_3$ of G the multiplicity is $k = 2t$, $t \geq 1$. We point out that these k-factorizations can be extended to each $n \geq 3$. Indeed, let $G = K \times L$, where K is of type $(2, 2, 2, 2, 2t)$ and define A_4 by $H_4 = L$ and $d_4 \in K \setminus \{e\}$. Clearly, A_4 is not a subgroup of G and $A_1 A_2 A_3 A_4$ is a k-factorization of G.

Let $n = 4$ and let G be of type $(4, 4, 2, t)$, $t \geq 3$ with basis x, y, z, u. Define A_1, A_2, A_3, A_4 by as shown in Table 5.

Let χ be a nonprincipal character of G.

If $\chi(x^2) = 1$, then $\chi(\overline{H}_1) \neq 0$ only if $\chi(y) = 1$. Now $\chi(y^2) = 1$ and $\chi(\overline{H}_2) \neq 0$ only if $\chi(u) = 1$. Then $\chi(\overline{H}_4) \neq 0$ only if $\chi(x) = 1$. Since χ is not the principal

i	H_i	d_i
1	$\langle y, u \rangle$	x
2	$\langle x, v \rangle$	y
3	$\langle x, w \rangle$	xy

Table 4

i	H_i	d_i
1	$\langle y \rangle$	x^2
2	$\langle u \rangle$	y^2
3	$\langle z, u \rangle$	$x^2 y^2$
4	$\langle x \rangle$	u

Table 5

character, $\chi(z) \neq 1$ and this gives $\chi(\overline{H}_3) = 0$. We see that $\chi(d_3) = \chi(x^2 y^2) = 1$ also holds.

If $\chi(y^2) = 1$, then $\chi(\overline{H}_2) \neq 0$ only if $\chi(u) = 1$. Further $\chi(\overline{H}_4) \neq 0$ only if $\chi(x) = 1$, which reduces the problem to the previous case.

If $\chi(x^2 y^2) = 1$, then $\chi(\overline{H}_3) \neq 0$ only if $\chi(z) = 1$ and $\chi(u) = 1$. Now $\chi(\overline{H}_4) \neq 0$ only if $\chi(x) = 1$, and this reduces the problem to the first case.

In this k-factorization $A_1 A_2 A_3 A_4$ the multiplicity $k = t$ and $t \geq 3$. As we have already seen, this factorization can be extended to each $n \geq 4$.

This completes the proof.

Exercise 4.2.2. Let p be a prime and let ρ and σ be (p)th roots of unity. The $\rho = 1$ and $\sigma = 1$ cases are included. Prove that at least one of

$$\sigma, \ \rho, \ \rho\sigma, \ \rho\sigma^2, \ldots, \rho\sigma^{p-1}$$

is always equal to 1. (Hint: Use the pigeon hole principle to show that the list contains two equal elements. Then distinguish two cases depending on whether $\sigma = \rho\sigma^i$ or $\rho\sigma^i = \rho\sigma^j$.)

4.3 Multiple factorization by cyclic subsets

Hajós's theorem is related to a geometric conjecture of H. Minkowski stated about 1900. In algebraic form Minkowski's conjecture reads that if a finite abelian group is factored into cyclic subsets, then at least one of the factors must be a subgroup of the group. In 1936, before Minkowski's conjecture had been proven, Ph. Furtwängler formulated a more general conjecture which in algebraic form states that in each multiple factorization of a finite abelian group by cyclic subsets at least one of the factors must be a subgroup no matter what is the multiplicity. In 1938, Hajós refuted Furtwängler's conjecture. Exercises 4.3.1 and 4.3.2 are about counterexamples.

Exercise 4.3.1. Let $G = \langle x \rangle \times \langle y \rangle \times \langle z \rangle$, $|x| = |y| = 2$, $|z| = 3$ and let

$$A_1 = \{e, xz\}, \qquad\qquad A_2 = \{e, yz\},$$
$$A_3 = \{e, xz, (xz)^2\}, \qquad A_4 = \{e, yz, (yz)^2\},$$
$$A_5 = \{e, xyz, (xyz)^2\}.$$

Verify that the product $A_1 \cdots A_5$ is a multiple factorization of G with multiplicity 9. Further check that none of the cyclic subsets $A_1, \ldots, A_5$ is a subgroup of G.

Exercise 4.3.2. Let $G = \langle x \rangle \times \langle y \rangle$, $|x| = |y| = 4$ and

$$A_1 = \{e, x\}, \qquad A_2 = \{e, xy\},$$
$$A_3 = \{e, y\}, \qquad A_4 = \{e, xy^2\},$$
$$A_5 = \{e, xy^3\}, \qquad A_6 = \{e, x^2 y\}.$$

Check that $A_1 \cdots A_6$ is a 4-factorization of G without subgroup factors.

We will characterize those finite abelian groups for which Furtwängler's conjecture holds. The result can be summarized in the next theorem.

Theorem 4.3.1. *Let G be a finite cyclic group or a group of type $(p^\alpha, p, \ldots, p)$, where p is a prime. In each multiple factorization of G into cyclic subsets at least one of the factors must be a subgroup of G. If G is a finite abelian group which is not cyclic and not of type $(p^\alpha, p, \ldots, p)$, then G has a multiple factorization by nonsubgroup cyclic subsets.*

Proof. The proof is divided into five steps. The outline is the following.

(1) We adopt the character test to study multiple factorizations by cyclic subsets.

(2) A cyclic set can always be replaced by a product of cyclic subsets of prime cardinality. So we may restrict our attention to products of cyclic subsets of prime cardinality. It is not clear a priori that these prime cardinalities divide the order of the group. We show that we may focus our attention on such situation when this is the case.

(3) We show that if G is cyclic or of type $(p^\alpha, p, \ldots, p)$, then each multiple factorization of G by cyclic subsets must contain at least one subgroup among the factors.

(4) Groups of types (p^2, p^2) and (p, p, q), where p and q are distinct primes, admit factorizations by nonsubgroup cyclic subsets.

(5) The above factorizations can be extended for each finite abelian group that is not cyclic and is not of type $(p^\alpha, p, \ldots, p)$.

Let us turn to the details.

(1) Of course characters of the finite abelian group G can be exploited to study multiple factorizations of G. Let $A_1, \ldots, A_n$ be subsets of G. The product

$A_1 \cdots A_n$ which is not necessarily a set but rather a multiset of G is an m-fold factorization of G if for each character χ of G,

$$
\begin{aligned}
m\chi(G) &= \chi(A_1 \cdots A_n) \\
&= \chi(A_1) \cdots \chi(A_n).
\end{aligned}
$$

For the principal character χ of G this is simply $m|G| = |A_1| \cdots |A_n|$. For a non-principal character this gives $0 = \chi(A_1) \cdots \chi(A_n)$. Thus the product $A_1 \cdots A_n$ is a multiple factorization of G if $\mathrm{Ann}(A_1) \cup \cdots \cup \mathrm{Ann}(A_n)$ contains each nonprincipal character χ of G. The multiplicity is $m = (|A_1| \cdots |A_n|)/|G|$. The annihilator of the cyclic subset

$$
A_i = \{e, a_i, a_i^2, \ldots, a_i^{r(i)-1}\} \tag{1}
$$

can be described easily. Since

$$
\begin{aligned}
\chi(A_i) &= \chi(e) + \chi(a_i) + \chi(a_i^2) + \cdots + \chi(a_i^{r(i)-1}) \\
&= 1 + \chi(a_i) + \big(\chi(a_i)\big)^2 + \cdots + \big(\chi(a_i)\big)^{r(i)-1},
\end{aligned}
$$

it follows that $\chi(A_i) = r(i)$ if $\chi(a_i) = 1$ and

$$
\chi(A_i) = \frac{1 - \chi(a_i^{r(i)})}{1 - \chi(a_i)}
$$

if $\chi(a_i) \neq 1$. Thus $\mathrm{Ann}(A_i)$ consists of all characters χ of G for which $\chi(a_i) \neq 1$ and $\chi(a_i^{r(i)}) = 1$.

It is convenient to code the cyclic subsets $A_1, \ldots, A_n$ of the finite abelian group G by the generator elements $a_1, \ldots, a_n$ and the cardinalities $r(1), \ldots, r(n)$ respectively. The product $A_1 \cdots A_n$ is a multiple factoring of G if and only if

(i) the product $r(1) \cdots r(n)$ is a multiple of $|G|$,

(ii) for each nonprincipal character χ of G there is an i, $1 \leq i \leq n$ with $\chi(a_i) \neq 1$ and $\chi(a_i^{r(i)}) = 1$.

If in addition $a_i^{r(i)} \neq e$, for each i, $1 \leq i \leq n$, then there is no subgroup among the subsets $A_1, \ldots, A_n$.

(2) Suppose that $A_1, \ldots, A_n$ are cyclic subsets of the finite abelian group G such that the cardinalities $|A_1| = r(1), \ldots, |A_n| = r(n)$ are primes and the product $A_1 \cdots A_n$ is a multiple factorization of G. Assume that one of the primes $r(1), \ldots, r(n)$, say $r(1)$, does not divide $|G|$. Note that there is no character χ of G for which

$$
\chi(a_1) \neq 1 \quad \text{and} \quad \chi(a_1^{r(1)}) = \big(\chi(a_1)\big)^{r(1)} = 1. \tag{2}
$$

To prove the claim assume the contrary, that there is a character χ of G satisfying (2). Now from the equation $\big(\chi(a_1)\big)^{r(1)} = 1$ it follows that the order of $\chi(a_1)$ divides $r(1)$. From $\chi(a_1) \neq 1$ it follows that the order of $\chi(a_1)$ is $r(1)$. Further the order of $\chi(a_1)$ divides the order of a_1 which divides $|G|$. This gives the contradiction that r_1 divides $|G|$. Therefore there is no character χ of G satisfying (2), that is, $\mathrm{Ann}(A_1)$ is empty. Consequently, the cyclic subset A_1 can be cancelled from the multiple factorization $A_1 \cdots A_n$. Of course, after cancellation the multiplicity of the factorization changes from m to $m/r(1)$.

(3) Let G be a finite cyclic group and $A_1, \ldots, A_n$ cyclic subsets of G such that $A_1 \cdots A_n$ is a multiple factorization of G. Since G is cyclic there is a character χ of G which defines an isomorphism between G and the cyclic subgroup of the multiplicative group of roots of unity of order $|G|$. Such a character χ is called *faithful* and can be defined by $\chi(g) = \rho$, where g is a generator of G and ρ is a root of unity of order $|G|$. There must be an i, $1 \le i \le n$ with

$$\chi(a_i) \neq 1 \quad \text{and} \quad \chi(a_i^{r(i)}) = \big(\chi(a_i)\big)^{r(i)} = 1.$$

But from $\big(\chi(a_i)\big)^{r(i)} = 1$ it follows that $a_i^{r(i)} = e$, that is, the cyclic subset A_i is a subgroup of G.

Let G be a finite group of type $(p^\alpha, p, \ldots, p)$ and $A_1, \ldots, A_n$ cyclic subsets of G such that $A_1 \cdots A_n$ is a multiple factorization of G. By observation (2) we may assume that the cardinalities of the cyclic subsets are all equal to p, that is, $r(1) = \cdots = r(n) = p$. The group G is the direct product of subgroups H and K of types (p^α) and $(p, \ldots, p)$ respectively. Consider a character χ of G which is faithful on H and which is principal on K. There is an i, $1 \le i \le n$ with

$$\chi(a_i) \neq 1 \quad \text{and} \quad \chi(a_i^p) = \big(\chi(a_i)\big)^p = 1.$$

Write a_i in the form $a_i = hk$, $h \in H$, $k \in K$. Now

$$1 = \big(\chi(a_i)\big)^p = \big(\chi(h)\big)^p \big(\chi(k)\big)^p = \big(\chi(h)\big)^p,$$

hence $h^p = e$. Therefore $(hk)^p = a_i^p = e$ and so A_i is a subgroup of G.

(4) Let p and q be distinct primes and let $G = \langle x \rangle \times \langle y \rangle \times \langle z \rangle$, $|x| = |y| = p$, $|z| = q$. Thus G is of type (p, p, q).

We show that G has a multiple factorization by nonsubgroup cyclic subsets $A_1, \ldots, A_n$. The number of the cyclic subsets will be $n = p+3$ and the multiplicity is $m = q^p$. We code the cyclic subsets by the generator elements $a_1, \ldots, a_n$ and the cardinalities $r(1), \ldots, r(n)$. Table 1 summarizes the data.

An inspection shows that $a_i^{r(i)} \neq e$ for each i, $1 \le i \le n$. In order to prove that the product $A_1 \cdots A_n$ is a multiple factorization of G consider a nonprincipal character χ of G. Note that one of

$$\chi(y^q), \ \chi(x^q), \ \chi(x^q y^q), \ \chi(x^q y^{2q}), \ldots, \chi(x^q y^{(p-1)q})$$

is equal to 1. (Exercise 4.2.2 completes this part of the argument.)

If $\chi(y^q) = 1$, then condition (ii) is met unless $\chi(yz) = 1$. As q is relatively prime to p, from $\chi(y^q) = 1$ it follows that $\chi(y) = 1$. This and $\chi(yz) = 1$ give

i	a_i	$r(i)$	$a_i^{r(i)}$
1	xz	p	z^p
2	yz	p	z^p
3	yz	q	y^q
4	xz	q	x^q
5	xyz	q	$x^q y^q$
$\vdots$	$\vdots$	$\vdots$	$\vdots$
n	$xy^{p-1}z$	q	$x^q y^{(p-1)q}$

Table 1

that $\chi(z) = 1$ and so that $\chi(z^p) = 1$. Now condition (ii) is met unless $\chi(xz) = \chi(yz) = 1$. But from these $\chi(x) = \chi(y) = 1$ follow and so χ would be the principal character of G, which is not the case.

If $\chi(x^q y^{iq}) = 1$, then condition (ii) is satisfied unless $\chi(xy^i z) = 1$. From this again $\chi(z) = 1$ follows and we can finish the proof as before.

Turn to the case of groups of type (p^2, p^2). Let p be a prime and let $G = \langle x \rangle \times \langle y \rangle$, $|x| = |y| = p^2$.

We show that G has a multiple factorization by nonsubgroup cyclic subsets $A_1, \ldots, A_n$. The number of the cyclic subsets will be $n = 2p+2$ and the multiplicity will be $m = p^{2p-2}$. We code the cyclic subsets by the generator elements $a_1, \ldots, a_n$ and the cardinalities $r(1), \ldots, r(n)$. Table 2 summarizes the information we need.

An inspection shows that $a_i^{r(i)} \neq e$ for each i, $1 \le i \le n$. In order to prove that the product $A_1 \cdots A_n$ is a multiple factorization of G consider a nonprincipal character χ of G. Start with the last column of the table. Let us observe that one of

$$\chi(y^p),\ \chi(x^p),\ \chi(x^p y^p),\ \chi(x^p y^{2p}), \ldots, \chi(x^p y^{(p-1)p})$$

is equal to 1.

Assume first that $\chi(y^p) = 1$. Condition (ii) is not satisfied only if $\chi(y) = \chi(x^p y) = 1$. From this it follows that $\chi(x^p) = 1$. Again condition (ii) is not satisfied only if $\chi(x) = 1$. But in this last case we have the contradiction $\chi(x) = \chi(y) = 1$.

Assume secondly that $\chi(x^p y^{ip}) = 1$. Condition (ii) is not satisfied only if $\chi(xy^i) = \chi(xy^{p+i}) = 1$. Now $\chi(y^p) = 1$ and the problem reduces to the previous case.

(5) In this step we will show that the above multiple factorizations can be extended to all finite abelian groups except cyclic groups and groups of type $(p^\alpha, p, \ldots, p)$.

Let G be a finite abelian group and let H be a subgroup of G such that the

i	a_i	$r(i)$	$a_i^{r(i)}$
1	x	p	x^p
2	xy	p	$x^p y^p$
3	xy^2	p	$x^p y^{2p}$
$\vdots$	$\vdots$	$\vdots$	$\vdots$
p	xy^{p-1}	p	$x^p y^{(p-1)p}$
$p+1$	y	p	y^p
$p+2$	xy^p	p	x^p
$p+3$	xy^{p+1}	p	$x^p y^p$
$p+4$	xy^{p+2}	p	$x^p y^{2p}$
$\vdots$	$\vdots$	$\vdots$	$\vdots$
$n-1$	xy^{p+p-1}	p	$x^p y^{(p-1)p}$
n	$x^p y$	p	y^p

Table 2

index $|G : H| = r$ is a prime. If H is an m-fold product of nonsubgroup cyclic subsets, then so is G. By the fundamental theorem of the finite abelian groups H is a direct product of cyclic subgroups of prime power orders. In other words H is of type $(q_1, \ldots, q_s)$, where $q_1, \ldots, q_s$ are prime powers. Since H is an m-fold product of nonsubgroup cyclic subsets it follows that $(q_1, \ldots, q_s)$ is not equal to $(p^\alpha, p, \ldots, p)$ and H is not a cyclic subgroup. As $|G : H| = r$, G is of type $(q_1, \ldots, q_s, r)$ or $(q_1 r, q_2, \ldots, q_s)$, where q_1 is a power of r.

Assume first that G is of type $(q_1, \ldots, q_s, r)$. In this case G is a direct product of subgroups K and L of types $(q_1, \ldots, q_s)$ and (r) respectively. Let x be a basis element of L. Clearly, $G = H[x, r] = H[xh, r]$ is a simple factorization of G for each $h \in H$. Here $[x, r]$ stands for the cyclic subset $\{e, x, x^2, \ldots, x^{r-1}\}$. If $(xh)^r = e$ for each $h \in H$, then H is of type $(p, \ldots, p)$. This is not the case so $(xh)^r \neq e$ for some $h \in H$ hence $[xh, r]$ is not a subgroup of G for some $h \in H$.

Assume that G is of type $(q_1 r, q_2, \ldots, q_s)$. Now G is a direct product of the subgroups K and L of type $(q_1 r)$ and $(q_2, \ldots, q_s)$ respectively. Let x be a generator element of K. Clearly, $K = [x, r]\langle x^r \rangle$ is a simple factorization of K and $[x, r]$ is not a subgroup of K. Hence $G = [x, r]\langle x^r \rangle L$. The subgroup $\langle x^r \rangle L$ is not cyclic and is not of type $(p^\alpha, p, \ldots, p)$ and so it can be represented as an m-fold product of nonsubgroup cyclic subsets.

Using steps (3), (4), and (5) we can now complete the proof of the theorem.

Let G be a group of type

$$\left(p(1)^{\alpha(1,1)}, \ldots, p(1)^{\alpha(1,s(1))}, \ldots, p(u)^{\alpha(u,1)}, \ldots, p(u)^{\alpha(u,s(u))}\right),$$

where $p(1), \ldots, p(u)$ are distinct primes and in addition we assume that

$$\alpha(i,1) \geq \cdots \geq \alpha(i,s(i)) \geq 1 \quad \text{for} \quad 1 \leq i \leq u.$$

First assume that $u = 1$. If $s(1) = 1$ or $s(1) \geq 2$ but $\alpha(1,2) = \cdots = \alpha(1, s(1)) = 1$, then by step (3), G cannot have a multiple factorization by nonsubgroup cyclic subsets. In the remaining cases $\alpha(1,2) \geq 2$ and so G has a subgroup H whose type is $((p(1))^2, (p(1))^2)$. Hence by step (4) H can be multiple factored into nonsubgroup cyclic subsets and then by step (5) so can be G.

Secondly assume that $u \geq 2$. If $s(1) = \cdots = s(u) = 1$, then G is cyclic and so G cannot have a multiple factorization by nonsubgroup cyclic subsets. In the remaining cases $s(i) \neq 1$ for some i, $1 \leq i \leq u$. We assume that $s(1) \geq 2$. Now G has a subgroup H of type $(p(1), p(1), p(2))$ and so by step (4), H and then by step (5), G has a multiple factorization by nonsubgroup cyclic subsets.

This completes the proof.

Theorem 4.3.2. *In a multiple factorization of a finite abelian group by cyclic subsets, there always occurs a subgroup among the factors if the multiplicity is relatively prime to the order of the group.*

Proof. Let $A_1, \ldots, A_n$ be cyclic subsets of the finite abelian group G and suppose that the product $A_1 \cdots A_n$ is an m-factoriza-tion of G, where m is relatively prime to $|G|$. As we have seen in the previous proof we may assume that each $|A_i|$ is a prime which divides $|G|$. If m has a prime divisor, say p, then from $m|G| = |A_1| \cdots |A_n|$ it follows that $p \mid |A_i|$ for some i, $1 \leq i \leq n$ and hence $p \mid |G|$. This is not possible. Thus $m = 1$. Therefore by Hajós's theorem, one of the factors $A_1, \ldots, A_n$ is a subgroup of G.

This completes the proof.

Hajós's theorem holds for multiple factorizations provided the multiplicity of the factorization is relatively prime to the order of the group. The next constructions exhibit examples to show that Rédei's theorem cannot be extended in the above way.

Let G be a group of type $\left(p(1), \ldots, p(s), q\right)$, where $p(1), \ldots, p(s)$, and q are primes. Let $x_1, \ldots, x_s, y$ be basis elements of G with orders $p(1), \ldots, p(s), q$ respectively. Set

$$A_1 = \left\{e, x_1, x_1^2, \ldots, x_1^{p(1)-2}, x_1^{p(1)-1} y\right\},$$

$$\vdots$$

$$A_s = \left\{e, x_s, x_s^2, \ldots, x_s^{p(s)-2}, x_s^{p(s)-1} y\right\},$$
$$B = \left\{e, y, y^2, \ldots, y^{q-2}, x_1 y^{q-1}\right\},$$
$$C = \langle x_1 \rangle \cup x_2 \langle y \rangle.$$

We claim that the product $A_1 \cdots A_s BC$ is a $(p(1)+q)$-fold factorization of G. To prove the claim it is enough to show that $\chi(A_1 \cdots A_s BC) = 0$ for each nonprincipal character χ of G. Let χ be a nonprincipal character of G. If $\chi(y) = 1$, then $\chi(x_i) \neq 1$ for some i, $1 \leq i \leq s$. Now $\chi(A_i) = 0$. Thus we may suppose that $\chi(y) \neq 1$. If $\chi(x_1) = 1$, then $\chi(B) = 0$. If $\chi(x_1) \neq 1$, then $\chi(C) = 0$.

Choosing $p(1)$ to be 2 and $2+q$ to be a prime r provides an r-fold factorization of G in which each factor has a prime number of elements.

Exercise 4.3.3. Let $G = \langle x \rangle \times \langle y \rangle \times \langle z \rangle$, $|x| = |y| = 2$, $|z| = 3$ and

$$A_1 = \{e, xz\}, \qquad A_2 = \{e, yz\},$$
$$A_3 = \{e, xz, z^2\}, \qquad A_4 = \{e, x, y, yz, yz^2\}.$$

Verify that the product $A_1 \cdots A_4$ is a multiple factorization of G with multiplicity 5. Show that none of the factors is a subgroup of G.

In the next construction the group is cyclic. Let G be a group of type (p^α, q), where p and q are distinct primes. Let x, y be basis elements of G such that $|x| = p^\alpha$, $|y| = q$. Let

$$A_1 = \{e, x, x^2, \ldots, x^{p-1}\},$$
$$A_2 = \{e, x^p, x^{2p}, \ldots, x^{(p-1)p}\},$$
$$\vdots$$
$$A_{\alpha-1} = \{e, x^{p^{\alpha-2}}, x^{2p^{\alpha-2}}, \ldots, x^{(p-1)p^{\alpha-2}}\},$$
$$A_\alpha = \{e, x^{p^{\alpha-1}}, x^{2p^{\alpha-1}}, \ldots, x^{(p-2)p^{\alpha-1}}, x^{(p-1)p^{\alpha-1}}y\},$$
$$B = \{e, y, y^2, \ldots, y^{q-2}, x^{p^{\alpha-1}}y^{q-1}\},$$
$$C = \langle x^{p^{\alpha-1}} \rangle \cup x\langle y \rangle.$$

We claim that the product $A_1 \cdots A_\alpha BC$ is a $(p+q)$-fold factorization of G. To prove the claim pick a nonprincipal character χ of G. Suppose $\chi(y) = 1$. As $\chi(x) \neq 1$, $\chi(x^{p^\alpha}) = 1$, it follows that $\chi(x^{p^{i-1}}) \neq 1$, $\chi(x^{p^i}) = 1$ for some i, $1 \leq i \leq \alpha$. Now $\chi(A_i) = 0$. Thus we may assume that $\chi(y) \neq 1$. If $\chi(x^{p^{\alpha-1}}) = 1$, then $\chi(B) = 0$. If $\chi(x^{p^{\alpha-1}}) \neq 1$, then $\chi(C) = 0$.

Choosing p to be 2 and $2+q$ to be a prime r provides an r-fold factorization of G in which each factor has a prime number of elements.

Exercise 4.3.4. Let $G = \langle x \rangle \times \langle y \rangle$, $|x| = 4$, $|y| = 3$ and

$$A_1 = \{e, xy\}, \qquad A_2 = \{e, x^2y\},$$
$$A_3 = \{e, x^2y, y^2\}, \qquad A_4 = \{e, x, x^2, xy, xy^2\}.$$

Check that the product $A_1 \cdots A_4$ is a 5-fold factorization of G without subgroup factors.

Theorem 4.3.1 can be used to give a partial answer to a problem of K. Corrádi. First we describe the problem. Let G be a finite abelian group and let $A_1, \ldots, A_n$ be cyclic subsets of G such that A_i is in form (1). If $G = BA_1 \cdots A_n$ is a factorization of G does it follow that $\langle B \rangle \cap \{a_1^{r(1)}, \ldots, a_n^{r(n)}\} \neq \emptyset$? This problem is an attempt to find an interpolation between Hajós's theorem and Keller's conjecture. By Keller's conjecture $B^{-1}B \cap \{a_1^{r(1)}, \ldots, a_n^{r(n)}\} \neq \emptyset$. Plainly $B^{-1}B \subset \langle B \rangle$ and Corrádi's problem is a weakened version of Keller's conjecture. On the other hand the $B = \{e\}$ special case of the problem gives $a_i^{r(i)} = e$ for some i, $1 \leq i \leq n$. In other words it would imply Hajós's theorem.

Theorem 4.3.3. *Let G be a finite cyclic group or a group of type $(p^\alpha, p, \ldots, p)$, where p is a prime. Let $A_1, \ldots, A_n$ be cyclic subsets of G such that A_i is in form (1). If $G = BA_1 \cdots A_n$ is a factorization of G, then $\langle B \rangle \cap \{a_1^{r(1)}, \ldots, a_n^{r(n)}\} \neq \emptyset$.*

Proof. If A_i has composite order, say $|A_i| = r(i) = uv$, then A_i is a direct product of cyclic subsets of order u and v. Namely,

$$A_i = \{e, a_i, a_i^2, \ldots, a_i^{u-1}\}\{e, a_i^u, a_i^{2u}, \ldots, a_i^{(v-1)u}\}.$$

Since $a_i^u \in \langle B \rangle$ implies $a_i^{uv} = a_i^{r(i)} \in \langle B \rangle$ we may assume that each cyclic subset A_i has a prime number of elements. Set $H = \langle B \rangle$. If $H = G$, then we are obviously done. Suppose $H \neq G$. We may assume that $a_1, \ldots, a_s \in H$ and $a_{s+1}, \ldots, a_n \notin H$ since this is only a matter of rearranging the factors $A_1, \ldots, A_n$. Considering the factor group G/H we can see that the product

$$(A_{s+1}H)/H \cdots (A_nH)/H$$

is a multiple factorization of G/H. Note that G/H is cyclic or is of type $(p^\alpha, p, \ldots, p)$. By Theorem 4.3.1, $(A_iH)/H$ is a subgroup of G/H for some i, $s + 1 \leq i \leq n$. It means that $a_i^{r(i)}H = H$, that is, $a_i^{r(i)} \in H$.

   This completes the proof.

4.4 Robinson's result

This is a long section about the complete characterization of dimensions and multiplicities for which Furtwängler's conjecture holds. We start with describing five constructions.

(1) Let G be of type $(2, 4, p)$ with basis elements x, y, z. Here $|x| = 2$, $|y| = 4$, $|z| = p$ and p is an odd prime. Set a_i, $r(i)$ as shown in Table 1. Notice that y^{2p}, x^p, $x^p y^{2p}$ are equal to y^2, x, xy^2 respectively. We claim that for each nonprincipal character χ of G there is an a_i such that

$$\chi(a_i) \neq 1, \quad \chi(a_i^{r(i)}) = 1, \quad a_i^{r(i)} \neq e.$$

One of $\chi(y^{2p})$, $\chi(x^p)$, $\chi(x^p y^{2p})$ is always 1. Suppose $\chi(y^{2p}) = \chi(y^2) = 1$. This is a bad situation only if $\chi(xyz) = 1$. This gives that $\chi(z) = 1$. Now $\chi(z^4) = 1$. This is bad only if $\chi(yz) = 1$, that is $\chi(y) = 1$. We have $\chi(z) = 1$, $\chi(y) = 1$, $\chi(xyz) = 1$ and they imply $\chi(x) = 1$. Therefore χ is the principal character of G.

Suppose $\chi(x^p) = \chi(x) = 1$. This is bad only if $\chi(xz) = 1$ which in turn gives that $\chi(z) = 1$ and then $\chi(z^4) = 1$. This is bad only if $\chi(yz) = 1$ hence $\chi(y) = 1$. Now χ is the principal character.

Suppose $\chi(x^p y^{2p}) = \chi(xy^2) = 1$. This is bad only if $\chi(xy^2 z) = 1$ and so $\chi(z) = 1$. Hence $\chi(z^4) = 1$. This is bad only if $\chi(yz) = 1$ and consequently $\chi(y) = 1$. Now $\chi(y^2) = 1$ which reduces the problem to the an earlier case.

The multiplicity of the factorization is

$$k = \frac{4 \cdot 2 \cdot 2p \cdot p \cdot p}{2 \cdot 4 \cdot p} = p^2$$

and the number of the factors is $n = 4$.

(2) Let G be of type $(2, 4, p)$ with basis elements x, y, z, where $|x| = 2$, $|y| = 4$, $|z| = p$ and p is an odd prime. Set a_i, $r(i)$ by Table 2. Note that x^p, $x^p y^{2p}$, y^2 are equal to x, xy^2, y^2 respectively. Let χ be a nonprincipal character of G. One of $\chi(x^p)$, $\chi(x^p y^{2p})$, $\chi(y^2)$ is always 1. Suppose $\chi(x^p) = \chi(x) = 1$. This is bad only if $\chi(xz) = 1$, that is, $\chi(z) = 1$. $\chi(z) = 1$ implies $\chi(z^2) = 1$. This is bad only if $\chi(y^2 z) = 1$, that is, $\chi(y^2) = 1$. This is bad only if $\chi(y) = 1$. But in this case χ is the principal character of G.

Suppose $\chi(y^2) = 1$. This is bad only if $\chi(y) = 1$ and $\chi(xy) = 1$, that is, $\chi(x) = 1$. This reduces the problem to the previous case.

Suppose $\chi(x^p y^{2p}) = \chi(xy^2) = 1$. This is bad only if $\chi(xy^2 z) = 1$, that is, $\chi(z) = 1$. $\chi(z) = 1$ implies $\chi(z^2) = 1$. This is bad only if $\chi(y^2 z) = 1$, that is $\chi(y^2) = 1$. This reduces the problem to the earlier case.

The multiplicity of the factoring is

$$k = \frac{p \cdot p \cdot 2 \cdot 2 \cdot 2}{2 \cdot 4 \cdot p} = p$$

and the number of factors is $n = 5$.

i	a_i	$r(i)$	$a_i^{r(i)}$
1	yz	4	z^4
2	xyz	$2p$	y^{2p}
3	xz	p	x^p
4	xy^2z	p	$x^p y^{2p}$

Table 1

i	a_i	$r(i)$	$a_i^{r(i)}$
1	xz	p	x^p
2	xy^2z	p	$x^p y^{2p}$
3	y	2	y^2
4	xy	2	y^2
5	y^2z	2	z^2

Table 2

i	a_i	$r(i)$	$a_i^{r(i)}$
1	xy^2z^3	2	y^4z^6
2	xy^2z^6	2	y^4z^3
3	xy^2	2	y^4
4	xz	6	z^6
5	xyz	9	x^9

Table 3

i	a_i	$r(i)$	$a_i^{r(i)}$
1	xz	2	z^2
2	xy	2	y^2
3	xyz	2	y^2z^2
4	xyz^2	2	y^2
5	xyz^3	2	y^2z^2
6	xy^2z	2	z^2

Table 4

(3) Let G be of type $(2,3,9)$ with basis elements x, y, z, where $|x| = 2$, $|y| = 3$, $|z| = 9$. Set a_i, $r(i)$ as given in Table 3.

Note that y^4z^6, y^4z^3, y^4, z^6 are equal to $y(z^3)^2$, yz^3, y, $(z^3)^2$ respectively. One of $\chi(y^4z^6)$, $\chi(y^4z^3)$, $\chi(y^4)$, $\chi(z^6)$ is always 1. Suppose $\chi(y^4) = \chi(y) = 1$. This is bad only if $\chi(xy^2) = 1$, that is $\chi(x) = 1$. $\chi(x) = 1$ implies $\chi(x^9) = 1$. this is bad only if $\chi(xyz) = 1$. Now $\chi(z) = 1$. So χ is the principal character of G.

Suppose $\chi(z^6) = \chi(z^3) = 1$. This is bad only if $\chi(xz) = 1$. Now $\chi(x) = 1$ and consequently $\chi(x^9) = 1$. This is bad if $\chi(xyz) = 1$ which in turn implies $\chi(y) = 1$. Then $\chi(y^4) = 1$ and this reduces the problem to the previous case.

Suppose $\chi(y^4z^3) = 1$. This is bad only if $\chi(xy^2z^3) = 1$ from which it follows that $\chi(x) = 1$ and so $\chi(x^9) = 1$. This is bad only if $\chi(xyz) = 1$. This implies $\chi(yz) = 1$. Therefore $\chi(y^2z^2) = 1$. Then $\chi(z) = 1$ and so $\chi(z^6) = 1$ which reduces the problem to the earlier case.

Suppose $\chi(y^4z^6) = 1$. This is bad only if $\chi(xy^2z^3) = 1$. This implies $\chi(x) = 1$. Then $\chi(x^9) = 1$. This is bad only if $\chi(xyz) = 1$. We get $\chi(yz) = 1$. From this $\chi(y^4z^4) = 1$. Comparing this with $\chi(y^4z^6) = 1$ results in $\chi(z^2) = 1$. Then $\chi(z) = 1$ and $\chi(z^6) = 1$ reduces the problem to an earlier case.

The multiplicity of the factorization is

$$k = \frac{2 \cdot 2 \cdot 2 \cdot 6 \cdot 9}{2 \cdot 3 \cdot 9} = 8$$

and the number of the factors is $n = 5$.

(4) Let G be of type $(2,4,4)$ with basis elements x, y, z, where $|x| = 2$, $|y| = 4$, $|z| = 4$. Set a_i, $r(i)$ by Table 4. Let χ be a nonprincipal character of G. One of $\chi(y^2)$, $\chi(z^2)$, $\chi(y^2z^2)$ is always 1. Suppose $\chi(y^2) = 1$. This is bad only if $\chi(xy) = 1$ and $\chi(xyz^2) = 1$. These give $\chi(z^2) = 1$. This is bad only if $\chi(xz) = 1$ and $\chi(xy^2z) = 1$. $\chi(y^2) = 1$ and $\chi(z^2) = 1$ imply $\chi(y^2z^2) = 1$. This is bad only if $\chi(xyz) = 1$ and $\chi(xyz^3) = 1$. Now $\chi(xy) = 1$, $\chi(xyz) = 1$ give $\chi(z) = 1$. Then from $\chi(xz) = 1$ we get $\chi(x) = 1$. Then from $\chi(xy) = 1$ we get $\chi(y) = 1$. Therefore χ is the principal character of G.

Suppose $\chi(z^2) = 1$. This is bad only if $\chi(xz) = 1$ and $\chi(xy^2z) = 1$. These give $\chi(y^2) = 1$ which reduces the problem to the previous case.

Suppose $\chi(y^2z^2) = 1$. This is bad only if $\chi(xyz) = 1$ and $\chi(xyz^3) = 1$. From these it follows that $\chi(z^2) = 1$ which reduces the problem to the earlier case.

The multiplicity of the factorization is

$$k = \frac{2^6}{2 \cdot 4 \cdot 4} = 2$$

and the number of the factors is $n = 6$.

(5) We have seen in Section 4.3 that a group of type (p^2, p^2) has a multiple factoring by nonsubgroup cyclic subsets, where the multiplicity is $k = p^{2p-2}$ and the number of the factors is $n = 2p + 2$. In the $p = 2$ special case this provides a multiple factoring with $k = 4$, $n = 6$.

Construction (1) shows that $(4, p^2)$ is an admissible (n, k) pair for each odd prime p. We will show that the multiplicity k can be extended for p^2t, where $t \geq 2$ is an integer. Write t as a product of primes, say $t = q(1) \cdots q(s)$. Note that $\left|a_1^{r(1)}\right| = p$ and $\left|a_4^{r(4)}\right| = 2$. When $q(1) = p$, then $q(1) \neq 2$. Replace $r(4)$ by $r'(4) = r(4)q$. The cyclic subset $A_4 = \{e, a_4, a_4^2, \ldots, a_4^{r(4)-1}\}$ will be replaced by $A_4' = \{e, a_4, a_4^2, \ldots, a_4^{r(4)q(1)-1}\}$. It is possible that A_4' has elements with multiplicity when $r(4)q(1)$ is greater than the order of a_4. In this case A_4' is a multiset. We claim that A_4' is not an integer multiple of a subgroup of G and that $A_1 A_2 A_3 A_4'$ is a $pq(1)$-fold factorization of G. Indeed,

$$\chi(a_4) \neq 1, \quad \chi\big(a_4^{r(4)}\big) = 1, \quad a_4^{r(4)} \neq e$$

implies

$$\chi(a_4) \neq 1, \quad \chi\big(a_4^{r(4)q(1)}\big) = 1, \quad a_4^{r(4)q(1)} \neq e.$$

When $q(1) = 2$, then $q(1) \neq p$. In this case replace $r(1)$ by $r'(1) = r(1)q(1)$ and $A_1 A_2 A_3 A_4'$ is a $pq(1)$-fold factorization of G. Continuing in this way we get a (pt)-fold factorization of G by nonsubgroup cyclic multisets. In short $(4, p^2t)$ is an admissible pair. For $n = 4$, only when k is not a multiple of an odd prime has the pair (n, k) a chance to be inadmissible.

Construction (2) shows that $(5, p)$ is an admissible pair for each odd prime p. Using that $\left|a_1^{r(1)}\right| = 2$, $\left|a_5^{r(5)}\right| = p$ we can conclude that $(5, pt)$ is an admissible pair. By construction (3), $(5, 8)$ is an admissible pair where $\left|a_1^{r(1)}\right| = 3$, $\left|a_5^{r(5)}\right| = 2$. From this it follows that $(5, 8t)$ is an admissible pair for each positive integer t. For $n = 5$, only $k = 1, 2, 4$ has the pair (n, k) a chance to be inadmissible.

Let H be a subgroup of G such that the factor group G/H is cyclic with basis element bH. The elements of G/H are

$$H, bH, b^2 H, \ldots, b^{r-1} H.$$

Clearly, $G = BH$ is a 1-fold factorization of G, where $B = \{e, b, b^2, \ldots, b^{r-1}\}$ is a cyclic subset. If H is given by its type $(t_1, \ldots, t_s)$, then we can choose G such that

B is not a subgroup. For example we can choose G to be of type $(t_1 r, t_2, \ldots, t_s)$ with basis elements $x_1, \ldots, x_s$. The set $b = x_1$, $H = \langle x_1^r, x_2, \ldots, x_s \rangle$. Suppose that the product of nonsubgroup cyclic multisets $A_1 \cdots A_5$ is a k-fold factorization of H. Then the product $A_1 \cdots A_5 B$ is a k-fold factorization of G. In short $(5, k)$ is an admissible pair, then so is $(6, k)$. In general, (n, k) is admissible for each $n \geq 5$. By constructions (4) and (5), $(6, 2)$, $(6, 4)$ are admissible pairs. This shows that for $n \geq 6$ only $k = 1$ has the pair (n, k) a chance to be inadmissible. This shows that for $n \geq 6$ only when $k = 1$ has the pair (n, k) a chance to be inadmissible.

We will need five lemmas.

Let $b_1, \ldots, b_n$ be elements of a finite abelian group G. The vanishing product

$$(e - b_1) \cdots (e - b_n) = 0 \tag{1}$$

in the group ring $Z(G)$ is called *primitive* if $n \geq 2$ and no factor can be cancelled from (1) without destroying the equation.

Lemma 4.4.1. *If (1) is primitive, then*
(i) $\langle b_i \rangle \not\subset \langle b_j \rangle$ for $i \neq j$ for each i, j, $1 \leq i, j \leq n$, $i \neq j$,
(ii) $|b_i| \leq 2^{n-2}$ for each i, $1 \leq i \leq n$.

Proof. In order to prove (i) assume that $\langle b_2 \rangle \subset \langle b_1 \rangle$. From (1) it follows that

$$[1 - \chi(b_1)][1 - \chi(b_2)] \cdots [1 - \chi(b_n)] = 0$$

for each character χ of G. We get then

$$[1 - \chi(b_2)] \cdots [1 - \chi(b_n)] = 0.$$

It is clear when $\chi(b_i) = 1$ for some i, $2 \leq i \leq n$. If $\chi(b_1) = 1$, then it implies $\chi(b_2) = 1$. Thus

$$\chi\big((e - b_2) \cdots (e - b_n)\big) = 0$$

holds for each character χ of G and so $(e - b_2) \cdots (e - b_n) = 0$, showing that (1) is not primitive.

To prove (ii) multiply out the product

$$(e - b_1) \cdots (e - b_{n-1}).$$

We get 2^{n-1} products in the form $c_1 \cdots c_{n-1}$, where each c_i is either e or a b_j. Half of the products are with positive sign, half of the products are with negative sign. The number of times an element g of G appears on the positive list is denoted by $\lambda(g)$. The number of times g appears on the negative list is designated by $\mu(g)$. Set

$$U = \sum_{0 < \lambda(g) - \mu(g)} g, \quad V = \sum_{\lambda(g) - \mu(g) < 0} g.$$

Clearly,

$$(e - b_1) \cdots (e - b_{n-1}) = U - V,$$

U, V have nonnegative coefficients and U, V have no terms in common. The set of elements of G that appear in U with nonzero coefficients is called the support of U and we will denote it by $\{U\}$. The equation (1) is equivalent to $(U-V)(e-b_n) = 0$ and then

$$U + V b_n = U b_n + V.$$

As U, V have no common terms, it follows that

$$\{V\} \subset \{V b_n\} = \{V\} b_n, \quad \{U\} \subset \{U b_n\} = \{U\} b_n.$$

Cardinalities show that in fact

$$\{V\} = \{V\} b_n, \quad \{U\} = \{U\} b_n,$$

that is, $\{U\}, \{V\}$ are periodic with period b_n. Multiplying elements of $\{U\}$ by b_n permutes elements of $\{U\}$. The permutation consists of cycles whose common length is $|b_n|$. Thus $|b_n|$ divides the cardinality of $\{U\}$. This gives $|b_n| \leq 2^{n-2}$.

The proof is complete.

Note that if (1) is primitive, then $n = 2$ is not possible. Indeed, by Lemma 4.4.1, $|b_1| = |b_2| = 1$. Thus (1) is not primitive.

Also note that if (1) is primitive, $n = 3$, then $H = \langle b_1, b_2, b_3 \rangle$ is of type $(2, 2)$ and $\langle b_1 \rangle, \langle b_2 \rangle, \langle b_3 \rangle$ are all the maximal subgroups of H. Indeed, by Lemma 4.4.1, $|b_1| = |b_2| = |b_3| = 2$. Since $|b_3| = 2$, $\{U\}, \{V\}$ must have two elements. Now $\{e, b_1 b_2\}, \{b_1, b_2\}$ are periodic with period b_3. This gives

$$b_1 b_2 b_3 = e, \quad b_1 b_3 = b_2, \quad b_2 b_3 = b_1$$

and so $\langle b_1, b_2 \rangle$ is a group of type $(2, 2)$.

Similarly, if (1) is primitive, $n = 4$ and $|b_4| = 4$, then $H = \langle b_1, \ldots, b_4 \rangle$ is of type $(2, 4)$ and $\langle b_1 \rangle, \ldots, \langle b_4 \rangle$ are all the maximal subgroups of H. The reason is the following. As $|b_4| = 4$, both $\{U\}, \{V\}$ must have four elements. Now

$$\{e, b_1 b_2, b_1 b_3, b_2 b_3\}, \quad \{b_1, b_2, b_3, b_1 b_2 b_3\}$$

are periodic with period b_4. We may assume that

$$(b_1 b_2 b_3) b_4 = b_1, \quad (b_1) b_4 = b_3, \quad (b_3) b_4 = b_2, \quad (b_2) b_4 = b_1 b_2 b_3$$

since this is only a matter of indexing the elements b_1, b_2, b_3. It follows that $b_1^2 = e$, $b_2 = b_1 b_4^2$, $b_3 = b_1 b_4$. The group $\langle b_1, b_4 \rangle$ is of type $(2, 4)$.

We carry out a more detailed analysis for $n = 4$. Our main tool will be Lemma 3.3.1. We will refer to this result as the first zero divisor lemma.

Lemma 4.4.2. *If (1) is primitive and $n = 4$, then the type of $H = \langle b_1, \ldots, b_4 \rangle$ is either (i) $(2, 4)$ or (ii) $(3, 3)$ and $\langle b_1 \rangle, \ldots, \langle b_4 \rangle$ are all the maximal subgroups of H.*

Proof. We may assume that $|b_1| \leq \cdots \leq |b_4|$. By Lemma 4.4.1, $|b_4| \leq 4$. We have seen that if $|b_4| = 4$, then we have (i). We may assume that $|b_3| \leq 3$. We claim that if $|b_4| = 3$, then $|b_i| = 4$ for each i, $1 \le i \le 4$. In order to verify this claim suppose that, say $|b_1| = |b_2| = 2$, $|b_3| = |b_4| = 3$. From (1) with the $B = (e - b_1)(e - b_2)$ choice by the first zero divisor lemma, it follows that $\langle b_1, b_2 \rangle$ contains a nonidentity element of $\langle b_3, b_4 \rangle$. This is not possible because of the order of the elements. Therefore if $|b_4| = 3$, then $H = \langle b_1, \ldots, b_4 \rangle$ is one of the types

$$(3), \ (3,3), \ (3,3,3), \ (3,3,3,3).$$

If H is type (3), then $\langle b_1 \rangle = \langle b_2 \rangle$ which is not possible by Lemma 4.4.1. From (1) the first zero divisor lemma with the $B = (e - b_1)(e - b_2)(e - b_3)$ choice gives that $b_4 \in \langle b_1, b_2, b_3 \rangle$. This means that H cannot be of type $(3,3,3,3)$. Assume that H is of type $(3,3,3)$ with basis elements b_1, b_2, b_3 and $b_4 = b_1^u b_2^v b_3^w$. An inspection shows that for any choice of $u \le v \le w$ there is a character χ of G such that

$$[1 - \chi(b_1)][1 - \chi(b_2)][1 - \chi(b_3)] \ne 0.$$

This gives the contradiction that (1) does not hold.

Finally we show that $|b_4| = 2$ is not possible. If $|b_1| = \cdots = |b_4| = 2$, then the type of $H = \langle b_1, \ldots, b_4 \rangle$ is one of

$$(2), \ (2,2), \ (2,2,2), \ (2,2,2,2).$$

From (1) with the $B = (e - b_1)(e - b_2)(e - b_3)$ choice the first zero divisor lemma gives that $b_4 \in \langle b_1, b_2, b_3 \rangle$ which sorts out the $(2,2,2,2)$ type. If H is of type (2) or $(2,2)$, then there are $i \ne j$ with $\langle b_i \rangle = \langle b_j \rangle$. Assume that H is of type $(2,2,2)$. We may assume that b_1, b_2, b_3 are the basis elements of H. If b_4 is a product of two basis elements, say $b_4 = b_1 b_2$, then $(e - b_1)(e - b_2)(e - b_4) = 0$ since $[1 - \chi(b_1)][1 - \chi(b_2)][1 - \chi(b_1 b_2)] = 0$ holds for each character χ of $\langle b_1, b_2 \rangle$. Thus (1) is not primitive. Therefore $b_4 = b_1 b_2 b_3$. Now

$$[1 - \chi(b_1)][1 - \chi(b_2)][1 - \chi(b_3)][1 - \chi(b_1 b_2 b_3)] \ne 0$$

for the character χ of H defined by $\chi(b_1) = \chi(b_2) = \chi(b_3) = -1$.

This completes the proof.

Lemma 4.4.3. *If (1) is primitive and $n = 5$, then there is a permutation $c_1, \ldots, c_5$ of $b_1, \ldots, b_5$ such that $|c_i| \mid |c_5|$ for each i, $1 \le i \le 5$.*

Proof. By Lemma 4.4.1, $|b_i| \le 2^3$, for each i, $1 \le i \le 5$. We claim that if $|b_i| = 7$ for one i, then $|b_i| = 7$ for all i. To verify the claim let, say b_4, b_5 be all the elements among $b_1, \ldots, b_5$ whose orders are 7. The first zero divisor lemma with the $B = (e - b_1)(e - b_2)(e - b_3)$ choice gives that a nonidentity element of $\langle b_4, b_5 \rangle$ is in $\langle b_1, b_2, b_3 \rangle$. This is not possible because of the orders of the elements. We can argue similarly if one of $b_1, \ldots, b_5$ has order 5, then all have order 5. We can sort out the case when each of $|b_1|, \ldots, |b_5|$ is either 3 or 2^α using the same idea. We are left with the case when at least one 6 and one 4 or 8 occur among $|b_1|, \ldots, |b_5|$.

Assume that exactly one 6 occurs, say $|b_5| = 6$. Multiply (1) by $(e + b_5)$ to get

$$(e - b_1) \cdots (e - b_4)(e - b_5^2) = 0. \tag{2}$$

Here $|b_5^2| = 3$. If (2) is primitive, then the zero divisor argument gives that $|b_1| = \cdots = |b_4| = 3$. As $|b_5| = 6$ we are done. We may assume that (2) is not primitive and factors can be cancelled. Surely, $(e - b_5^2)$ cannot be cancelled. As $|b_5^2| = 3$ only one factor can be cancelled, say $(e - b_1)$. By Lemma 4.4.2, it follows that $|b_2| = |b_3| = |b_4| = 3$. In the $|b_1| = 2$ case there is nothing to prove. So we may assume that $|b_1| = 4$ or $|b_1| = 8$. From (1) with the $B = (e - b_2) \cdots (e - b_5)$ choice the first zero divisor lemma gives that $b_1 \in \langle b_2, \ldots, b_5 \rangle$. This is not possible by the orders.

Assume next that exactly two of $b_1, \ldots, b_5$ have order 6, say $|b_4| = |b_5| = 6$. Consider (2). If it is primitive, then the problem reduces to the previous case. We may assume that (2) is not primitive and one of the factors can be cancelled. Clearly $(e - b_5^2)$ cannot be cancelled. If $(e - b_1)$ can be cancelled, then $|b_2| = |b_3| = |b_4| = 3$. This is not possible as $|b_4| = 6$. Thus $(e - b_4)$ can be cancelled and $|b_1| = |b_2| = |b_3| = 3$. As $|b_4| = |b_5| = 6$ we are done. We can continue in this way to settle the cases when exactly three elements have order 6 and when exactly four elements have order 6.

This completes the proof.

Lemma 4.4.4. *Let $A = a + a^2 + \cdots + a^{r-1}$ and assume that $|b_i| \mid |a^r|$ for each i, $1 \leq i \leq n$. Then $(e - b_1) \cdots (e - b_n)A = 0$ implies $(e - b_1) \cdots (e - b_n) = 0$.*

Proof. We will show that $(e - b_1) \cdots (e - b_n) \neq 0$ implies $(e - b_1) \cdots (e - b_n)A \neq 0$ We embark on a proof that there is a character χ of G for which

$$\chi(b_1) \neq 1, \ldots, \chi(b_n) \neq 1, \ \chi(a^r) = 1 \ \text{ implies } \ \chi(a) = 1.$$

Set $H = \langle b_1, \ldots, b_n, a^r \rangle$, $|a^r| = s$. Assume first that $a \in H$. From $|b_i| \mid s$ we get $b_i^s = e$. Computing a^s we can see that $a^s = e$. This means that the order of a divides the order of a^r. Therefore $|a| = |a^r|$ and so $\langle a \rangle = \langle a^r \rangle$. Thus $\chi(a^r) = 1$ implies $\chi(a) = 1$.

Assume next that $a \notin H$. Let t be the smallest positive integer for which $a^t \in H$. Divide r by t with remainder

$$r = tq + u, \quad 0 \leq u \leq t - 1$$

and compute a^r. We get

$$a^r = a^{tq+u} = (a^t)^q a^u.$$

From $a^r \in H$, $a^t \in H$ it follows that $a^u \in H$. The minimality of t gives that $u = 0$ and so $a^r = (a^t)^q \in \langle a^t \rangle$. By the earlier case there is a character χ of H such that

$$\chi(b_1) \neq 1, \ldots, \chi(b_n) \neq 1, \ \chi\big((a^t)^q\big) = 1 \ \text{ implies } \ \chi(a^t) = 1.$$

Now χ can be extended to a character χ' of G with $\chi'(a) = 1$. Each g in G is uniquely expressible in the form $g = a^i h$, $0 \le i \le t - 1$, $h \in H$. Define χ' by $\chi'(a^i h) = \chi(h)$. One can check that χ' is well defined and extends χ.

This completes the proof.

For the remaining part of the section we fix a standard notation. Let

$$A_i = e + a_i + a_i^2 + \cdots + a_i^{r(i)-1}$$

and set

$$b_i = a_i^{r(i)}, \quad |b_i| = \alpha(i), \quad H_i = \langle a_i \rangle,$$

$$B_i = e + a_i + \cdots + a_i^{\alpha(i)r(i)-1}.$$

Clearly $|H_i| = |a_i|$ divides $\alpha(i)r(i)$ and $B_i = t\overline{H}_i$, with some positive integer t.

Lemma 4.4.5. *Assume $A_1 \cdots A_n = k\overline{G}$ and $|b_1|, \ldots, |b_{u-1}|$ are relatively prime to $|b_u|$. Further $b_u \in H_1 \cap \cdots \cap H_{u-1}$, then $|b_u|^{u-2}$ divides k.*

Proof. From $b_u \in H_i$, it follows that $\alpha(u) \mid |a_i|$ and so $\alpha(u) \mid \alpha(i)r(i)$. As $\alpha(u)$ is relatively prime to $\alpha(i)$, there is a $\beta(i)$ such that $\alpha(i)r(i) = \alpha(i)\alpha(u)\beta(i)$. As $\langle b_u \rangle$ and $\langle a_i^{\alpha(i)\beta(i)} \rangle$ are subgroups of the cyclic group H_i and $\left| a_i^{\alpha(i)\beta(i)} \right|$ divides $\alpha(u)$ it follows that $\langle a_i^{\alpha(i)\beta(i)} \rangle \subset \langle b_u \rangle$ and therefore $a_i^{\alpha(i)\beta(i)} \in H_i$. Multiply $A_1 \cdots A_n = k\overline{G}$ by

$$\left(e + b_1 + \cdots + b_1^{\alpha(1)-1}\right) \cdots \left(e + b_u + \cdots + b_u^{\alpha(u)-1}\right)$$

to get

$$B_1 \cdots B_u A_{u+1} \cdots A_n = \alpha(1) \cdots \alpha(u) k\overline{G}.$$

Set

$$C_i = e + a_i + \cdots + a_i^{\alpha(i)\beta(i)-1},$$

$$D_i = e + a_i^{\alpha(i)\beta(i)} + \cdots + a_i^{[\alpha(u)-1]\alpha(i)\beta(i)}.$$

Clearly $B_i = C_i D_i$. Note that $a_i^{\alpha(i)\beta(i)} H_u = H_u$. This gives $D_i B_u = \alpha(u) B_u$. We get that

$$[\alpha(u)]^{u-1} C_1 \cdots C_{u-1} B_u A_{u+1} \cdots A_n = \alpha(1) \cdots \alpha(u) k\overline{G}.$$

Comparing the coefficients of e on the two sides gives that $\alpha(1) \cdots \alpha(u)k$ is a multiple of $[\alpha(u)]^{u-1}$.

This completes the proof.

Consider an n-dimensional cube tiling without twins for $1 \le n \le 3$. Suppose $A_1 \cdots A_n = k\overline{G}$ and $a_i^{r(i)} \ne e$ for $1 \le i \le n$. we will show that this is not possible. Multiplying by $(e - a_1) \cdots (e - a_n)$ we get $(e - b_1) \cdots (e - b_n) = 0$. For $n \le 2$, this implies the $b_i = e$ contradiction. So we may assume that $n = 3$. Then we may assume that $|b_1|, |b_2| \mid |b_3|$. Multiplying by $(e - a_1)(e - a_2)$ we get

$(e - b_1)(e - b_2)A_3 = 0$. Now Lemma 4.4.4 gives that $(e - b_1)(e - b_2) = 0$ which again leads to a contradiction.

Next we will deal with 4-dimensional tilings.

Suppose that

$$A_1 A_2 A_3 A_4 = k\overline{G}. \tag{3}$$

and $a_i^{r(i)} \neq e$, $1 \leq i \leq 4$. We want to show that $p^2 \mid k$ for some odd prime p.

Assume that $a_i^{r(i)} \neq e$, $1 \leq i \leq 4$. Multiplying (3) by $(e - a_1) \cdots (e - a_4)$ gives $(e - b_1) \cdots (e - b_4) = 0$, where $b_i = a_i^{r(i)}$. If this equation is primitive then by Lemma 4.4.2, there is a permutation $c_1, \ldots, c_4$ of $b_1, \ldots, b_4$ such that $|c_i| \mid |c_4|$. We may assume that $c_1 = b_1, \ldots, c_4 = b_4$. From the equation $(e - b_1)(e - b_2)(e - b_3)A_4 = 0$ by Lemma 4.4.4, it follows that

$$(e - b_1)(e - b_2)(e - b_3) = 0. \tag{4}$$

If one of the factors, say $(e - b_3)$ can be cancelled, then from $(e - b_1)(e - b_2) = 0$ we get the $b_1 = e$, $b_2 = e$ contradiction. Thus (4) is primitive. We may choose the basis elements of $H = \langle b_1, b_2, b_3 \rangle$ such that

$$b_1^2 = b_2^2 = b_3^2 = b_1 b_2 b_3 = e.$$

We claim that $b_i \notin H_j$ for each i, j, $1 \leq i, j \leq 3$, $i \neq j$. To prove the claim assume the contrary that $b_i \in H_j$. Since $b_1^2 = e$ we get the contradiction that $\langle b_i \rangle$ is equal to the unique second order subgroup of H_j which is $\langle b_j \rangle$. Next we claim that $b_4 \in H_i$ for each i, $1 \leq i \leq 3$. In order to verify, say $b_4 \in H_1$ consider the equation

$$A_1(e - b_2)(e - b_3)(e - b_4) = 0$$

and write it in the forms

$$A_1(e + b_2 b_3 + b_2 b_4 + b_3 b_4) = A_1(b_2 + b_3 + b_4 + b_2 b_3 b_4),$$

$$B_1(e + b_2 b_4) = B_1(b_2 + b_4).$$

The coefficient of a_1 on the left side is not zero so one of the following must hold.

$$\begin{aligned}
a_1 \in \{B_1 b_2\} \quad &\text{implies} \quad b_2 \in H_1, \\
a_1 \in \{B_1 b_4\} \quad &\text{implies} \quad b_4 \in H_1.
\end{aligned}$$

The first conclusion is not possible, therefore $b_4 \in H_1$.

Let $s = \alpha(4) = |b_4|$. If $s = 2t$, then $(b_4^t)^2 = e$, $b_4^t \neq e$. On the other hand $b_4^t \in \langle b_1 \rangle \cap \langle b_2 \rangle = \{e\}$ gives that $b_4^t = e$. But this is not the case. Thus s is odd. Lemma 4.4.5 gives that $s^2 \mid k$. Thus $p^2 \mid k$ for some odd prime p.

Finally we turn to 5-dimensional tilings.

Suppose

$$A_1 \cdots A_5 = k\overline{G} \tag{5}$$

and $a_i^{r(i)} \neq e$ for each i, $1 \leq i \leq 5$. We want to deduce that $k \neq 1, 2, 4$. Multiplying (5) by $(e - a_1) \cdots (e - a_5)$ gives that $(e - b_1) \cdots (e - b_5) = 0$. If this equation is primitive, then by Lemma 4.4.2, there is a permutation $c_1, \ldots, c_5$ of $b_1, \ldots, b_5$ such that $|c_i| \mid |c_5|$ for each i, $1 \leq i \leq 5$. We assume that $c_1 = b_1, \ldots, c_5 = b_5$. From $(e - b_1) \cdots (e - b_4)A_5 = 0$ by Lemma 4.4.4, it follows that

$$(e - b_1) \cdots (e - b_4) = 0. \tag{6}$$

Either (6) is primitive or one of the factors, say $(e - b_4)$ can be cancelled and

$$(e - b_1)(e - b_2)(e - b_3) = 0 \tag{7}$$

is primitive. We consider three cases.

Case 1: (6) is primitive and $H = \langle b_1, \ldots, b_4 \rangle$ is of type $(3, 3)$.

Case 2: (6) is primitive and $H = \langle b_1, \ldots, b_4 \rangle$ is of type $(2, 4)$.

Case 3: (7) is primitive.

Let us turn to Case 1. We may choose the basis elements of $H = \langle b_1, \ldots, b_4 \rangle$ such that

$$b_1^3 = b_2^3 = b_3^3 = b_4^3 = b_1 b_2 b_3 = e,$$

$$b_1 b_4 = b_2, \quad b_2 b_4 = b_3, \quad b_3 b_4 = b_1.$$

We claim that $b_i \notin H_j$ for each i, j, $1 \leq i, j \leq 4$, $i \neq j$. To verify the claim assume the contrary, that $b_i \in H_j$. As $b_i^3 = e$, we get that $\langle b_i \rangle$ is equal to $\langle b_j \rangle$ which is the unique subgroup of order 3 of H_j. But by Lemma 4.4.1, $\langle b_i \rangle = \langle b_j \rangle$ is not possible. Next we claim that $b_5 \in H_i$ for each i, $1 \leq i \leq 4$. We verify only one case, say $b_5 \in H_1$. Start with $A_1(e - b_2) \cdots (e - b_5) = 0$ and write it in the form

$$A_1(e + b_2 b_3 + b_2 b_4 + b_2 b_5 + b_3 b_4 + b_3 b_5 + b_4 b_5 + b_2 b_3 b_4 b_5) =$$

$$A_1(b_2 + b_3 + b_4 + b_5 + b_2 b_3 b_4 + b_2 b_3 b_5 + b_2 b_4 b_5 + b_3 b_4 b_5).$$

Then

$$A_1(e + b_1 b_2 b_3)(e + b_2 b_5) = A_1(e + b_1 + b_2 b_3)(b_2 + b_5).$$

After multiplying by b_1 we get

$$B_1(e + b_2 b_5) = B_1(b_2 + b_5).$$

The coefficient of a_1 on the left side is not zero. One of the following must hold.

$$a_1 \in \{B_1 b_2\} \quad \text{implies} \quad b_2 \in H_1,$$
$$a_1 \in \{B_1 b_5\} \quad \text{implies} \quad b_5 \in H_1.$$

The first is not possible therefore $b_5 \in H_1$.

Let $s = \alpha(5) = |b_5|$. If $s = 3t$, then $(b_5^t)^3 = e$, $b_5^t \neq e$. On the other hand $b_5^t \in \langle b_1 \rangle \cap \langle b_2 \rangle = \{e\}$ provides the $b_5^t = e$ contradiction. Therefore $3 \nmid s$. By Lemma 4.4.5, $s^3 \mid k$. Consequently there is a prime p, $p \neq 3$ for which $p^3 \mid k$.

Turn to Case 2 when (6) is primitive and $H = \langle b_1, \ldots, b_4 \rangle$ is of type $(2,4)$. We may choose the basis for H such that

$$b_1^2 = b_2^2 = b_3^4 = b_4^4 = b_2 b_3 b_4 = e,$$

$$b_3 b_4 = b_2, \quad b_1 b_3 = b_4, \quad b_1 b_4 = b_3.$$

We claim that $b_i \notin H_j$ for each i, j, $1 \leq i, j \leq 4$, $i \neq j$. To prove the claim assume the contrary, that $b_i \in H_j$. If $|b_i| = |b_j|$, then we get the $\langle b_i \rangle = \langle b_j \rangle$ contradiction. If $|b_i| = 2$, $|b_j| = 4$, then $\langle b_i \rangle = \langle b_j^2 \rangle$ and we get the $\langle b_i \rangle \subset \langle b_j \rangle$ contradiction. If $|b_i| = 4$, $|b_j| = 2$, then $\langle b_i^2 \rangle = \langle b_j \rangle$ which leads to the $\langle b_j \rangle \subset \langle b_i \rangle$ contradiction.

We claim that $b_5 \in H_i$ for each i, $1 \leq i \leq 4$. We check only two typical cases, say $b_5 \in H_1$ and $b_5 \in H_3$. In order to prove $b_5 \in H_1$ start with $A_1(e - b_2) \cdots (e - b_5) = 0$ and write it in the form

$$A_1(b_3 + b_4)(b_2 + b_5) = A_1(b_3 + b_4)(e + b_2 b_5).$$

Then multiply by b_3^{-1} to get

$$B_1(e + b_2 b_5) = B_1(b_2 + b_5).$$

One of the following must hold.

$$a_1 \in \{B_1 b_2\} \quad \text{implies} \quad b_2 \in H_1,$$
$$a_1 \in \{B_1 b_5\} \quad \text{implies} \quad b_5 \in H_1.$$

The first one is not possible. To check $b_5 \in H_3$ start with

$$(e - b_1)(e - b_2) A_3 (e - b_4)(e - b_5) = 0$$

and convert it to

$$A_3(e + b_3)(e + b_1 b_3 + b_1 b_5 + b_2 b_5)$$
$$= A_3(e + b_3)(b_1 + b_2 + b_5 + b_1 b_2 b_5)$$

and then to

$$B_3(e + b_1 b_5) = B_3(b_1 + b_5).$$

One of the following must hold.

$$a_3 \in \{B_3 b_1\} \quad \text{implies} \quad b_1 \in H_3,$$
$$a_3 \in \{B_3 b_5\} \quad \text{implies} \quad b_5 \in H_3.$$

The first one is not possible. Let $s = \alpha(5) = |b_5|$. If $s = 2t$, then $(b_5^t)^2 = e$, $b_5^t \neq e$. Then $b_5^t = \langle b_1 \rangle \cap \langle b_2 \rangle = \{e\}$ gives the $b_5^t = e$ contradiction. Therefore s is odd. Using Lemma 4.4.5, it follows that $s^3 \mid k$. Therefore $p^3 \mid k$ for some odd prime p.

Let us turn to Case 3 when (6) is primitive. We will choose a basis for $H = \langle b_1, b_2, b_3 \rangle$ such that

$$b_1^2 = b_2^2 = b_3^2 = b_1 b_2 b_3 = e.$$

We claim that $b_i \notin H_j$ for each i, j, $1 \leq i, j \leq 3$, $i \neq j$. To prove the claim assume the contrary that $b_i \in H_j$. Using $b_i^2 = e$ we get that $\langle b_i \rangle = \langle b_j \rangle$. But this is not possible by Lemma 4.4.1. Next we claim that

$$\left[b_4 \in H_1\right] \ \text{ or } \ \left[b_5 \in H_1\right] \ \text{ or } \ \left[b_4^2, b_5^2, b_2 b_4 b_5, b_3 b_4 b_5 \in H_1\right],$$
$$\left[b_4 \in H_2\right] \ \text{ or } \ \left[b_5 \in H_2\right] \ \text{ or } \ \left[b_4^2, b_5^2, b_3 b_4 b_5, b_1 b_4 b_5 \in H_2\right],$$
$$\left[b_4 \in H_3\right] \ \text{ or } \ \left[b_5 \in H_3\right] \ \text{ or } \ \left[b_4^2, b_5^2, b_1 b_4 b_5, b_2 b_4 b_5 \in H_3\right].$$

In order to verify the first line of the claim start with $A_1(e - b_2) \cdots (e - b_5) = 0$ and write it in the form

$$B_1(e + b_2 b_4 + b_2 b_5 + b_4 b_5) = B_1(b_2 + b_4 + b_5 + b_2 b_4 b_5). \tag{8}$$

One of the following must hold

$$
\begin{array}{lll}
a_1 \in \{B_1 b_2\} & \text{implies} & b_2 \in H_1, \\
a_1 \in \{B_1 b_4\} & \text{implies} & b_4 \in H_1, \\
a_1 \in \{B_1 b_5\} & \text{implies} & b_5 \in H_1, \\
a_1 \in \{B_1 b_2 b_4 b_5\} & \text{implies} & b_2 b_4 b_5 \in H_1.
\end{array}
$$

Suppose $b_2 b_4 b_5 \in H_1$, then $B_1 = B_1 b_2 b_4 b_5$, $B_1 b_4 b_5 = B_1 b_2$ and (8) reduces to

$$B_1(b_2 b_4 + b_2 b_5) = B_1(b_4 + b_5). \tag{9}$$

Multiplying (9) by b_5 gives

$$B_1(b_2 b_4 b_5 + b_2 b_5^2) = B_1(b_4 b_5 + b_5^2),$$

$$B_1(e + b_2 b_5^2) = B_1(b_4 b_5 + b_5^2).$$

One of the following must hold.

$$
\begin{array}{lll}
a_1 \in \{B_1 b_4 b_5\} = \{B_1 b_2\} & \text{implies} & b_2 \in H_1, \\
a_1 \in \{B_1 b_5^2\} & \text{implies} & b_5^2 \in H_1.
\end{array}
$$

Only the second one is possible. Multiplying (9) by b_4 gives

$$B_1(b_2 b_4^2 + b_2 b_4 b_5) = B_1(b_4^2 + b_4 b_5),$$

$$B_1(e + b_2 b_4^2) = B_1(b_4^2 + b_4 b_5).$$

One of the following must hold.

$$a_1 \in \{B_1 b_4^2\} \qquad \text{implies} \qquad b_4^2 \in H_1,$$
$$a_1 \in \{B_1 b_4 b_5\} = \{B_1 b_2\} \qquad \text{implies} \qquad b_2 \in H_1.$$

The second one is not possible.

We claim $b_4^2 = b_5^2 = e$ is not possible. To prove the claim assume that $b_4^2 = b_5^2 = e$. Now $b_4^2 \in H_1, H_2, H_3$ and we get that

$$b_2 b_4 b_5, b_3 b_4 b_5 \in H_1,$$

$$b_3 b_4 b_5, b_1 b_4 b_5 \in H_2,$$

$$b_1 b_4 b_5, b_2 b_4 b_5 \in H_3.$$

As $(b_1 b_4 b_5)^2 = (b_2 b_4 b_5)^2 = (b_3 b_4 b_5)^2 = e$ it follows that

$$\langle b_2 b_4 b_5 \rangle = \langle b_3 b_4 b_5 \rangle = \langle b_1 \rangle$$

$$\langle b_3 b_4 b_5 \rangle = \langle b_1 b_4 b_5 \rangle = \langle b_2 \rangle$$
$$\langle b_1 b_4 b_5 \rangle = \langle b_2 b_4 b_5 \rangle = \langle b_3 \rangle$$

This implies the contradiction $\langle b_1 \rangle = \langle b_2 \rangle$.

As $b_4^2 = b_5^2 = e$ is not possible we may assume that $b_5^2 \neq e$. We know that

$$\begin{array}{ccc} \left[b_4 \in H_1 \right] & \text{or} & \left[b_5^2 \in H_1 \right], \\ \left[b_4 \in H_2 \right] & \text{or} & \left[b_5^2 \in H_2 \right], \\ \left[b_4 \in H_3 \right] & \text{or} & \left[b_5^2 \in H_3 \right]. \end{array}$$

(This is a particular case of an earlier result.) We may assume that either

$$b_4 \in H_1, \ b_4 \in H_2, \ b_5^2 \in H_3 \tag{10}$$

or

$$b_5^2 \in H_1, \ b_5^2 \in H_2, \ b_4 \in H_3. \tag{11}$$

Let us deal with possibility (10). Let $s = |b_4|$. If $s = 2t$, then $(b_4^t)^2 = e$, $b_4^t \neq e$. On the other hand $b_4^t \in \langle b_1 \rangle \cap \langle b_2 \rangle = \{e\}$. Thus s is odd. This implies $s^2 \mid 4sk$ and then $s \mid k$. In other words k is divisible by an odd prime.

Turn to possibility (11). Multiply $A_1 \cdots A_5 = k\overline{G}$ by $(e + b_5)$ to get $A_1 \cdots A_4 A_5' = 2k\overline{G}$, where

$$A_5' = A_5(e + b_5) = e + a_5 + \cdots + a_5^{2r(5)-1}.$$

Set $b_5' = b_5^2$, $r'(5) = 2r(5)$. The earlier argument gives that $2k$ is divisible by an odd prime.

Chapter 5

Applying the machinery

5.1   The size of an annihilator

In 1987, K. Corrádi conjectured that in any factorization of a finite abelian group by its subsets the annihilator of one of the factors must span the whole character group.

We verify this conjecture in three special cases. In the first case the finite abelian group is a p-group generated (or spanned) by two elements. In the second case the number of the factors in the factorization is not greater than four. In the third case $p \geq n$, where p is the least prime factor of the order of the group and n is the number of the factors in the factorization.

Let G be a finite abelian group and let $\mathcal{G}$ be its character group. The identity element of $\mathcal{G}$ is the character which takes the value 1 on each element of G. We call this character the principal character of G and we will denote it by ε. For a character χ of G we use the notation $\mathrm{Ker}\chi$ to denote the set of elements of G on which χ takes the value 1. $\mathrm{Ker}\chi$ is a subgroup of G. It is the largest subgroup of G on which the restriction of χ is principal.

To a subgroup H of G we assign the subgroup $\mathcal{H} = \{\chi : H \subset \mathrm{Ker}\chi\}$ of $\mathcal{G}$. The mapping $H \to \mathcal{H}$ is a bijection between the subgroups of G and $\mathcal{G}$. This map has the following properties.

If $H \to \mathcal{H}$, $K \to \mathcal{K}$, then $HK \to \mathcal{H} \cap \mathcal{K}$ and $H \cap K \to \mathcal{H}\mathcal{K}$.

$\{e\} \to \mathcal{G}$.

$G \to \{\varepsilon\}$.

In short, this map is an antiisomorphism between the subgroup lattices of G and $\mathcal{G}$ and it is called the Dirichlet correspondence. To a subgroup $\mathcal{H}$ of $\mathcal{G}$ the inverse map assigns the subgroup

$$H = \bigcap_{\chi \in \mathcal{H}} \mathrm{Ker}\chi$$

of G.

Exercise 5.1.1. Verify Corrádi's conjecture for finite cyclic groups. (Hint: Let G be a finite cyclic group and let χ be a character of G with $\mathrm{Ker}\chi = \{e\}$. Then apply χ to a factorization $G = A_1 \cdots A_n$.)

In the next few lines we describe two more cases when Corrádi's conjecture holds. Let $G = A_1 \cdots A_n$ be a factorization of the finite abelian group G, where each A_i is a cyclic subset of G. We will show that Hajós's theorem implies Corrádi's conjecture and Corrádi's conjecture implies Hajós's theorem. Let

$$A_i = \{e, a_i, a_i^2, \ldots, a_i^{r(i)-1}\}.$$

We recall that the annihilator of A_i is a difference of two subgroups of $\mathcal{G}$. Namely, $\mathrm{Ann}(A_i) = \mathcal{L}_i \setminus \mathcal{M}_i$, where

$$\mathcal{L}_i = \left\{\chi : \big(\chi(a_i)\big)^{r(i)} = 1\right\}, \quad \mathcal{M}_i = \{\chi : \chi(a_i) = 1\}.$$

The product of the subsets $A_1, \ldots, A_n$ of a finite abelian group G is direct and gives G if and only if $|A_1| \cdots |A_n| = |G|$ and

$$\mathrm{Ann}(A_1) \cup \cdots \cup \mathrm{Ann}(A_n) = \mathcal{G} \setminus \{\varepsilon\}.$$

By Hajós's theorem there is an i, $1 \le i \le n$ such that A_i is a subgroup of G. This means that $a_i^{r(i)} = e$. Now $\big(\chi(a_i)\big)^{r(i)} = 1$ for each character χ of G and so $\mathcal{L}_i = \mathcal{G}$. Since $a_i \ne e$, $\mathcal{M}_i \ne \mathcal{G}$ and so $|\mathcal{M}_i| \le |\mathcal{G}|/2$. Hence $\langle \mathrm{Ann}(A_i)\rangle = \mathcal{G}$ as $|\mathcal{G} \setminus \mathcal{M}_i| \ge |\mathcal{G}|/2$. (Exercise 1.2.1 helps to complete the argument.) Thus Hajós's theorem implies Corrádi's conjecture.

Conversely, assume that Corrádi's conjecture holds, that is, $\langle \mathrm{Ann}(A_i)\rangle = \mathcal{G}$ for some i, $1 \le i \le n$. From this it follows that $\mathcal{L}_i$ must be $\mathcal{G}$. This gives that $\big(\chi(a_i)\big)^{r(i)} = 1$ for each character χ of G. Therefore $a_i^{r(i)} = e$ and so A_i is a subgroup of G. Thus Corrádi's conjecture implies Hajós's theorem.

Let $G = A_1 \cdots A_n$ be a normalized factorization of the finite abelian group G, where each A_i is a simulated subset. We will show that Theorem 1.2.1 implies Corrádi's conjecture and Corrádi's conjecture implies Theorem 1.2.1. Let

$$A_i = \big(H_i \setminus \{h_i\}\big) \cup \{h_i d_i\},$$

where H_i is a subgroup of G,

$$h_i \in H_i, \quad h_i \notin A_i, \quad a_i \notin H_i, \quad a_i \in A_i, \quad a_i = h_i d_i, \quad d_i \in G.$$

We know that $\mathrm{Ann}(A_i) = \mathcal{L}_i \setminus \mathcal{M}_i$, where

$$\mathcal{L}_i = \{\chi : \chi(d_i) = 1\}, \quad \mathcal{M}_i = \{\chi : \chi(h) = 1 \ \text{ for each } \ h \in H_i\}.$$

By Theorem 1.2.1, $d_i = e$ holds for some i, $1 \le i \le n$. It follows that $\mathcal{L}_i = \mathcal{G}$. As $\mathcal{M}_i$ is a subgroup of $\mathcal{G}$ and $\mathcal{M}_i \ne \mathcal{G}$ we get that $\langle \mathrm{Ann}(A_i)\rangle = \mathcal{G}$.

Next suppose that Corrádi's conjecture holds. We have that $\langle \mathrm{Ann}(A_i) \rangle = \mathcal{G}$ for some i, $1 \leq i \leq n$. It follows that $\mathcal{L}_i = \mathcal{G}$ and so $\chi(d_i) = \chi(e)$ for each character χ of G. Therefore $d_i = e$. Thus Corrádi's conjecture implies Theorem 1.2.1.

We need the next lemma to verify the conjecture for finite abelian p-groups generated by two elements.

Lemma 5.1.1. *Let G be a finite abelian group and let $\mathcal{G}$ be its character group. If $a \in G$ and $\mathcal{H}$ is a subgroup of $\mathcal{G}$, then*

$$\sum_{\chi \in \mathcal{H}} \chi(a) = \begin{cases} |\mathcal{H}|, & \text{if } a \in K, \\ 0, & \text{if } a \notin K, \end{cases}$$

where

$$K = \bigcap_{\chi \in \mathcal{H}} \mathrm{Ker}\chi.$$

Proof. If $a \in K$, then $\chi(a) = 1$ for each $\chi \in \mathcal{H}$ and so

$$\sum_{\chi \in \mathcal{H}} \chi(a) = |\mathcal{H}|.$$

If $a \notin K$, then there is a $\chi' \in \mathcal{H}$ for which $\chi'(a) \neq 1$. Multiplying the elements of $\mathcal{H}$ by χ' permutes the elements of $\mathcal{H}$. Hence

$$\sum_{\chi \in \mathcal{H}} \chi(a) = \sum_{\chi \in \mathcal{H}} (\chi'\chi)(a) = \sum_{\chi \in \mathcal{H}} \chi'(a)\chi(a),$$

from which we get

$$0 = \left(1 - \chi'(a)\right) \sum_{\chi \in \mathcal{H}} \chi(a).$$

As $\chi'(a) \neq 1$ it follows that

$$\sum_{\chi \in \mathcal{H}} \chi(a) = 0.$$

This completes the proof.

Theorem 5.1.1. *Let p be a prime and let G be a finite abelian p-group generated by two elements. If $G = A_1 \cdots A_n$ is a factorization of G, then there is an i, $1 \leq i \leq n$ such that $\langle \mathrm{Ann}(A_i) \rangle = \mathcal{G}$, where $\mathcal{G}$ is the character group of G.*

Proof. First of all note that in proving the theorem we may restrict our attention to normalized factorizations. Indeed, let $a_1 \in A_1, \ldots, a_n \in A_n$. Multiplying the factorization $G = A_1 \cdots A_n$ by $a = a_1^{-1} \cdots a_n^{-1}$ leads to the normalized factorization $G = aG = (a_1^{-1}A_1) \cdots (a_n^{-1}A_n)$. In addition, it is clear that $\mathrm{Ann}(A_i) = \mathrm{Ann}(a_i^{-1}A_i)$.

In order to prove the theorem assume the contrary that

$$\langle \operatorname{Ann}(A_i) \rangle \neq \mathcal{G}$$

for each i, $1 \leq i \leq n$. Further assume that n is minimal with this property. As $\langle \operatorname{Ann}(A_i) \rangle \neq \mathcal{G}$ for each i, $1 \leq i \leq n$ there is a maximal subgroup $\mathcal{M}_i$ of $\mathcal{G}$ such that $\operatorname{Ann}(A_i) \subset \mathcal{M}_i$. By the character test for factorization $\operatorname{Ann}(A_1) \cup \cdots \cup \operatorname{Ann}(A_n) = \mathcal{G} \setminus \{\varepsilon\}$ and consequently $\mathcal{M}_1 \cup \cdots \cup \mathcal{M}_n = \mathcal{G}$. The minimality of n in our counterexample implies that the subgroups $\mathcal{M}_1, \ldots, \mathcal{M}_n$ are distinct. To keep the notational difficulties to a minimum we prove that $\mathcal{M}_{n-1} \neq \mathcal{M}_n$. From the factorization $G = A_1 \cdots A_{n-2}(A_{n-1}A_n)$ by the minimality of n it follows that $\langle \operatorname{Ann}(A_{n-1}A_n) \rangle = \mathcal{G}$. If we assume that $\mathcal{M}_{n-1} = \mathcal{M}_n$, then we get the contradiction

$$
\begin{aligned}
\operatorname{Ann}(A_{n-1}A_n) \quad &\subset \quad \operatorname{Ann}(A_{n-1}) \cup \operatorname{Ann}(A_n) \\
&\subset \quad \mathcal{M}_{n-1} \cup \mathcal{M}_n \\
&= \quad \mathcal{M}_{n-1}.
\end{aligned}
$$

Thus $\mathcal{M}_{n-1} \neq \mathcal{M}_n$ or in general $\mathcal{M}_i \neq \mathcal{M}_j$ if $i \neq j$, $1 \leq i, j \leq n$.

As G and $\mathcal{G}$ are isomorphic, $\mathcal{G}$ is a finite abelian p-group generated by two elements. Let $\Phi(\mathcal{G})$ be the Frattini subgroup of $\mathcal{G}$ and consider the factor group $\mathcal{G}/\Phi(\mathcal{G})$. Clearly $\mathcal{G}/\Phi(\mathcal{G})$ is an elementary abelian group of order p^2. We know that

$$\big(\mathcal{M}_1\Phi(\mathcal{G})\big)/\Phi(\mathcal{G}), \ldots, \big(\mathcal{M}_n\Phi(\mathcal{G})\big)/\Phi(\mathcal{G})$$

are distinct subgroups of order p and

$$\mathcal{G}/\Phi(\mathcal{G}) = \big(\mathcal{M}_1\Phi(\mathcal{G})\big)/\Phi(\mathcal{G}) \cup \cdots \cup \big(\mathcal{M}_n\Phi(\mathcal{G})\big)/\Phi(\mathcal{G}).$$

This gives that $\big(\mathcal{M}_1\Phi(\mathcal{G})\big)/\Phi(\mathcal{G}), \ldots, \big(\mathcal{M}_n\Phi(\mathcal{G})\big)/\Phi(\mathcal{G})$ are all the subgroups of $\mathcal{G}/\Phi(\mathcal{G})$ of order p and so $n = p + 1$.

We claim that $\mathcal{M}_i \cap \mathcal{M}_j = \Phi(\mathcal{G})$ for each $i \neq j$, $1 \leq i, j \leq n$. By definition $\Phi(\mathcal{G}) = \mathcal{M}_1 \cap \cdots \cap \mathcal{M}_n$. Hence $\Phi(\mathcal{G}) \subset \mathcal{M}_i \cap \mathcal{M}_j$ and so it is enough to establish that $|\mathcal{G} : \Phi(\mathcal{G})| \leq |\mathcal{G} : (\mathcal{M}_i \cap \mathcal{M}_j)|$. As $\mathcal{G}$ is a p-group generated by 2 elements, $|\mathcal{G} : \Phi(\mathcal{G})| = p^2$. As $\mathcal{M}_i$ is a maximal subgroup of $\mathcal{G}$, $|\mathcal{G} : \mathcal{M}_i| = p$. From $\mathcal{M}_i \neq \mathcal{M}_j$ it follows that $\mathcal{M}_i \cap \mathcal{M}_j \neq \mathcal{M}_i$ and so from $\Phi(\mathcal{G}) \subset \mathcal{M}_i \cap \mathcal{M}_j \neq \mathcal{M}_i$ it follows that $|\mathcal{G} : (\mathcal{M}_i \cap \mathcal{M}_j)| \geq p^2$.

Next we claim that $\mathcal{M}_i \setminus \Phi(\mathcal{G}) \subset \operatorname{Ann}(A_i)$ for each i, $1 \leq i \leq n$. Or equivalently, that from $\tau_i \in \mathcal{M}_i \setminus \Phi(\mathcal{G})$ it follows that $\tau_i\Phi(\mathcal{G}) \subset \operatorname{Ann}(A_i)$ for each i, $1 \leq i \leq n$. To prove this claim assume the contrary, that there is a $\xi \in \tau_i\Phi(\mathcal{G})$ with $\xi \notin \operatorname{Ann}(A_i)$. Since $\tau_i\Phi(\mathcal{G}) \cap \Phi(\mathcal{G}) = \emptyset$, $\xi \neq \varepsilon$ and so from

$$\operatorname{Ann}(A_1) \cup \cdots \cup \operatorname{Ann}(A_n) = \mathcal{G} \setminus \{\varepsilon\}$$

it follows that $\xi \in \operatorname{Ann}(A_j)$ for some j, $j \neq i$. But $\operatorname{Ann}(A_j) \subset \mathcal{M}_j$ which leads to the contradiction $\xi \in \mathcal{M}_i \cap \mathcal{M}_j = \Phi(\mathcal{G})$.

Relying on these results we may argue in the following way. For each $\tau_i \in \mathcal{M}_i \setminus \Phi(\mathcal{G})$, $\tau_i \Phi(\mathcal{G}) \subset \mathrm{Ann}(A_i)$, that is, for each $\xi \in \tau_i \Phi(\mathcal{G})$, $\xi(A_i) = 0$ and so

$$
\begin{aligned}
0 &= \sum_{\xi \in \tau_i \Phi(\mathcal{G})} \xi(A_i) \\
&= \sum_{\xi \in \tau_i \Phi(\mathcal{G})} \left(\sum_{a \in A_i} \xi(a) \right) \\
&= \sum_{a \in A_i} \left(\sum_{\xi \in \tau_i \Phi(\mathcal{G})} \xi(a) \right).
\end{aligned}
$$

Each character ξ in $\tau_i \Phi(\mathcal{G})$ is uniquely expressible in the form $\xi = \tau_i \chi$, where $\chi \in \Phi(\mathcal{G})$. Hence

$$
\begin{aligned}
0 &= \sum_{a \in A_i} \left(\sum_{\chi \in \Phi(\mathcal{G})} (\tau_i \chi)(a) \right) \\
&= \sum_{a \in A_i} \left(\sum_{\chi \in \Phi(\mathcal{G})} \tau_i(a) \chi(a) \right) \\
&= \sum_{a \in A_i} \tau_i(a) \left(\sum_{\chi \in \Phi(\mathcal{G})} \chi(a) \right).
\end{aligned}
$$

Consider the Dirichlet correspondence between the subgroup lattices of G and $\mathcal{G}$. Let K be the image of $\Phi(\mathcal{G})$. By Lemma 5.1.1,

$$
\sum_{\chi \in \Phi(\mathcal{G})} \chi(a) = \begin{cases} |\Phi(\mathcal{G})|, & \text{if } a \in K, \\ 0, & \text{if } a \notin K, \end{cases}
$$

and hence

$$
0 = \sum_{a \in K \cap A_i} \tau_i(a).
$$

As $e \in K \cap A_i$, the summation is not empty. The terms are roots of unity of p-power orders. Thus by Lemma 3.1.1, $|K \cap A_i| \geq p$ for each i, $1 \leq i \leq n$.

Now we can draw two conclusions about $|K|$. On one hand the Dirichlet correspondence gives $|K| = |\mathcal{G} : \Phi(\mathcal{G})| = p^2$. On the other hand note that the product $(K \cap A_1) \cdots (K \cap A_n)$ is direct and is part of K. From this and from $|K \cap A_i| \geq p$ it follows that $|K| \geq p^n$. Thus $p^2 = |K| \geq p^n = p^{p+1}$ which lands on the $2 \geq p + 1$ contradiction.

This completes the proof.

We reformulate Corrádi's conjecture. Let G be a finite abelian group and let $\mathcal{G}$ be its character group. To a subset A of G we assign the subgroup

$$
K = \bigcap_{\chi(A)=0} \mathrm{Ker}\chi
$$

of G. Using the Dirichlet correspondence we can verify that if $\langle \mathrm{Ann}(A) \rangle \neq \mathcal{G}$, then $K \neq \{e\}$ and conversely if $K \neq \{e\}$, then $\langle \mathrm{Ann}(A) \rangle \neq \mathcal{G}$. The new version of Corrádi's conjecture now reads as follows.

Let $G = A_1 \cdots A_n$ be a factorization of the finite abelian group G and let $K_1, \ldots, K_n$ be the subgroups assigned to the factors $A_1, \ldots, A_n$ respectively. Then there is an i, $1 \leq i \leq n$ such that $K_i = \{e\}$.

Lemma 5.1.2. *Let G be a finite abelian group and let $\mathcal{G}$ be its character group. If A is a periodic subset of G, then $\langle \mathrm{Ann}(A) \rangle = \mathcal{G}$.*

Proof. As A is periodic, it admits a factorization of the form $A = BH$, where B is a subset and H is a subgroup of G, where the nonidentity elements of H are the periods of A and so $H \neq \{e\}$. Let $\mathcal{M}$ be the subgroup of $\mathcal{G}$ containing each character χ of G which is principal on H. As $H \neq \{e\}$, $\mathcal{M} \neq \mathcal{G}$. By Lemma 3.2.6, $\mathcal{G} \setminus \mathcal{M} \subset \mathrm{Ann}(A)$. Now from $|\mathcal{G} \setminus \mathcal{M}| \geq |\mathcal{G}|/2$ it follows that $\langle \mathrm{Ann}(A) \rangle = \mathcal{G}$.

This completes the proof.

Lemma 5.1.3. *Let A, B, C be subsets of an abelian group with $e \in B$ and $e \in C$. If the product ABC is direct, then $AB \cap AC = A$.*

Proof. Let $d \in AB \cap AC$. Since $d \in AB$, there are $a \in A$ and $b \in B$ such that $d = ab$. Since $d \in AC$, there are $a' \in A$ and $c \in C$ such that $d = a'c$. Now

$$d = \underbrace{(a)}_{\in A} \underbrace{(b)}_{\in B} \underbrace{(e)}_{\in C} = \underbrace{(a')}_{\in A} \underbrace{(e)}_{\in B} \underbrace{(c)}_{\in C}.$$

By the directness of the product ABC it follows that $a = a'$, $b = e$, $e = c$, that is, $d \in A$. Hence $AB \cap AC \subset A$. The containment $A \subset AB \cap AC$ is a consequence of $e \in B$ and $e \in C$.

This completes the proof.

Theorem 5.1.2. *Let $G = A_1 \cdots A_n$ be a factorization of the finite abelian group G and let $K_1, \ldots, K_n$ be the subgroups assigned to the factors $A_1, \ldots, A_n$ respectively. If $n \leq 4$, then there is an i, $1 \leq i \leq n$ such that $K_i = \{e\}$.*

Proof. We assume to the contrary that $K_i \neq \{e\}$ for each i, $1 \leq i \leq n$. From this assumption we will draw the conclusion that one of the factors $A_1, \ldots, A_n$ is periodic. By Lemma 5.1.2, this is a contradiction.

Case $n = 1$. Now $G = A_1$ and so A_1 is clearly periodic.

Case $n = 2$. Now $G = A_1 A_2$ and $\chi(A_1) = 0$ for each character χ of G for which $\chi(K_2) = 0$. By Lemma 3.2.6, this means that A_1 is periodic.

Case $n = 3$. Now $G = A_1 A_2 A_3$ and $\chi(A_1) = 0$ for each character χ of G for which $\chi(K_2) = 0$ and $\chi(K_3) = 0$. By Lemma 3.2.7, there are subsets X_2, X_3 of G such that $A_1 = X_2 K_2 \cup X_3 K_3$, where the union is disjoint and the products $X_2 K_2$ and $X_3 K_3$ are direct. If $X_2 = \emptyset$ or $X_3 = \emptyset$, then A_1 is periodic. So we may assume that $X_2 \neq \emptyset$ and $X_3 \neq \emptyset$. Similarly $\chi(A_2) = 0$ for each character χ of

G for which $\chi(K_1) = 0$ and $\chi(K_3) = 0$. Thus there are subsets Y_1, Y_3 of G such that $A_2 = Y_1 K_1 \cup Y_3 K_3$, where the union is disjoint and the products $Y_1 K_1$ and $Y_3 K_3$ are direct. If $Y_1 = \emptyset$ or $Y_3 = \emptyset$, then A_2 is periodic. So we may assume that $Y_1 \neq \emptyset$ and $Y_3 \neq \emptyset$. As $X_3 \neq \emptyset$ and $Y_3 \neq \emptyset$, there are elements $x_3 \in X_3$ and $y_3 \in Y_3$. Multiplying the factorization $G = A_1 A_2 A_3$ by $g = x_3^{-1} y_3^{-1}$ we get the factorization

$$G = gG = (x_3^{-1} A_1)(y_3^{-1} A_2) A_3$$
$$= \left[x_3^{-1}(X_2 K_2 \cup X_3 K_3) \right] \left[y_3^{-1}(Y_1 K_1 \cup Y_3 K_3) \right] A_3.$$

Here $K_3 \subset x_3^{-1} A_1$ and $K_3 \subset y_3^{-1} A_2$. This contradicts the definition of factorization as K_3 contains a nonidentity element.

Case $n = 4$. Assume that $A_1 A_2$ is periodic and the periods together with e form the subgroup L of G. By Lemma 3.2.6, $\chi(A_1 A_2) = 0$ for each χ for which $\chi(L) = 0$. So $\chi(A_1) = 0$ for each χ with $\chi(K_2) = 0$ and $\chi(L) = 0$. There are $U_2, U \subset G$ such that $A_1 = U_2 K_2 \cup UL$, where the union is disjoint and the products are direct. Similarly, there are $V_1, V \subset G$ such that $A_2 = V_1 K_1 \cup VL$. The directness of the product $A_1 A_2$ implies that $U = \emptyset$ or $V = \emptyset$ and so A_1 or A_2 is periodic. In the remaining part we will show that $A_1 A_2$ is periodic.

First we prove that after a suitable relabelling of the factors $K_3 \subset A_4$. From the factorization $G = (A_1 A_2) A_3 A_4$ it follows that there are $X_3, X_4 \subset G$ such that $A_1 A_2 = X_3 K_3 \cup X_4 K_4$, where the union is disjoint and the products are direct. As $e \in A_1 A_2$, one of $K_3 \subset A_1 A_2$, $K_4 \subset A_1 A_2$ holds. In general, for each $\{k, l\} \subset \{1, 2, 3, 4\}$ there is an $i \notin \{k, l\}$ such that $K_i \subset A_k A_l$. There are six choices for $\{k, l\}$ and i ranges over four values so by the pigeon hole principle there are i, $\{k, l\}$, $\{k', l'\}$ such that $K_i \subset A_k A_l$ and $K_i \subset A_{k'} A_{l'}$. If $\{k, l\}$ and $\{k', l'\}$ are disjoint, then $i \notin \{k, l\} \cup \{k', l'\} = \{1, 2, 3, 4\}$ is a contradiction. Thus we may assume that $k = k'$. Now $K_i \subset A_k A_l \cap A_k A_{l'} = A_k$. By relabelling we may assume that $K_3 \subset A_4$.

Let $a \in A_1 A_2$ and consider the factorization

$$G = (a^{-1} A_1 A_2) A_3 A_4.$$

From this it follows that there are $X_3, X_4 \subset G$ such that

$$a^{-1} A_1 A_2 = X_3 K_3 \cup X_4 K_4,$$

where the union is disjoint and the products are direct. Consequently

$$K_3 \subset (a^{-1} A_1 A_2) \quad \text{or} \quad K_4 \subset (a^{-1} A_1 A_2).$$

If $K_3 \subset (a^{-1} A_1 A_2)$, then $K_3 \subset (a^{-1} A_1 A_2) \cap A_4 = \{e\}$ is a contradiction and so $K_4 \subset (a^{-1} A_1 A_2)$ for each $a \in A_1 A_2$. By Lemma 1.2.1, $A_1 A_2$ is periodic. This completes the proof.

Theorem 5.1.3. *Let $G = A_1 \cdots A_n$ be a factorization of the finite abelian group G and let $K_1, \ldots, K_n$ be the subgroups assigned to $A_1, \ldots, A_n$ respectively. If $p \geq n$, where p is the least prime divisor of $|G|$, then $K_i = \{e\}$ for some i, $1 \leq i \leq n$.*

Proof. There is nothing to prove when $n = 1$. So we assume that $n \geq 2$. Let $\mathcal{G}$ be the character group of G. To a subgroup K of G we assign the subgroup

$$\mathcal{K} = \{\chi \in \mathcal{G} : K \subset \mathrm{Ker}\chi\}$$

of $\mathcal{G}$. We will use the fact that the factor group G/K is isomorphic to $\mathcal{K}$. In order to prove the theorem assume the contrary, that $K_i \neq \{e\}$ and consequently $p \leq |K_i|$ for each i, $1 \leq i \leq n$. We claim that

$$\mathcal{G} \setminus \{\varepsilon\} = \bigcup_{i=1}^{n} (\mathcal{K}_i \setminus \{\varepsilon\}).$$

Indeed, if χ is not the principal character of G, then from

$$0 = \chi(G) = \chi(A_1 \cdots A_n) = \chi(A_1) \cdots \chi(A_n)$$

it follows that $\chi(A_i) = 0$ for some i, $1 \leq i \leq n$. It means that $K_i \subset \mathrm{Ker}\chi$ and so $\chi \in \mathcal{K}_i$. Using the properties of the Dirichlet correspondence we get

$$|G| = |\mathcal{G}| = n - 1 + \left| \bigcup_{i=1}^{n} \mathcal{K}_i \right| \leq \sum_{i=1}^{n} |\mathcal{K}_i| = \sum_{i=1}^{n} |G|/|K_i|.$$

Using $n \geq 2$ it follows that

$$1 < \sum_{i=1}^{n} \frac{1}{|K_i|}$$

and consequently $|K_i| < n$ for some i, $1 \leq i \leq n$. This leads to the $p \leq |K_i| < n$ contradiction.

The proof is complete.

Next we describe an observation that opens up a possibility for extending Hajós's theorem for noncommutative groups. Let G be a finite abelian group. We say that the subgroups $L_1, \ldots, L_n, M_1, \ldots, M_n$ of G form a *covering system* if
(1) $M_1 \subset L_1, \ldots, M_n \subset L_n$,
(2) $G \setminus \{e\} = (L_1 \setminus M_1) \cup \cdots \cup (L_n \setminus M_n)$,
(3) $|G| = |L_1 : M_1| \cdots |L_n : M_n|$.
We say that this covering system is *regular* if $G = L_i$ holds for some i, $1 \leq i \leq n$.

Theorem 5.1.4. *If for each finite abelian group each covering system is regular, then Hajós's theorem holds.*

Proof. Assume that for each finite abelian group each covering system is regular. We will show that Hajós's theorem holds. Choose a finite abelian group G and consider a factorization $G = A_1 \cdots A_n$, where

$$A_i = \left\{ e, a_i, a_i^2, \ldots, a_i^{r(i)-1} \right\}.$$

Let $\mathcal{G}$ be the character group of G and let

$$\mathcal{L}_i = \{\chi \in \mathcal{G} : \chi(a_i^{r(i)}) = 1\}, \quad \mathcal{M}_i = \{\chi \in \mathcal{G} : \chi(a_i) = 1\}.$$

We know that $\mathrm{Ann}(A_i) = \mathcal{L}_i \setminus \mathcal{M}_i$ and so

$$\mathcal{G} \setminus \{\varepsilon\} = (\mathcal{L}_1 \setminus \mathcal{M}_1) \cup \cdots \cup (\mathcal{L}_n \setminus \mathcal{M}_n).$$

Note that $|\mathcal{L}_i : \mathcal{M}_i| = r(i)$. This gives that

$$|\mathcal{G}| = |G| = r(1)\cdots r(n) = |\mathcal{L}_1 : \mathcal{M}_1| \cdots |\mathcal{L}_n : \mathcal{M}_n|.$$

Hence $\mathcal{L}_1, \ldots, \mathcal{L}_n, \mathcal{M}_1, \ldots, \mathcal{M}_n$ is a covering system for $\mathcal{G}$. By our assumption $\mathcal{G} = \mathcal{L}_i$ for some i, $1 \le i \le n$. Now $\chi(a_i^{r(i)}) = \chi(e)$ holds for each character χ of G. Therefore $a_i^{r(i)} = e$ and A_i is a subgroup of G.

This completes the proof.

Problem 5.1.1. Is it true that each covering system of a not necessarily commutative finite group is regular?

Let G be a group of type $(4,4)$ with basis elements x, y and let

$$\begin{aligned}
L_1 &= \langle x^2, y \rangle, & M_1 &= \langle x^2 \rangle, \\
L_2 &= \langle x, y^2 \rangle, & M_2 &= \langle y^2 \rangle, \\
L_3 &= \langle xy \rangle, & M_3 &= \langle x^2 y^2 \rangle, \\
L_4 &= \langle xy^3 \rangle, & M_4 &= \langle x^2 y^2 \rangle.
\end{aligned}$$

Exercise 5.1.2. Verify that

$$M_1 \subset L_1, \ldots, M_4 \subset L_4,$$

$$G \setminus \{e\} = (L_1 \setminus M_1) \cup \cdots \cup (L_4 \setminus M_4),$$

$$|G| \ne |L_1 : M_1| \cdots |L_4 : M_4|$$

and $G = L_i$ does not hold for $1 \le i \le 4$.

Exercise 5.1.2 shows that condition (3) is essential in the covering problem.

5.2 Cyclic type subsets

This section returns to the subject of Section 2.1. We take advantage of a larger stock of available tools. The map $f : G \to G$ defined by $f(g) = ga$ permutes the elements of G and so it consists of cycles of equal length. The elements of a given subset A of G are distributed in some way on the cycles. Consecutive elements of A form blocks or sections. If there is an element a of G such that the elements of A form one single section on a cycle, then A is essentially a cyclic subset. In general the fewer sections a subset consists of the closer is the subset to being a cyclic subset. If there is an element a of $G \setminus \{e\}$ such that the elements of A fill complete cycles of the permutation f, then A is periodic with period a.

Two types of subsets will appear in this section. A subset in form the

$$C\langle a \rangle \cup g[a, r],$$

where $C \subset G$, the product $C\langle a \rangle$ is direct and the union is disjoint was defined to be weakly periodic in Section 2.1. The part $C\langle a \rangle$ is a union of complete cycles of the permutation f and $[a, r]$ is an abbreviation for the cyclic subset $\{e, a, a^2, \ldots, a^{r-1}\}$. The general form of a subset we deal with in this section is

$$g_1[a, r(1)] \cup \cdots \cup g_s[a, r(s)],$$

which is a disjoint union of s sections. In case the subset is normalized we generally assume that $g_1 = e$. Clearly every subset can be written in the above form for each given element a of G. An upper bound will be placed on the number of the components s. In the special case when each of the s section is situated on the same cycle of f, that is, when $g_1, \ldots, g_s$ are powers of a we call the subset a *lacunary cyclic* subset. The next theorem and Theorem 2.1.1 overlap. Here no restriction is placed on the 2-component of G. In Section 2.1 a larger variety of factors has appeared.

Theorem 5.2.1. *Let $G = A_1 \cdots A_n$ be a normalized factorization of the finite abelian group G, where each A_i is a weakly periodic subset. Then at least one of the factors $A_1, \ldots, A_n$ is periodic.*

Proof. We prove the result by induction on n. If $n = 1$, then $G = A_1$ and so A_1 is periodic. Assume that $n \geq 2$. Each A_i is of one of the forms

$$C_i\langle a_i \rangle, \quad g_i[a_i, r(i)], \quad C_i\langle a_i \rangle \cup g_i[a_i, r(i)]. \qquad (1, 2, 3)$$

Here $C_i \subset G$, $a_i, g_i \in G$, $1 \leq r(i) \leq |a_i| - 1$ and the unions are disjoint. If there is a factor of form (1) then we are done. If A_i is of form (2) for each i, $1 \leq i \leq n$, then by Hajós's theorem at least one of the factors is a subgroup and so we are done. Thus we assume that A_i is of form (3) for some i, $1 \leq i \leq n$. We may assume $i = n$ since this is only a matter of indexing the factors $A_1, \ldots, A_n$. Let $B = A_1 \cdots A_{n-1}$. By Lemma 2.1.1, B is periodic. From the proof of the lemma we can read off that B is periodic with period $a_n^{r(n)}$. Thus B is a union of complete cosets modulo $H = \langle a_n^{r(n)} \rangle$. Since B is normalized, it contains H and so $a_n^{r(n)} \in B$. As $A_n = C_n\langle a_n \rangle \cup g_n[a_n, r(n)]$, A_n is normalized and $C_n \neq \emptyset$, it follows that $a_n^{r(n)} \in A_n$. Now

$$a_n^{r(n)} = \underbrace{(\, a_n^{r(n)} \,)}_{\in B} \underbrace{(\, e \,)}_{\in A_n} = \underbrace{(\, e \,)}_{\in B} \underbrace{(\, a_n^{r(n)} \,)}_{\in A_n}$$

violates the factorization $G = BA_n$. This reduces the proof to the case when each factor is of form (1) or (2).

This completes the proof.

We can prove Theorem 5.2.1 using replacement arguments as well. This approach will pay off when we try to extend Theorem 5.2.1 for multiple factorizations. Also it makes possible to extend some results from factorization of a group to factorization of a periodic subset in Section 5.4.

Second proof of Theorem 5.2.1. Consider the normalized factorization $G = A_1 \cdots A_n$ of the finite abelian group G, where each A_i is of one of the forms (1), (2), or (3). If the factor A_i is of form (1), then A_i is periodic and so we are done. So we may assume that A_i is of form (2) or (3) for each i, $1 \leq i \leq n$. If A_i is of form (2), then it is a cyclic subset. Using the character test we show that if A_i is of form (3), then it is replaceable by $B_i = [a_i, s(i)]$, where $s(i) = |A_i|$. We would like to point out that B_i is not necessarily a subset of G. In general B_i contains elements with multiplicity distinct from 0 and 1, that is, in general B_i is a multiset. Namely this is the case when $|A_i|$ is larger than the order of a_i.

By Lemma 4.2.1, $\chi(B_i) = 0$ if and only if $\chi(a_i) \neq 1$ and $\chi(a_i^{s(i)}) = 1$. In order to prove that $\mathrm{Ann}(A_i) \subset \mathrm{Ann}(B_i)$ it is sufficient to show that from $\chi(A_i) = 0$ it follows that (i) $\chi(a_i) \neq 1$ and (ii) $\chi(a_i^{s(i)}) = 1$.

To prove (i) assume the contrary that χ is a character of G for which $\chi(A_i) = 0$ and $\chi(a_i) = 1$. Now $0 = \chi(A_i) = \chi(C_i)t(i) + r(i)$, where $t(i) = |a_i|$. Hence $\chi(C_i) = -(r(i)/t(i))$. The left-hand side is an algebraic integer of the cyclotomic field over the rationals, the right-hand side is a rational number. Therefore both are integers. But this is impossible since $-1 < -(r(i)/t(i)) < 0$.

To prove (ii) consider a character χ of G with $\chi(A_i) = 0$. Now $\chi(A_i)\chi(a_i) = \chi(A_ia_i)$. From $\chi(A_i) = \chi(A_ia_i)$ after canceling we have that $\chi(g_i) = \chi(g_ia_i^{r(i)})$. Thus $\chi(a_i^{s(i)}) = \chi(a_i^{|C_i|t(i)+r(i)}) = 1$.

Suppose that A_i is of form (3) for some i, $1 \leq i \leq n$. Now B_i contains at least one element with multiplicity larger than 1. Replace A_i by B_i in the factorization $G = A_1 \cdots A_n$. The product $A_1 \cdots A_{i-1}B_iA_{i+1} \cdots A_n$ is equal to G. But this is a contradiction since the product contains at least one element with multiplicity larger than 1 while each multiplicity is 1 in G. Thus we may assume that A_i is of form (2) for each i, $1 \leq i \leq n$. Therefore by Hajós's theorem at least one of the factors must be a subgroup of G.

This completes the proof.

Next we might attempt to extend Theorem 5.2.1 for multiple factorizations. But Theorem 5.2.1 cannot be extended for multiple factorization even in the special case when each factor is cyclic, that is, Hajós's theorem does not extend to multiple factorizations in general. However, Theorem 5.2.1 can be extended for multiple factorization imposing the extra condition that the multiplicity is relatively prime to the order of the factored group.

Theorem 5.2.2. *Let $A_1 \cdots A_n$ be a k-fold factorization of the finite abelian group G, where k is relatively prime to $|G|$ and each A_i is a weakly periodic subset of G. Then at least one of the factors $A_1, \ldots, A_n$ is periodic.*

Proof. Replace each A_i by the corresponding cyclic multiset $B_i = [a_i, s(i)]$ in the same way as in the second proof of Theorem 5.2.1. The product $B_1 \cdots B_n$ is a k-fold factorization of G. As k is relatively prime to $|G|$ by Theorem 4.3.2, $a_i^{s(i)} = e$ for some i, $1 \le i \le n$. Hence $s(i)$ is a multiple of the order of a_i and so A_i is a union of complete cosets modulo $\langle a_i \rangle$. Therefore A_i is periodic.

      This completes the proof.

Theorem 5.2.3. *Let G be a finite abelian group and let p be the least prime factor of $|G|$. Let $A_1, \ldots, A_n$ be lacunary cyclic subsets of G with gaps $s_1 - 1, \ldots, s_n - 1$ respectively such that $s_1 < p, \ldots, s_n < p$. Then in the following two cases A_i is a subgroup of G for some i, $1 \le i \le n$.*

     (a) The product $A_1 \cdots A_n$ is direct giving a periodic subset A of G such that $G = AB$ is a factorization of G.

     (b) The product $A_1 \cdots A_n$ forms a k-fold product of G, where k is a positive integer relatively prime to $|G|$.

Proof. First we show that if $s < p$, then the lacunary cyclic subset

$$b_1[a, r(1)] \cup \cdots \cup b_s[a, r(s)]$$

can be replaced by the cyclic subset $B = [a, t]$, where $t = r(1) + \cdots + r(s)$. We know that $\chi(B) = 0$ if and only if $\chi(a) \ne 1$ and $\chi(a^t) = 1$. We will show that from $\chi(A) = 0$ it follows that (i) $\chi(a) \ne 1$ and (ii) $\chi(a^t) = 1$. If $\chi(a) = 1$, then $\chi(A) = t \ne 0$ which proves (i). Multiplying $\chi(A) = 0$ by $\chi(a)$ we get $0 = \chi(A)\chi(a) = \chi(Aa)$. From $\chi(A) = \chi(Aa)$ it follows that

$$\chi(b_1) + \cdots + \chi(b_s) = \chi(b_1 a^{r(1)}) + \cdots + \chi(b_s a^{r(s)}). \tag{4}$$

Each term in (4) is a power of a fixed uth root of unity, where u is a divisor of $|G|$ and hence the least prime divisor of u is not smaller than p. As $s < p$, Lemma 3.1.5 applies to (4) and we can conclude that $\chi(b_1), \ldots, \chi(b_s)$ is a rearrangement of $\chi(b_1 a^{r(1)}), \ldots, \chi(b_s a^{r(s)})$ and so

$$\chi(b_1) \cdots \chi(b_s) = \chi(b_1 a^{r(1)}) \cdots \chi(b_s a^{r(s)}).$$

After canceling each factor on the left-hand side against some factor on the right-hand side we get the sought $1 = \chi(a^{r(1) + \cdots + r(s)}) = \chi(a^t)$ relation which proves (ii).

     In the product $A_1 \cdots A_n$ replace the lacunary cyclic subset

$$A_i = b_{i1}[a_i, r_{i1}] \cup \cdots \cup b_{is}[a_i, r_{is}]$$

by the corresponding cyclic set $B_i = [a_i, t_i]$, where $t_i = r_{i1} + \cdots + r_{is}$ for each i, $1 \le i \le n$. By Hajós's theorem, in the product $B_1 \cdots B_n$ at least one of the factors is a subgroup of G, say B_i is a subgroup of G. Note that B_i is a subgroup of G if and only if $|a_i| = t_i$. From this it follows that $B_i = A_i$ and so A_i is a subgroup of G.

This completes the proof.

In the following results we will consider subsets of form $[a, r] \cup g[a, s]$, where the union is disjoint. We would like to show that if a finite abelian group is factored into the above type of subsets, then at least one of the factors must be a subgroup. We are able to verify this fact in the special case when the order of the finite abelian group is odd. The next exercise provides an example that shows that the result does not extend to abelian groups of even order.

Exercise 5.2.1. Let G be the direct product of two cyclic groups of order four, say $G = \langle x \rangle \times \langle y \rangle$, where $|x| = |y| = 4$. Choose the subsets A and B to be $A = [x, 2] \cup y^2[x, 2]$, $B = [y, 2] \cup x^2 y[y, 2]$. Verify that $G = AB$ is a factorization of G and none of the factors A and B is a subgroup of G.

Lemma 5.2.1. *Let G be a finite abelian group of odd order and let A be a subset of G such that $A = [a, r] \cup g[a, s]$, where the union is disjoint and $r + s$ is odd. Then $\mathrm{Ann}(A) \subset \mathrm{Ann}([a, r + s])$.*

Proof. Let $B = [a, r + s]$. As it is known $\mathrm{Ann}(B)$ consists of each character χ of G for which $\chi(a) \neq 1$ and $\chi(a^{r+s}) = 1$. Thus it is enough to verify that from $\chi(A) = 0$ it follows that (i) $\chi(a) \neq 1$ and (ii) $\chi(a^{r+s}) = 1$.

To prove (i) assume the contrary, that χ is a character of G for which $\chi(A) = 0$ and $\chi(a) = 1$. Now $0 = \chi(A) = r + \chi(g)s$ or equivalently $\chi(g) = -(r/s)$. Taking the absolute values of both sides we have $s = r$. Hence $r + s$ would be even, which is not the case.

To prove (ii) consider a character χ of G with $\chi(A) = 0$. Now $0 = \chi(A)\chi(a) = \chi(Aa)$. From $\chi(A) = \chi(Aa)$ after canceling we get

$$\chi(e) + \chi(g) = \chi(a^r) + \chi(ga^s).$$

The roots of unity occurring are of odd order by Lemma 3.1.7, and $\chi(e)$, $\chi(g)$ is a rearrangement of $\chi(a^r)$, $\chi(ga^s)$. Hence $\chi(e)\chi(g) = \chi(a^r)\chi(ga^s)$, which is equivalent to $1 = \chi(a^{r+s})$.

This completes the proof.

If $G = AC$ is a factorization of the finite abelian group G, where $A = [a, r] \cup g[a, s]$, then by Lemma 5.2.1, A can be replaced by $B = [a, r + s]$ to get factorization $G = BC$. Now B must contain $r + s$ elements and so $|a| \geq r + s$. Thus when A is a factor of a factorization, then $|a| \geq r + s$ holds. We would like to point out that this is not the case in general. Let $G = \langle x \rangle \times \langle y \rangle$, $|x| = 5$, $|y| = 3$ and $A = [x, 3] \cup y[x, 4]$. Now $|A| = 7$ and $|x| < 7$.

Lemma 5.2.2. *Let G be a finite abelian group of odd order and let A be a subset of G such that $A = [a, r] \cup g[a, s]$, where the union is disjoint and $r + s$ is odd. If A is periodic and $|a| \geq r + s$, then $A = \langle a \rangle$.*

Proof. As $|a| \geq r + s$, it is enough to prove that (i) $a^{r+s} = e$ and (ii) $g = a^r$.

If $\chi(a^{r+s}) = 1$ for each character χ of G, then $a^{r+s} = 1$. So to prove (i) we consider $\mathcal{C} = \{\chi : \chi(a^{r+s}) = 1\}$ and we show that $\mathcal{C}$ in fact coincides with the character group $\mathcal{G}$ of G. Note that $\mathcal{C}$ is a subgroup of $\mathcal{G}$ and $\mathrm{Ann}(A) \subset \mathcal{C}$.

Let x be a period of A. By Lemma 3.2.6, $\chi(A) = 0$ whenever $\chi(x) \neq 1$. Counting the number of characters χ of G for which $\chi(x) \neq 1$ we get a lower bound for $|\mathrm{Ann}(A)|$.

$$
\begin{aligned}
\left|\mathrm{Ann}(A)\right| &\geq |\mathcal{G}| - |G : \langle x \rangle| \\
&= |G| - |G : \langle x \rangle| \\
&= |G|\left(1 - \frac{1}{|x|}\right) \\
&\geq |G|\left(1 - \frac{1}{p}\right) \\
&> |G|(1/2) \\
&= (1/2)|\mathcal{G}|.
\end{aligned}
$$

Here p is the smallest prime divisor of $|G|$. As $|\mathrm{Ann}(A)| > (1/2)|\mathcal{G}|$, $\mathrm{Ann}(A)$ generates $\mathcal{G}$ and consequently $\mathcal{C} = \mathcal{G}$.

To prove $g = a^r$ assume the contrary, that $g \neq a^r$. Let χ be a character of G for which $\chi(A) = 0$. Applying χ to $g \neq a^r$ we face two possibilities, (a) $\chi(g) = \chi(a^r)$ and (b) $\chi(g) \neq \chi(a^r)$. We establish an upper bound for $|\mathrm{Ann}(A)|$. If $\chi(g) = \chi(a^r)$, then $\chi(ga^{-r}) = 1$ and the number of these characters is

$$
\left|G : \langle ga^{-r} \rangle\right| - 1 \leq (|G|/p) - 1
$$

since $\chi(A) = |A| \neq 0$ for the principal character χ of G. Turn to the case when $\chi(g) \neq \chi(a^r)$ and let

$$
B = [a, r] \cup a^r[a, s] = [a, r + s].
$$

By Lemma 5.2.1, from $\chi(A) = 0$ it follows that $\chi(B) = 0$ and so

$$
\begin{aligned}
0 &= \chi(A) - \chi(B) \\
&= \chi([a, r]) + \chi(g)\chi([a, s]) - \chi([a, r]) - \chi(a^r)\chi([a, s]) \\
&= \chi([a, s])\left(\chi(g) - \chi(a^r)\right).
\end{aligned}
$$

Hence $\chi([a, s]) = 0$ and consequently $\chi([a, r]) = 0$. Therefore $\chi(a) \neq 1$, $\chi(a^s) = 1$, $\chi(a^r) = 1$. If t is the greatest common divisor of s and r, then $\chi(a^t) = 1$. The number of these characters is

$$
\left|G : \langle a^t \rangle\right| - 1 \leq (|G|/p) - 1.
$$

We now combine the lower and upper bounds for $|\mathrm{Ann}(A)|$ together.

$$|G|\left(1 - \frac{1}{p}\right) \leq |\mathrm{Ann}(A)|$$
$$\leq \left(\frac{|G|}{p}\right) - 1 + \left(\frac{|G|}{p}\right) - 1$$
$$< |G|(2/p).$$

Canceling $|G|$ we get $1 - (1/p) < (2/p)$ or equivalently $p < 3$ which is not the case. This completes the proof.

Theorem 5.2.4. *Let G be a finite abelian group of odd order and $A_1, \ldots, A_n$ be subsets of G such that $A_i = [a_i, r(i)] \cup g_i[a_i, s(i)]$. If $G = A_1 \cdots A_n$ is a factorization of G, then A_i is a subgroup of G for some i, $1 \leq i \leq n$.*

Proof. For $n = 1$ the theorem holds and so we proceed by induction on n. Replace the factor A_i by $B_i = [a_i, r(i) + s(i)]$ for each i, $1 \leq i \leq n$ in the factorization $G = A_1 \cdots A_n$ to get the factorization $G = B_1 \cdots B_n$. By Lemma 5.2.1, this can be done. From the factorization $G = B_1 \cdots B_n$, by Hajós's theorem it follows that at least one of the factors B_i is a subgroup of G. We may assume that $B_1 = H$ is a subgroup of G since this is only a matter of indexing the factors. In the factorization $G = A_1 A_2 \cdots A_n$ replace A_1 by $B_1 = H$ to get the factorization $G = H A_2 \cdots A_n$. From this we get the factorization

$$G/H = (A_2 H)/H \cdots (A_n H)/H$$

of the factor group G/H. By the inductive assumption some of the factors $(A_i H)/H$, say $(A_2 H)/H$, is a subgroup of G/H, that is, $H A_2$ is a subgroup of G. Continuing in this way we have that

$$H, \ H A_2, \ H A_2 A_3, \ldots, H A_2 \cdots A_n$$

are subgroups of G. Let $K = H A_2 \cdots A_{n-1}$. If $g_1 \in K$, then $A_1 \subset K$ and so $K = A_1 A_2 \cdots A_{n-1}$ is a factorization of K. By the inductive assumption one of the factors is a subgroup of K and so of G.

In the remaining part of the proof we assume that $g_1 \notin K$. Let $b \in A_n$. From the factorizations $G = A_1 A_2 \cdots A_n$ and $G = H A_2 \cdots A_n$ by multiplying with b^{-1} we have that

$$G = A_1 A_2 \cdots A_{n-1}\left(b^{-1} A_n\right),$$
$$G = H A_2 \cdots A_{n-1}\left(b^{-1} A_n\right),$$
$$G = K\left(b^{-1} A_n\right)$$

are also factorizations of G. We use now the Corrádi trick. As $G = K\left(b^{-1} A_n\right)$ is a factorization of G, $b^{-1} A_n$ is a complete set of representatives of G modulo K. There is an element t_b in $b^{-1} A_n$ such that $t_b^{-1} K$ contains g_1. Now $g_1 t_b \in K$. Let

$$C_b = [a_1, r(1)] \cup [a_1, s(1)] g_1 t_b.$$

We claim that $K = C_b A_2 \cdots A_{n-1}$ is a factorization of K. Indeed, products coming from $C_b A_2 \cdots A_{n-1}$ occur among the product coming from $A_1 A_2 \cdots A_{n-1} (b^{-1} A_n)$. But these latter ones are distinct as $G = A_1 A_2 \cdots A_{n-1} (b^{-1} A_n)$ is a factorization of G. From $K = C_b A_2 \cdots A_{n-1}$ by the inductive assumption we have that one of the factors is a subgroup of K. If this is not C_b, then we are done. Thus we suppose that C_b is a subgroup of K. Now C_b is periodic and so by Lemma 5.2.2, $C_b = \langle a_1 \rangle$. Consequently $g_1 t_b = a_1^{r(1)}$ or equivalently $t_b = g_1^{-1} a_1^{r(1)} \in b^{-1} A_n$. If $t_b = e$ for some $b \in A_n$, then $g_1 = a_1^{r(1)}$ and $A_1 = \langle a_1 \rangle$. If $t_b \neq e$ for each $b \in A_n$, then

$$e \neq t_b = g_1^{-1} a_1^{r(1)} \in \bigcap_{b \in A_n} b^{-1} A_n.$$

We can test the periodicity of A_n by its translates. By Lemma 1.2.1, A_n is periodic. Now by Lemma 5.2.2, $A_n = \langle a_n \rangle$.

This completes the proof.

Consider the subset

$$A = \{ e, b^u, b^{u+v}, b^{u+2v}, \ldots, b^{u+tv} \}$$

of the finite abelian group G. The exponents

$$u, u + v, u + 2v, \ldots, u + tv$$

form an arithmetic progression and hence we call A an *arithmetic* subset of G.

Exercise 5.2.2. Let $G = A_1 \cdots A_n$ be a factorization of the finite abelian group G whose order is odd and assume that each A_i is arithmetic. Show that A_i is a subgroup of G for some i, $1 \leq i \leq n$. (Hint: Note that A can be written in the form $[b^v, 1] \cup b^u [b^v, t + 1]$ then apply Theorem 5.2.4.)

5.3   Factoring by simulated subsets

Let A be a subset of an elementary p-group G such that $e \in A$ and $|A| = p$. Let $A = \{ e, a_1, a_2, \ldots, a_{p-1} \}$. For each i, $2 \leq i \leq p - 1$ there is a $b_i \in G$ such that $a_i = a_1^i b_i$. In other words A can be written in the form

$$A = \{ e, a_1, a_1^2 b_2, \ldots, a_1^{p-1} b_{p-1} \},$$

that is, A differs from the subgroup $H = \langle a_1 \rangle$ in at most $p - 2$ elements. In general, a subset A of a finite abelian group G is called k-*simulated* (or simply simulated) if there is an integer k and a subgroup H of G such that

$$|A| = |H| \leq |A \cap H| + k.$$

If $k = 0$, then the subset A is equal to the subgroup H. The smaller is k, the closer is A to be a subgroup of G. Each subset A of G is k-simulated for some k if $|A|$ divides $|G|$.

Consider a factorization $G = A_1 \cdots A_n$ of the finite abelian group G, where A_i is a k_i-simulated subset of G. For which $k_1, \ldots, k_n$ does it follow that at least one of the factors $A_1, \ldots, A_n$ is a subgroup of G?

In Section 1.2 we showed that if $G = A_1 \cdots A_n$ is a normalized factorization of the finite abelian group G, where $|A_1| \geq 3, \ldots, |A_n| \geq 3$ and $k_1 = \cdots = k_n = 1$, then $A_i = H_i$ for some i.

We extend this theorem, proving that the same conclusion holds if $k_1 = p_1 - 2, \ldots, k_n = p_n - 2$, where p_i is the least prime divisor of $|H_i|$. The next result is about multiple factorization of a group of type (p, p).

Lemma 5.3.1. *Let p be a prime and let G be an elementary abelian group of order p^2. If $rsG = AB$ and there are subgroups H, K of G such that*

$$rp = r|H| = |A| \leq |A \cap rH| + p - 2$$

and

$$sp = s|K| = |B| \leq |B \cap sK| + p - 2,$$

then $A = rH$ or $B = sK$.

Proof. If $r = s = 1$, then we have an ordinary factorization $G = AB$ and from Lemma 1.4.5 it follows that either A or B is a subgroup. Then, if A is a subgroup of order p, from $|A \cap H| \geq 2$ it follows that $A = H$. Similarly B being a subgroup implies that $B = K$.

Let c be a generator of H with $c \in A$ and let d be a generator of K with $d \in B$. Then c, d generate G. Let ρ be a pth primitive root of unity. We may assume that p is odd since, for $p = 2$, we have

$$|A| \leq |A \cap rH|, \quad |B| \leq |B \cap sK|$$

and so $A = rH$, $B = sK$. Let $\chi(c) = \rho^k$, $\chi(d) = \rho^l$ define a character χ of G. If $\chi(A) = 0$, then this will remain true upon replacing ρ by any other pth primitive root of unity ρ^u, $2 \leq u \leq p - 1$. Thus while there are $p^2 - 1$ nonidentity characters χ of G we need only consider $p + 1$ essentially distinct cases in the annihilators of A and of B. We shall set

$$\chi_y(c) = \rho^y, \quad \chi_y(d) = \rho, \quad 0 \leq y \leq p - 1$$

and $\chi(c) = \rho$, $\chi(d) = 1$ to obtain one representative from each of these $p + 1$ classes of characters. By symmetry we may assume that at least $(p+1)/2$ of these characters belong to the annihilator of A. There are two cases to consider. Firstly where there are $(p + 1)/2$ values of y with $\chi_y(A) = 0$ and secondly where there are only $(p - 1)/2$ such values of y but also $\chi(A) = 0$.

Let

$$A = \{c^{u(i)} d^{v(i)} : 1 \leq i \leq rp\}, \quad 0 \leq u(i), v(i) \leq p - 1.$$

Since $|A| - |A \cap rH| \leq p - 2$ it follows that $v(i) = 0$ for at least $(r - 1)p + 2$ values of i. Since A is a multiset the pairs $(u(i), v(i))$, $1 \leq i \leq rp$, are not necessarily distinct, but it is more convenient here to list all these pairs, rather than to use multiplicities.

Assume first that $\chi_y(A) = 0$ for at least $(p+1)/2$ values of y. Then

$$\sum_{i=1}^{rp} \rho^{u(i)y+v(i)} = 0$$

for these values of y and hence the numbers $u(i)y + v(i)$, taken modulo p, form the multiset $r\{0, 1, \ldots, p-1\}$, that is, each number here occurs precisely r times. We define polynomials as follows, with coefficients in the Galois field $GF(p)$ of p elements:

$$D(x,y) = \prod_{i=1}^{rp}(x - u(i)y - v(i)) = \sum_{i=0}^{rp} d_i(y)x^i,$$

$$D(x,0) = \prod_{i=1}^{rp}(x - v(i)) = x^{(r-1)p}E(x),$$

$$E(x) = \sum_{i=0}^{p} e_i x^i, \quad G(x) = \prod(x - v(i)),$$

where this product is taken over only the distinct numbers $v(i)$ occurring in the presentation of A. Since $v(i) = 0$ for at least $(r-1)p+2$ values of i it follows that x^2 divides $E(x)$. The remaining factors $(x - v(i))$, $v(i) \neq 0$, also occur in $E(x)$ and so $G(x)$ divides $E(x)$ over $GF(p)$. When y is replaced by any of the $(p+1)/2$ values which give $\chi_y(A) = 0$, then

$$D(x,y) = \left[x(x-1)\cdots(x-(p-1))\right]^r = (x^p - x)^r.$$

So whenever the coefficient of x^i in $(x^p - x)^r$ is zero we have $d_i(y) = 0$ for $(p+1)/2$ values of y. If the degree of $d_i(y)$, as a polynomial in y, is less than $(p+1)/2$ it follows that $d_i(y) = 0$ for all y. Now

$$d_{pr-j}(y) = (-1)^j S_j\big(u(i)y + v(i)\big),$$

where S_j denotes the jth elementary symmetric function. This has degree at most j. This implies that $d_{pr-j}(y) = 0$ for all y for $1 \leq j \leq (p-1)/2$. In particular this is true for $y = 0$. Now

$$D(x,0) = \sum_{i=0}^{rp} d_i(0)x^i = x^{(r-1)p}\sum_{i=0}^{p} e_i x^i.$$

It follows that $e_i = 0$ for $(p+1)/2 \leq i \leq p-1$. We also have $e_p = 1$. Hence $E(x) - (x^p - x)$ has degree $\leq (p-1)/2$ but is not the zero polynomial, as x^2 divides $E(x)$. Since $G(x)$ divides $E(x)$ and also $x^p - x$ it follows that $G(x)$ has degree $\leq (p-1)/2$. Hence $E(x)/G(x)$ has degree $\geq (p+1)/2$. Now $E(x)/G(x)$

contain only the repeated linear factors of $E(x)$ and to a lower power than their occurrence in $E(x)$. Therefore $E(x)/G(x)$ divides $E'(x)$. Since $px^{p-1} = 0$ either $E'(x) = 0$ or else degree $E'(x) < (p-1)/2$. The second case is not possible from consideration of the degree of $E(x)/G(x)$ and so we must have $E'(x) = 0$. This gives $E(x) = x^p$ and so all $v(i) = 0$. It follows that $A = rH$.

We turn to the case where $\chi_y(A) = 0$ for only $(p-1)/2$ values of y but where $\chi(A) = 0$ also. With the polynomials defined as above we obtain $e_i = 0$ but now only for $(p-1)/2 \le i \le p-1$. Now

$$\chi(A) = \sum \rho^{u(i)} = 0$$

implies that the numbers $u(i)$ are equal to the multiset

$$r\{0, 1, \ldots, p-1\}$$

when taken modulo p. Let us consider

$$d_{pr-j}(y) = (-1)^j S_j\big(u(i)y + v(i)\big)$$

with $j = (p-1)/2$. The coefficient of y^j is $(-1)^j S_j\big(u(i)\big)$. Now

$$\prod_{i=1}^{rp} (x - u_i) = (x^p - x)^r$$

and $(-1)^j S_j\big(u(i)\big)$ is the coefficient of x^{pr-j} here. For $j = (p-1)/2$ this coefficient is 0. Hence $d_{pr-j}(y)$ with $j = (p-1)/2$ has degree less than $(p-1)/2$ but is equal to zero for $(p-1)/2$ values of y. This leads as before to $d_{pr-j}(0) = 0$ and so to $e_i = 0$ for $i = p - (p-1)/2$. The proof that $A = rH$ now follows, as before.

This completes the proof.

Lemma 5.3.2. *Let A and H be a subset and a subgroup of the finite abelian group G such that $|H| = |A| \le |A \cap H| + p - 2$, where p is the least prime factor of $|A|$. Then A can be replaced by H.*

Proof. First deal with the case when $|H|$ is composite. In this case $|A| = |H| < 2|A \cap H|$ since $|A \cap H| < p$. Let χ be a character of G for which $\chi \in \mathrm{Ann}(A)$. We want to prove that $\chi \in \mathrm{Ann}(H)$. If $\chi \notin \mathrm{Ann}(H)$, then since H is a subgroup $\chi(h) = 1$ for each $h \in H$. Then

$$
\begin{aligned}
0 = \chi(A) \; &= \; \sum_{a \in A} \chi(a) \\
&= \; \sum_{a \in A \cap H} \chi(a) + \sum_{a \in A \setminus H} \chi(a) \\
&= \; |A \cap H| + \sum_{a \in A \setminus H} \chi(a).
\end{aligned}
$$

Hence

$$\begin{aligned}
|A \cap H| &= \left| \sum_{a \in A \setminus H} \chi(a) \right| \\
&\leq \sum_{a \in A \setminus H} |\chi(a)| \\
&= |A \setminus H|.
\end{aligned}$$

This contradicts $|A| = |H| < 2|A \cap H|$. Thus $\chi \in \mathrm{Ann}(A)$.

Let us turn to the case when $|A| = |H| = p$. Let $H = \{e, h, h^2, \ldots, h^{p-1}\}$ and assume that $h \in A$. By Lemma 1.4.3, A can be replaced by $H = \langle h \rangle$.

This completes the proof.

The following result is a straightforward consequence of Lemma 3.1.5 and Lemma 5.3.2.

Lemma 5.3.3. *Let*

$$H = \{h_1, \ldots, h_u, h_{u+1}, \ldots, h_{u+v}\}$$

and

$$A = \{h_1, \ldots, h_u, a_1, \ldots, a_v\}$$

be a subgroup and a subset of the finite abelian group G such that $a_1, \ldots, a_v \notin H$. Assume that $v \leq p-2$, where p is the least prime divisor of $|G|$. Let χ be a character of G for which $\chi(A) = 0$. If $|H|$ is composite or $|H| = p$, then $\chi(h_{u+1}), \ldots, \chi(h_{u+v})$ is a rearrangement of $\chi(a_1), \ldots, \chi(a_v)$.

Proof. First consider the case when $|H|$ is composite. Then $|A| = |H| < 2|A \cap H|$ and hence by the proof of Lemma 5.3.2, from $\chi(A) = 0$ it follows that $\chi(H) = 0$. Therefore $\chi(A) = \chi(H)$. After canceling the identical terms from both sides of this equation we obtain equal sums of roots of unity

$$\chi(h_{u+1}) + \cdots + \chi(h_{u+v}) = \chi(a_1) + \cdots + \chi(a_v).$$

By Lemma 3.1.5 of, $\chi(a_1), \ldots, \chi(a_v)$ is a rearrangement of $\chi(h_{u+1}), \ldots, \chi(h_{u+v})$.

Secondly consider the case when $|H| = p$. If $\chi(H) = 0$ for each character χ of G for which $\chi(A) = 0$, then we can proceed as before. So assume that $\chi(H) \neq 0$ for some character χ of G with $\chi(A) = 0$. Then χ is principal over H. As $A = \{h_1, h_2, a_1, \ldots, a_{p-2}\}$ and as χ is principal over H, from $\chi(A) = 0$ we obtain the vanishing sum of roots of unity

$$1 + 1 + \chi(a_1) + \cdots + \chi(a_{p-2}) = 0.$$

This equation contradicts the result of Exercise 3.1.3.

This completes the proof.

Under certain conditions if a simulated subset is periodic, then it is a subgroup. This is the content of the next lemma.

Lemma 5.3.4. *Let G be a finite abelian group and let A and H be a subset and a subgroup of G respectively. Suppose that $|H| = |A| \leq |A \cap H| + p - 2$, where p is the least prime divisor of $|H|$. If A is periodic, then $A = H$.*

Proof. The result is clear for $p = 2$ so we assume that p is odd. Let g be a period of G, that is, let $gA = A$. We may suppose that $|g| = r$ is a prime. The permutation $a \to ga$, $a \in A$ consists of cycles of length r. Therefore r is a divisor of $|A|$ and so $p \leq r$. Consider a cycle of the permutation. There must be two elements a and b in this cycle contained by H. Since $g^t a = b$ for some t, $1 \leq t \leq r - 1$, it follows that $g \in H$. Let c be an element of the cycle. Now, $g^s a = c$ for some s, $0 \leq s \leq r - 1$. Thus $c \in H$ which gives that $A = H$.

    This completes the proof.

    Now we are in position to prove the main results of the section. First we prove the result for factorization containing two factors. Then using this we turn to the general case.

Theorem 5.3.1. *Let G have odd order and let G be a direct product of subsets A and B such that there exist subgroups H and K of G with*

$$|A| = |H| \leq |A \cap H| + p - 2,$$
$$|B| = |K| \leq |B \cap K| + q - 2,$$

where p and q are the least prime factors of $|H|$ and $|K|$ respectively. Then either $A = H$ or $B = K$.

Proof. By Lemma 5.3.2, we can replace A by H and B by K in the factorization $G = AB$ to obtain $G = HB = AK = HK$. Let

$$H = \{h_1, \ldots, h_u, h_{u+1}, \ldots, h_{u+v}\},$$
$$A = \{h_1, \ldots, h_u, a_1, \ldots, a_v\},$$

where $a_i \notin H$. Since $a_i \in hK$ implies $a_i K = hK$ from $G = HK = AK$ it follows that we can order the elements a_i such that $a_i K = h_{u+i} K$. Thus there are elements $x_i \in K \setminus \{e\}$ with $a_i = h_{u+i} x_i$. Similarly we have

$$K = \{k_1, \ldots, k_s, k_{s+1}, \ldots, k_{s+t}\},$$
$$B = \{k_1, \ldots, k_s, b_1, \ldots, b_t\},$$

with $b_i = k_{s+i} y_i$, $1 \leq i \leq t$, $y_i \in H \setminus \{e\}$.

    If $|A| = p$ and $|B| = q$ then, by Rédei's theorem $A = H$ or $B = K$. Thus we may proceed by induction on $|G|$.

    If $v = 0$ then $A = H$. If $v = 1$ then, by Lemma 1.2.3, x_1 is a period of B and so, by Lemma 5.3.4, $B = K$. Thus we need only consider $v \geq 2$ and also $t \geq 2$.

    Let $X = \langle x_1, \ldots, x_v \rangle$ and $Y = \langle y_1, \ldots, y_t \rangle$. Then X is a non-identity subgroup of K and Y of H. Hence $|X| \geq q$ and $|Y| \geq p$. Suppose that $Y \neq H$. Let

$d_1,\ldots,d_w$ be a set of coset representatives for H modulo Y and so also for G modulo YK. If, for each i, $1 \leq i \leq w$, $A \cap d_i YK \subset H$, then $A = H$. So we may suppose that $A \cap d_i YK \not\subset H$ holds for some i, $1 \leq i \leq w$. As $B \subset YK$ we have $(A \cap d_i YK)d_i^{-1}B \subset YK$. Let $g \in YK$. Then $d_i g = ab$ for some $a \in A$, $b \in B$. Then $a = d_i gb^{-1} \in (A \cap d_i YK)$ and so $g = ad_i^{-1}b \in (A \cap d_i YK)d_i^{-1}B$. Hence $YK = (A \cap d_i YK)d_i^{-1}B$. Thus $|Y| = |(A \cap d_i YK)d_i^{-1}|$. Note that

$$|Y \cap (A \cap d_i YK)d_i^{-1}| = |d_i Y \cap A \cap d_i YK| = |A \cap d_i Y|.$$

Since H is a disjoint union of the cosets $d_i Y$ and $|H| \leq |A \cap H| + p - 2$ it follows that $|Y| \leq |A \cap d_i Y| + p - 2$. So we may apply the inductive assumption to this factorization of YK to conclude that $B = K$, as $A \cap d_i YK \neq d_i Y \subset H$.

There is a similar argument if $X \neq K$. So we may assume that $Y = H$ and $X = K$.

We may suppose that $p \leq q$ and so that p is the least prime factor of $|G|$. Let χ be a character of G with $\chi(A) = 0$. By Lemma 5.3.3, the sequences of roots of unity $\chi(h_{u+1}),\ldots,\chi(h_{u+v})$ and $\chi(a_1),\ldots,\chi(a_v)$ are rearrangements of each other.

Suppose that q does not divide $|H|$. From $\chi(h_{u+i}) = \chi(a_j) = \chi(h_{u+j})\chi(x_j)$ it follows that $\chi(x_{j|q}) = 1$. By the replacement principle, A can be replaced by

$$A' = \{h_1,\ldots,h_u,h_{u+1}(x_{1|q})^{\alpha(1)}x_{1|q'},\ldots,h_{u+v}(x_{v|q})^{\alpha(v)}x_{v|q'}\}$$

for any exponents $\alpha(1),\ldots,\alpha(v)$. We define X' for A', as before. If $x_{j|q'} = e$ for each j, $1 \leq j \leq v$, then X' and so K are q-groups. By suitable choice of exponents $\alpha(j)$ we can make A' differ from H in one position only. Then $G = A'B$ implies $B = K$.

If, for some j, $x_{j|q'} \neq e$, then we may choose all $\alpha(j) = 0$. This gives an X' with q'-elements and so $X' \neq K$. As above we have $H = A'$ or $B = K$. However $x_{j|q'} \neq e$ implies $H \neq A'$.

Finally we have the case where q divides $|H|$. Then H, K contain subgroups T, S of index q and $G' = G/(TS)$ is an elementary group of order q^2. The factorization $G = AB$ leads to a multiple factorization $mnG' = A'B'$, where $|H| = mq$, $|K| = nq$ and A', B' are the multisets in G' obtained from A, B. Let $H' = (HS)/(TS)$, $K' = (TK)/(TS)$. Then

$$|A'| = m|H'| \leq |A' \cap mH'| + q - 2,$$
$$|B'| = n|K'| \leq |B' \cap nK'| + q - 2.$$

By Lemma 5.3.1, $A' = mH'$ or $B' = nK'$. $A' = mH'$ implies $A \subset HS$ and so $X \subset S \neq K$. Similarly $B' = nK'$ implies $Y \neq H$.

This contradiction completes the proof.

Theorem 5.3.2. *Let G be a direct product of subsets $A_1,\ldots,A_n$ and let there exist subgroups $H_1,\ldots,H_n$ such that $|H_i| = |A_i| \leq |A_i \cap H_i| + p_i - 2$, where p_i is the least prime factor of $|A_i|$. Then there exists an i such that $A_i = H_i$.*

Proof. For $p_i = 2$ we have $|A_i| = |A_i \cap H_i| = |H_i|$ and so $A_i = H_i$. Thus we may assume that p_i is odd for each i.

For $n = 1$ we have $A_1 = G = H_1$. For $n = 2$ by Theorem 5.3.1 it follows that $A_1 = H_1$ or $A_2 = H_2$.

We proceed by induction on n.

By Lemma 5.3.2, A_i can be replaced by H_i in the factorization

$$G = A_1 \cdots A_n \tag{1}$$

to get the factorization

$$G = A_1 \cdots A_{i-1} H_i A_{i+1} \cdots A_n. \tag{2}$$

Considering the factor group G/H_i from (2) we get a new factorization

$$G/H_i = (A_1 H_i)/H_i \cdots (A_{i-1} H_i)/H_i \cdot (A_{i+1} H_i)/H_i \cdots (A_n H_i)/H_i.$$

By the inductive assumption there exists $j = f(i) \neq i$ such that $A_j H_i = H_j H_i$. There must exist a set of indices, say $\{1, 2, \ldots, m\}$ permuted cyclically by f. Then for each i, $1 \leq i \leq m$, $A_i \subset H_1 \cdots H_m = K$ gives that

$$K = A_1 \cdots A_m$$

is a factorization of K.

If $m \leq n - 1$, then by the inductive assumption $A_i = H_i$ for some i. Thus we need to consider only the case $m = n$. Then $H_i A_{i+1} = H_i H_{i+1}$ for each i, $1 \leq i \leq n$. For each $a \in A_{i+1}$ we have $a = bc$, $b \in H_i$, $c \in H_{i+1}$. Hence $H_i a = H_i bc = H_i c$. It follows that as a varies over A_{i+1}, c must vary over H_{i+1}, each element of H_{i+1} occurring once.

Let $K = H_1 \cdots H_{n-1}$. For each i, $2 \leq i \leq n - 1$, $A_i \subset K$ and so

$$K = H_1 A_2 \cdots A_{n-1} \tag{3}$$

is a factorization of K.

Let $b \in A_n$. From factorizations (1) and (2), by multiplying by b^{-1} we have that

$$G = A_1 A_2 \cdots A_{n-1}(b^{-1} A_n), \tag{4}$$

$$G = H_1 A_2 \cdots A_{n-1}(b^{-1} A_n) \tag{5}$$

are factorizations of G.

Consider (3) and distinguish two cases depending on $A_1 \subset K$ or $A_1 \not\subset K$. If $A_1 \subset K$, then

$$K = A_1 A_2 \cdots A_{n-1}$$

is also a factorization of K. By the inductive assumption $A_i = H_i$ for some i, $1 \leq i \leq n - 1$.

Secondly suppose that $A_1 \not\subset K$. Let

$$H_1 = \{h_1, \ldots, h_u, h_{u+1}, \ldots, h_{u+v}\},$$
$$A_1 = \{h_1, \ldots, h_u, a_1, \ldots, a_v\}$$

such that $a_i \notin H_1$. There are elements $d_1, \ldots, d_v \in H_n \setminus \{e\}$ with $a_i = h_{u+i} d_i$. Since $A_1 \not\subset K$, one d_i, say d_1 is not in K. By (3) and (5) $G = K(b^{-1}A_n)$ is a factorization of G. Therefore, $b^{-1}A_n$ forms a complete set of representatives modulo K. We employ the Corrádi trick. There are elements $x_{1b}, \ldots, x_{vb} \in b^{-1}A_n$ such that the coset $x_{ib}K$ contains a_i^{-1}. Now $a_i x_{ib} \in K$.

Let
$$C_b = \{h_1, \ldots, h_u, a_1 x_{1b}, \ldots, a_v x_{vb}\}.$$

Note that

$$K = C_b A_2 \cdots A_{n-1} \tag{6}$$

is a factorization of K. To verify this first observe that $|K| = |C_b||A_2| \cdots |A_{n-1}|$. Secondly observe that products coming from $C_b A_2 \cdots A_{n-1}$ occur among the products coming from

$$A_1 A_2 \cdots A_{n-1}(b^{-1}A_n)$$

and these latter are distinct since (4) is a factorization of G.

From (6) by the inductive assumption it follows that $C_b = H_1$ or $A_i = H_i$ for some i, $2 \le i \le n-1$. We need to consider only the $C_b = H_1$ case. Then $a_1 x_{1b} \in H_1$. Clearly, $x_{1b} = t_b s_b$, where $t_b \in H_1$ and $s_b \in H_2 \cdots H_n$. From $a_1 x_{1b} = h_{u+1} d_1 t_b s_b \in H_1$ it follows that $d_1 s_b = e$, that is, $x_{1b} = d_1^{-1} t_b$. Therefore, $x_{1b} \in H_n H_1$. On the other hand $x_{1b} \in b^{-1}A_n \subset H_n H_{n-1}$. Thus $n = 2$ or $t_b = e$ for each $b \in A_n$. The case $n = 2$ has already been settled in Theorem 5.3.1. So we consider only the case when $t_b = e$ for each $b \in A_n$. Then

$$x_{1b} = d_1^{-1} \in \bigcap_{b \in A_n} b^{-1}A_n.$$

We can test the periodicity of A_n by means of its translates. By Lemma 1.2.1, A_n is periodic and so by Lemma 1.2.2, $A_n = H_n$.

This completes the proof.

In construction (1) in the proof of Lemma 2.3.2, factor A_1 differs from a subgroup in $p-1$ elements and factor A_2 differs from a subgroup in two elements. This shows that the conditions of Theorem 5.3.2 cannot be removed in the $n = 2$ case.

5.4   Factoring of periodic subsets

The nonidentity elements of a finite abelian group are periods of the group. Thus a finite abelian group which has at least two elements is a periodic set. So far we dealt with factorizations of finite abelian groups. In this section we will consider factorizations of a periodic subset. Our main tools will be Lemmas 3.3.1, 3.3.2, and 3.3.3. We will refer to these results as the zero divisor lemmas. We will see later how results about factoring a periodic subset can be applied to study ordinary factorizations.

Rédei's theorem for finite abelian p-groups states that if a finite abelian p-group is a direct product of normalized subsets of prime cardinality, then at least one of the factors must be a subgroup. The next theorem generalizes this result.

Theorem 5.4.1. *Let G be a finite abelian p-group and let A be a periodic subset of G such that*

$$A = A_1 \cdots A_n \tag{1}$$

is a normalized factorization of A. If for each i, $1 \le i \le n$ $|A_i|$ is a prime or A_i is a special simulated subset of G, then at least one of the factors must be a subgroup of G.

Proof. Let g be a period of A. The factorization (1) corresponds to the equation $\overline{A} = \overline{A}_1 \cdots \overline{A}_n$ in the group ring $Z(G)$. Multiplying this equation by $e - g$ we get

$$0 = \overline{A}_1 \cdots \overline{A}_n (e - g). \tag{2}$$

From this after the possible cancellations and reindexing if this is necessary, we have

$$0 = \overline{A}_1 \cdots \overline{A}_t (e - g). \tag{3}$$

Here $t \ge 1$ since $e - g \ne 0$ and $(e - g)$ cannot be cancelled from (3) as a partial product of $\overline{A}_1 \cdots \overline{A}_n$ cannot be zero.

If A_i is a special simulated subset of G, then there is a subgroup H_i of G such that $|H_i \cap A_i| + 1 \ge |A_i| = |H_i| \ge 3$. As H_i is a subgroup of G, it follows that $|A_i|$ is a power of p.

We claim that if $|A_i| = q$ is a prime and $q \ne p$, then $\mathrm{Ann}(A_i)$ is empty and consequently A_i is cancelable from (1). In order to prove the claim let χ be a character of G for which $\chi \in \mathrm{Ann}(A_i)$. Now $\chi(A_i)$ is a vanishing sum of powers of primitive (p^n)th roots of unity, where n is a nonnegative integer. The $n = 0$ assumption leads to the contradiction $0 = \chi(A_i) = |A_i| = q$. Thus $n \ge 1$. By Lemma 3.1.1, the number of the terms of the vanishing sum $0 = \chi(A_i)$ is a multiple of p. On the other hand $\chi(A_i)$ consists of q terms. This is a contradiction since p and q are distinct primes and so they are relatively primes too.

Thus we may assume that $|A_i|$ is a power of p for each i, $1 \le i \le n$. Hence the third zero divisor lemma is applicable to (3) with the choice $B = e$ to get

$$\gamma(A_1, \ldots, A_t, g) < m_1 + \cdots + m_t + 1$$

and then

$$\gamma(A_1, \ldots, A_t, g) \le m_1 + \cdots + m_t.$$

Hence

$$\gamma(A_1, \ldots, A_t) \leq m_1 + \cdots + m_t. \tag{4}$$

Set $l = m_1 + \cdots + m_t$ and $K = \langle A_1, \ldots, A_t \rangle$. As G is a p-group by (4) it follows that $|K| \leq p^l$. On the other hand

$$p^l \leq |A_1| \cdots |A_t| \leq p^l.$$

Therefore $A_1 \cdots A_t$ is a direct product giving the subgroup K. In this case by (an extended version of) Rédei's theorem, one of the factors $A_1, \ldots, A_t$ is a subgroup of K and so of G.

This completes the proof.

In certain applications it is guaranteed that the factored periodic subset itself is a factor of a factorization of a group. In the next theorem it will be assumed that the factored periodic subset has a complementary factor such that the product of the two subsets is direct and gives a group.

Theorem 5.4.2. *Let G be a finite abelian group and let $G = AB$ and*

$$A = A_1 \cdots A_n \tag{5}$$

be normalized factorizations where A is a periodic subset of G. Assume that for each i, $1 \leq i \leq n$ $|A_i| = p_i$ is a prime or A_i is a special simulated subset. Further assume that in case $|A_i| = p_i$ is a prime there is at most one prime divisor of $|\langle A_i \rangle|$ that is less than p_i. Then at least one of the factors must be a subgroup of G.

Proof. Let g be a period of A. First assume that each A_i for which $|A_i| = p_i$ is a prime contain only (p_i, q_i)-elements, where q_i is a prime distinct from p_i. The factorization (5) corresponds to the equation $\overline{A} = \overline{A}_1 \cdots \overline{A}_n$ in the group ring $Z(G)$. Multiplying this equation by $e - g$ results $0 = \overline{A}_1 \cdots \overline{A}_n(e - g)$. After the possible cancellations and reindexing if this is necessary, we get $0 = \overline{A}_1 \cdots \overline{A}_t(e-g)$, where $t \geq 1$ since $e - g \neq 0$ and $e - g$ cannot be deleted since a partial product of $\overline{A}_1 \cdots \overline{A}_n$ cannot be zero. The third zero divisor lemma is applicable to this equation and gives

$$\gamma(A_1, \ldots, A_t) \leq m_1 + \cdots + m_t. \tag{6}$$

Set $T = A_1 \cdots A_t$, $S = A_{t+1} \cdots A_n B$, and $K = \langle T \rangle$. Restricting the factorization $G = TS$ to the subgroup K results the factorization $K = K \cap G = T(S \cap K)$ of K. Hence $|T|$ divides $|K|$. As $|T|$ is the product of $m_1 + \cdots + m_t$ primes and as by (6) $|K|$ is a product of at most $m_1 + \cdots + m_t$ primes, it follows that the product $A_1 \cdots A_t$ is direct and gives the subgroup K. By Rédei's theorem, we get that at least one of the factors $A_1, \ldots, A_t$ is a subgroup of K and so of G.

Secondly assume that there is a factor A_i in the factorization (4) such that $|A_i| = p_i$ is a prime and A_i contains not only (p_i, q_i)-elements. For the sake of definiteness assume that $|A_1| = p$ is a prime $a, a' \in A_1$ and $a_{|r} \neq e$, $a'_{|q} \neq e$, where p, q, r are distinct primes. In this case we proceed by an induction on the quantity

$$h(A_1, \ldots, A_n) = \prod_{i=1}^{n} \prod_{a \in A_i} |a|.$$

By the hypotheses of the theorem there are at most one among the primes p, q, r that is less than p. Thus one of them, say q, must be larger than p. Set

$$A'_1 = \{a_{|q'} : a \in A_1\}.$$

By Lemma 3.2.3, $\mathrm{Ann}(A_1) \subset \mathrm{Ann}(A'_1)$ and hence A_1 is replaceable by A'_1 in the factorization $G = A_1 \cdots A_n B$ to get the factorization $G = A'_1 A_2 \cdots A_n B$. Let $A' = A'_1 A_2 \cdots A_n$. As

$$\mathrm{Ann}(\langle g \rangle) \subset \mathrm{Ann}(A) \subset \mathrm{Ann}(A'),$$

it follows that g is also a period of A'. Now

$$h(A_1, A_2, \ldots, A_n) < h(A'_1, A_2, \ldots, A_n).$$

Therefore by the induction assumption at least one of the factors $A'_1, A_2, \ldots, A_n$ must be a subgroup of G. If A'_1 is a subgroup of G, then A_1 contains only (p, q)-elements which is not the case. Thus one of $A_2, \ldots, A_n$ is a subgroup of G.

This completes the proof.

The above result can be viewed as a generalization of Hajós's and Rédei's theorems since if there were a counterexample for these theorems, then there were such ones in which the factors contain only (p, q)-elements. The next exercise exhibits a factorization of a periodic subset of a finite abelian group by nonsubgroup factors of prime cardinality. This shows that the extra condition that the factored periodic subset is itself a factor of a factorization of a whole finite abelian group is essential in Theorem 5.4.2.

Exercise 5.4.1. Let $G = \langle x \rangle \times \langle y \rangle \times \langle z \rangle$, where $|x| = 2$, $|y| = |z| = 3$ and define the subsets A and A_1, A_2, A_3, A_4 by

$$A = G \setminus \{xz^2, z^2\},$$

$$A_1 = \{e, xz\}, \qquad A_2 = \{e, xy\},$$
$$A_3 = \{e, xyz\}, \qquad A_4 = \{e, xyz^2\}.$$

Verify that the product $A_1 \cdots A_4$ is direct and gives the subset A. Check that A is periodic by x.

In the previous exercise each factor was a cyclic subset. In the next construction the factors are not cyclic subsets but the group we are working in is cyclic.

Let G be a group of type (p, q, r), where p, q, r are primes, $p < q$. Let x, y, z be basis elements of G with order p, q, r respectively. Set

$$A_1 = \langle y \rangle \cup z \langle x \rangle,$$

$$A_2 = \{e, xy, (xy)^2, \ldots, (xy)^{p-1}\},$$

$$A = \left(\langle y \rangle \cup z\{e, y, y^2, \ldots, y^{p-1}\} \right) \langle x \rangle.$$

We claim that the product $A_1 A_2$ is direct and is equal to A. It is enough to show that $\chi(A) = \chi(A_1 A_2)$ for each character χ of G. This holds when χ is the principal character of G since $\chi(A) = |A| = (q + p)p$, $\chi(A_1) = |A_1| = q + p$, $\chi(A_2) = |A_2| = p$. Pick a nonprincipal character χ of G. Suppose $\chi(x) \neq 1$. Now $\chi(A) = 0$. If $\chi(y) = 1$, then $\chi(A_2) = 0$. If $\chi(y) \neq 1$, then $\chi(A_1) = 0$. In either case $\chi(A_1 A_2) = 0$ and so $\chi(A) = \chi(A_1 A_2)$. Thus we may assume that $\chi(x) = 1$. Now

$$\chi(A) = \left[\chi(\langle y \rangle) + \chi(z)(1 + \chi(y) + \cdots + \chi(y^{p-1})) \right] p,$$

$$\chi(A_1) = \chi(\langle y \rangle) + p\chi(z),$$

$$\chi(A_2) = 1 + \chi(y) + \cdots + \chi(y^{p-1}).$$

We can see that $\chi(A) = \chi(A_1 A_2)$ holds.

Finally choose p to be 2 and $2 + q$ to be a prime.

Exercise 5.4.2. Let $G = \langle x \rangle \times \langle y \rangle \times \langle z \rangle$, where $|x| = 2$, $|y| = 3$, $|z| = 5$ and define the subsets A_1, A_2 and A by

$$A_1 = \{e, xyz, y, y^2, yz\}, \quad A_2 = \{e, xy^2\},$$

$$A = \{e, y, y^2, yz, z\}\{e, x\}.$$

It is clear from the definition of A that x is a period of A. Verify that the product $A_1 A_2$ is direct and gives A.

It is a special case of Theorem 2.1.1 that if G is a finite abelian group whose 2-component is cyclic and $G = A_1 \cdots A_n$ is a factorization of G, where each A_i is a distorted cyclic subset of G, then at least one of the factors must be periodic. Using the second zero divisor lemma we will show that the above result can be extended for factorizations of periodic subsets but we have to impose the extra condition that $|G|$ is relatively prime to 6.

We need the following lemma.

Lemma 5.4.1. *Let G be a finite abelian group whose order is relatively prime to 6. Let*

$$A = \{e, a, a^2, \ldots, a^{i-1}, a^i d, a^{i+1}, \ldots, a^{r-1}\}$$

be a distorted cyclic subset of G with $r \geq 2$ and let

$$B = \{e, a, a^2, \ldots, a^{r-1}\}$$

be the associated cyclic subset. Let $H = \langle d \rangle \times \langle a^r \rangle$, $K = \langle d \rangle \times \langle a \rangle$. Then
 (1) $K = AH$ is a factorization of K.
 (2) $\mathrm{Ann}(A) \subset \mathrm{Ann}(B)$.
 (3) $\chi(A) = 0$ implies $\chi(h) = 1$ for each $h \in H$ and each character χ of G.

Proof. We do not detail the proof of (1). To prove (2) assume the contrary, that there is a character χ of G with $\chi(A) = 0$ and $\chi(B) \neq 0$. Note that $\chi(B) \neq 0$ implies $\chi(a) = 1$ and so

$$0 = \chi(A) = 1 + \chi(a) + \cdots + \chi(a^{r-1}) + \chi(a^i d) - \chi(a^i) = r + \chi(d) - 1,$$

that is $|r - 1| = |-\chi(d)| = 1$. This is impossible as $r \geq 5$.

To prove (3) we show that $\chi(A) = 0$ implies $\chi(a^r) = 1$, $\chi(d) = 1$. We have seen that $\chi(A) = 0$ implies $\chi(a) \neq 1$. Hence

$$0 = \chi(A) = \frac{1 - \chi(a^r)}{1 - \chi(a)} + \chi(a^i d) - \chi(a^i).$$

This is equivalent to

$$1 + \chi(a^i d) + \chi(a^{i+1}) = \chi(a^i) + \chi(a^{i+1} d) + \chi(a^r). \tag{7}$$

By Lemma 3.1.5 it follows that the terms on the left side form a permutation of the terms on the right side. The product of the terms of the left side is equal to the product of the terms of the right side. This gives $\chi(a^{2i+1} d) = \chi(a^{2i+1} d a^r)$ and so $\chi(a^r) = 1$. Now (7) reduces to

$$\chi(a^i d) + \chi(a^{i+1}) = \chi(a^i) + \chi(a^{i+1} d)$$

and to

$$\chi(d) + \chi(a) = 1 + \chi(ad).$$

We can use again Lemma 3.1.5. As $\chi(a) = 1$ is impossible it follows that $\chi(d) = 1$. This completes the proof.

Theorem 5.4.3. *Let G be a finite abelian group such that the least prime divisor of $|G|$ is at least 5. Suppose that $G = AB$ is a normalized factorization of G, where A is a periodic subset of G and*

$$A = A_1 \cdots A_n \tag{8}$$

is a normalized factorization of A into distorted cyclic subsets. Then at least one of the factors $A_1, \ldots, A_n$ is periodic.

Proof. Let g be a period of A. The factorization (8) corresponds to the equation $\overline{A} = \overline{A}_1 \cdots \overline{A}_n$ in the group ring $Z(G)$. As $A = Ag$ we get $\overline{A}_1 \cdots \overline{A}_n(e-g) = 0$. After possible canceling and reordering we get an irreducible product $\overline{A}_1 \cdots \overline{A}_t(e - g) = 0$. Here $t \geq 1$ and $e-g$ is not cancelled. The second zero divisor lemma is applicable and gives that $\gamma(A_1, \ldots, A_t) \leq m_1 + \cdots + m_t$. It follows that the product $A_1 \cdots A_t$ is equal to a subgroup of G. By Theorem 2.1.1, one of the factors $A_1, \ldots, A_t$ is periodic.

This completes the proof.

It is well worth pointing out that in Theorem 5.4.3 above the periodic factor is in fact a subgroup. This follows from the next lemma. (The ideas in the proof are similar to those in the proof of Lemma 5.2.2.)

Lemma 5.4.2. *Let G be a finite abelian group whose order $|G|$ is relatively prime to 6 and let*

$$A = \{e, a, a^2, \ldots, a^{i-1} a^i d, a^{i+1}, \ldots, a^{r-1}\}$$

be a distorted cyclic subset of G such that $|a| \geq r$. If A is periodic then $A = \langle a \rangle$.

Proof. Assume that g is a period of A. It is enough to show that $d = e$ and $a^r = e$. In turn it is enough to verify that $\chi(d) = 1$ and $\chi(a^r) = 1$ for each character χ of G. Consider the subsets $\mathcal{D} = \{\chi : \chi(d) = 1\}$ and $\mathcal{A} = \{\chi : \chi(a^r) = 1\}$ of the character group $\mathcal{G}$ of G.

From $A = Ag$ it follows that $\chi(g) \neq 1$ implies $\chi(A) = 0$. Counting the characters χ of G with $\chi(g) \neq 1$ gives

$$|\mathrm{Ann}(A)| \geq |\mathcal{G}| - |G : \langle g \rangle| = |G|\left(1 - \frac{1}{|g|}\right) \geq (4/5)|G|.$$

Now

$$|\mathcal{A}| \geq \left|\{\chi : \chi(a) \neq 1, \chi(a^r) = 1, \chi(d) = 1\}\right|$$

$$= |\mathrm{Ann}(A)| \geq (4/5)|G|.$$

Note that $\mathcal{A}$ is a subgroup of $\mathcal{G}$. It follows that $\mathcal{A} = \mathcal{G}$ and so $a^r = e$. A similar argument gives that $\mathcal{D} = \mathcal{G}$ and consequently $d = e$.

This completes the proof.

Chapter 6

Cyclic prime component

6.1 Groups with cyclic p-component

In this section we present a further extension of Rédei's theorem allowing factors of prime power cardinality to occur in the factorization with the provision that in this case the corresponding prime component of the group is cyclic.

Lemma 6.1.1. *Let the finite abelian group G be the direct product of two of its subgroups H and K such that $|H|$ and $|K|$ are relatively prime. If $G = AB$ is a factorization of G, $|A|$ divides $|H|$, A' consists of the H-parts of elements of A, then A can be replaced by A'.*

Proof. Let $|A| = l$, $|H| = m$, $|K| = n$. The congruence $nt \equiv 1 \pmod{m}$ is solvable for t as n is prime to m. Replace A by A^{nt} in the factorization $G = AB$ to get the factorization $G = A^{nt}B$. This can be done since nt is prime to l. We claim that A^{nt} consists of the H-parts of the elements of A. Choose an $a \in A$. Let $a_{|H}$ be the H-part of a. Let $a_{|K}$ be the K-part of a and let s be the order of $a_{|H}$. Clearly $s \mid m$. Computing a^{nt} we get

$$a^{nt} = \left[(a_{|H})^{nt}(a_{|K})^{nt}\right] = (a_{|H})^{nt} = a_{|H}$$

since $nt \equiv 1 \pmod{s}$ holds too.

This completes the proof.

Theorem 6.1.1. *Let $G = AB$ be a factorization of the finite abelian group G whose p-component is cyclic, where $|A|$ is a power of p. Then either A or B is periodic.*

Proof. Let H be the p-component and let K be the p'-component of G. Let $H = \langle h \rangle$, where $|h| = p^{\alpha}$ and

$$A = \{h^{a(1)}k_1, \ldots, h^{a(s)}k_s\},$$

where $s = p^\beta$, $1 \leq \beta \leq \alpha$, $0 \leq a(i) \leq p^\alpha - 1$, $k_i \in K$. We denote the subgroup $\langle h^{p^{\alpha-1}} \rangle$ by L. Clearly $\mathrm{Ann}(L)$ consists of all the characters χ of G for which $\chi(h)$ is a primitive (p^α)th root of unity, say ρ. Since $\mathrm{Ann}(L) \subset \mathrm{Ann}(G)$, from the factorization $G = AB$ it follows that $\mathrm{Ann}(L) \subset \mathrm{Ann}(A) \cup \mathrm{Ann}(B)$.

If $\mathrm{Ann}(L) \subset \mathrm{Ann}(B)$, then by Lemma 3.2.6, B is periodic. We may assume that $\mathrm{Ann}(L) \not\subset \mathrm{Ann}(B)$, that is, $\mathrm{Ann}(L) \cap \mathrm{Ann}(A) \neq \emptyset$. If $\chi \in \mathrm{Ann}(L) \cap \mathrm{Ann}(A)$, then

$$0 = \chi(A) = \rho^{a(1)}\chi(k_1) + \cdots + \rho^{a(s)}\chi(k_s).$$

Replace ρ by x to get the polynomial

$$A(x) = x^{a(1)}\chi(k_1) + \cdots + x^{a(s)}\chi(k_s),$$

then replace $\chi(k_i)$ by 1 to get

$$A^*(x) = x^{a(1)} + \cdots + x^{a(s)}.$$

The (p^α)th cyclotomic polynomial

$$F_{p^\alpha}(x) = 1 + x^{p^{\alpha-1}} + x^{2p^{\alpha-1}} + \cdots + x^{(p-1)p^{\alpha-1}}$$

is irreducible over the (p')th cyclotomic field and $F(\rho) = A(\rho) = 0$. Hence $F(x)$ divides $A(x)$ and so $A(x) = F(x)D(x)$. By Lemma 6.1.1, the exponents $a(1), \ldots, a(s)$ are distinct and so $A(x)$ is not the zero polynomial. The $\deg F(x) = (p-1)p^{\alpha-1}$ and $0 \leq \deg A(x) \leq p^\alpha - 1$ relations give that $\deg D(x) \leq p^{\alpha-1} - 1$. Therefore the nonzero terms of $D(x)$ occur among the nonzero terms of $A(x)$. Hence the coefficients of $x^{a(u)}$ and $x^{a(v)}$ in $A(x)$ are equal if $a(u) \equiv a(v) \pmod{p^{\alpha-1}}$. Consequently $\chi(k_u) = \chi(k_v)$, that is $k_u k_v^{-1} \in \mathrm{Ker}\chi$ and $F(x) \mid A^*(x)$. Conversely, if $F(x) \mid A^*(x)$ and if $a(u) \equiv a(v) \pmod{p^{\alpha-1}}$ implies $k_u k_v^{-1} \in \mathrm{Ker}\chi$, then $\chi(A) = 0$.

Thus if $\chi \in \mathrm{Ann}(L) \cap \mathrm{Ann}(A)$, $\chi' \in \mathrm{Ann}(L)$ and $\mathrm{Ker}\chi \subset \mathrm{Ker}\chi'$, then $\chi' \in \mathrm{Ann}(A)$. Further if $\chi, \chi' \in \mathrm{Ann}(L) \cap \mathrm{Ann}(A)$, $\chi'' \in \mathrm{Ann}(A)$ and $\mathrm{Ker}\chi \cap \mathrm{Ker}\chi' \subset \mathrm{Ker}\chi''$, then $\chi'' \in \mathrm{Ann}(A)$. It follows taking the intersection of the kernels of all such characters, that there is a subgroup $M \subset K$ such that $\chi \in \mathrm{Ann}(L)$ annihilates A if and only if $M \subset \mathrm{Ker}\chi$.

If $M = \{e\}$, then $\mathrm{Ann}(L) \subset \mathrm{Ann}(A)$ and so A is periodic.

If $M \neq \{e\}$, then $\chi \in \mathrm{Ann}(L)$ annihilates B for each χ with $M \not\subset \mathrm{Ker}\chi$. This means that

$$\mathrm{Ann}(L) \cap \mathrm{Ann}(M) \subset \mathrm{Ann}(B).$$

By Lemma 3.2.7, B can be represented in the form

$$B = XL \cup YM, \tag{1}$$

where $X, Y \subset G$, the products are direct and the union is disjoint. Multiplying (1) by A we get

$$G = AB = AXL \cup AYM.$$

Since G and AYM are unions of cosets modulo M, so is AXL. Assume that $X \neq \emptyset$ and choose an $x \in X$. Then choose an element $h^{a(u)}k_u \in A$ and $l \in L$. The element $(h^{a(u)}k_u)(x)(l)$ belongs to a coset gM for some $g \in G$. Let $a(u) \equiv a(v)$ $(\mathrm{mod}\ p^{\alpha-1})$. Now $h^{a(u)-a(v)} \in L$ and $k_u k_v^{-1} \in M$ and so

$$(h^{a(v)}k_v)(x)(h^{a(u)-a(v)}l)(k_u k_v^{-1}) = (h^{a(u)}k_u)(x)(l) \in gM.$$

Hence $(h^{a(v)}k_v)(x)(h^{a(u)-a(v)}l) \in gM$. Given u we can find v in p ways (including $u = v$) with $a(u) \equiv a(v)$ $(\mathrm{mod}\ p^{\alpha-1})$. We found that elements belonging to gM occur in disjoint subsets of p elements. However this is impossible since $M \subset K$ and so $|M|$ is relatively prime to p. Therefore $X = \emptyset$ and by (1), B is periodic.

This completes the proof

The statement and the proof of the next lemma are similar to those of Lemma 2.1.3.

Lemma 6.1.2. *Let p be a prime and G a finite abelian group whose p-component is cyclic. If $G = AB$ is a normalized factorization, where $|B| = p^n$ and B is not periodic, then there are nonsubgroup cyclic subsets $C_1, \ldots, C_n$ for which $G = AC_1 \cdots C_n$ is also a factorization.*

Proof. We proceed by induction on the number of the prime factors of $|A|$. First let $|A| = q$ be a prime. From the factorization $G = AB$ by Theorem 6.1.1, it follows that A or B is periodic. By the hypothesis of the lemma, A is periodic. As $|A| = q$ is a prime, $A = H$ is a subgroup of G.

Note that G is a cyclic group of order $p^n q$. Indeed, if $p = q$, then G is a p-group and it is cyclic by the hypothesis of the lemma. If $p \neq q$, then the q-component of G is cyclic and the p-component of G is also cyclic. Hence G is cyclic.

Let $G = \langle a \rangle$. Clearly, $H = \langle a^{p^n} \rangle$. Now

$$G = HC_1 \cdots C_n = AC_1 \cdots C_n$$

is a factorization of G, where

$$C_i = \left\{ e, a^{p^{i-1}}, a^{2p^{i-1}}, \ldots, a^{(p-1)p^{i-1}} \right\}, \quad 1 \leq i \leq n.$$

Clearly C_i is not a subgroup since $(a^{p^{i-1}})^p = a^{p^i} \neq e$.

Secondly assume that $|A|$ is composite. From the factorization $G = AB$ as before it follows that A is periodic. There is a factorization $A = DH$, where H is a subgroup of G and $|H| = q$ is a prime. From the factorization $G = DHB$ we have the normalized factorization

$$G/H = (DH)/H \cdot (BH)/H.$$

If $(BH)/H$ is periodic, then there are subsets $B_1, K_1 \subset G$ such that $|K_1| = p$, $G = DHK_1 B_1$ is a factorization and $L_1 = HK_1$ is a subgroup of G. From the factorization $G = DL_1B_1$ we have the factorization

$$G/L_1 = (DL_1)/L_1 \cdot (B_1L_1)/L_1.$$

If $(B_1L_1)/L_1$ is periodic, then there are subsets $B_2, K_2 \subset G$ such that $|K_2| = p^2$, $G = DHK_2B_2$ is a factorization and $L_2 = HK_2$ is a subgroup of G. Continuing in this way we have that there are subsets $B_r, K_r \subset G$ such that $|K_r| = p^r$, $G = DHK_rB_r$ is a factorization and $L_r = HK_r$ is a subgroup of G.

From the factorization $G = DL_rB_r$ we have the factorization

$$G/L_r = (DL_r)/L_r \cdot (B_rL_r)/L_r. \tag{2}$$

Here we may assume that either $r = n$ or $(B_rL_r)/L_r$ is not periodic.

First assume that $(B_rL_r)/L_r$ is not periodic. From factorization (2) by the inductive assumption it follows that in (2) $(B_rL_r)/L_r$ can be replaced by a product of nonsubgroup cyclic subsets. There are cyclic subsets

$$C_i = \{e, c_i, c_i^2, \ldots, c_i^{p-1}\}, \qquad r+1 \le i \le n$$

of G such that

$$G/L_r = (DL_r)/L_r \cdot (C_{r+1}L_r)/L_r \cdots (C_nL_r)/L_r$$

is a factorization of G/L_r. It means that $DC_{r+1} \cdots C_n$ is a complete set of representatives of G modulo L_r. Therefore

$$\begin{aligned}
G &= DC_{r+1} \cdots C_n L_r \\
&= DL_rC_{r+1} \cdots C_n
\end{aligned}$$

are factorizations of G. Clearly C_i is not a subgroup of G since $c_i \notin L_r$.

Note that L_r is a cyclic subgroup of order $p^r q$. Let $L_r = \langle a \rangle$. Obviously $H = \langle a^{p^r} \rangle$. It is easy to verify that $L_r = HC_1 \cdots C_r$ is a factorization of L_r, where

$$C_i = \{e, a^{p^{i-1}}, a^{2p^{i-1}}, \ldots, a^{(p-1)p^{i-1}}\}, \quad 1 \le i \le r.$$

Since $(a^{p^{i-1}})^p = a^{p^i} \ne e$, C_i is not a subgroup of L_r and hence is not a subgroup of G. Finally

$$\begin{aligned}
G &= DL_rC_{r+1} \cdots C_n \\
&= DHC_1 \cdots C_rC_{r+1} \cdots C_r \\
&= AC_1 \cdots C_n
\end{aligned}$$

are factorizations of G and none of the cyclic factors is a subgroup of G. (In the $r = n$ case we need only the second part of the argument.)

This completes the proof.

The next two theorems show that, for groups with cyclic p-component, stronger versions of Rédei's theorem hold.

Theorem 6.1.2. *Let G be a finite abelian group and let R be a subset of the prime divisors of $|G|$. Suppose that the r-components of G are cyclic for each $r \in R$. If $G = A_1 \cdots A_n$ is a normalized factorization of G, where for each i, $1 \le i \le n$ either $|A_i|$ is a prime or $|A_i|$ is a power of the prime $r \in R$ or A_i is a simulated subset of G, then one of the factors A_i is periodic.*

Proof. If each factor is simulated or of prime cardinality, then we can use an argument similar to the proof of Theorem 3.3.2 to show that one of the factors is periodic. Assume that there is a factor, say A_n such that $|A_n| = r^\alpha$, $\alpha \ge 2$, $r \in R$. If one of the factors $A_1, \ldots, A_n$ is periodic we are done. Suppose that none of the factors $A_1, \ldots, A_n$ is periodic. Let $A = A_1 \cdots A_{n-1}$ and $B = A_n$. Therefore $G = AB$ is a factorization. By Lemma 6.1.3, B can be replaced by a product of nonsubgroup factors of cardinality r. Repeat this procedure in connection with each factor whose cardinality is r^α, $r \in R$. The resulting factorization contains factors of prime cardinality and none of these factors is a subgroup. This violates Rédei's theorem.

The proof is complete.

Theorem 6.1.3. *Let G be a finite abelian group whose p-component is cyclic. If $G = BA_1 \cdots A_n$ is a factorization of G, where each $|A_i|$ is a power of p, then one of the factors is periodic.*

Proof. Assume the contrary, that there is a factorization $G = BA_1 \cdots A_n$ in which none of the factors is periodic. From the factorization

$$G = (BA_1 \cdots A_{i-1}A_{i+1} \cdots A_n)A_i$$

by Lemma 6.1.2, we get that A_i can be replaced by a direct product of nonsubgroup cyclic subsets of prime order. A nonsubgroup cyclic subset of prime order can be replaced by a nonsubgroup cyclic subset containing only (p, q)-elements. Thus we may assume that A_i is a nonsubgroup cyclic subset containing only (p, q)-elements for each i, $1 \le i \le n$. From the factorization $G = B(A_1 \cdots A_n)$ by Theorem 6.1.1, it follows that B or $A_1 \cdots A_n$ is periodic. As B is not periodic, $A_1 \cdots A_n$ is periodic. Now Theorem 5.4.2 is applicable and gives the contradiction that one of the factors $A_1, \ldots, A_n$ is periodic.

This completes the proof.

6.2 Cyclic groups

A. D. Sands showed that for finite cyclic groups a stronger result than Rédei's theorem holds.

Theorem 6.2.1. *Let G be a finite cyclic group and let $G = A_1 \cdots A_n$ be factorization of G such that each $|A_i|$ is either a prime power or a product of two distinct primes. Then at least one of the factors is periodic.*

Proof. Assume the contrary, that there is a factorization $G = A_1 \cdots A_n$ such that none of the factors is periodic. We may assume that each factor is normalized. By Lemma 6.1.2, we may replace each factor of prime power cardinality by a product of nonsubgroup cyclic subsets of prime order. So there is a factorization $G = A_1 \cdots A_n$ such that each $|A_i|$ is either a product of two distinct primes or $|A_i|$ is a prime and A_i is a nonsubgroup cyclic subset of G. For convenience assume that $A_1, \ldots, A_s$ are the factors with composite order and $A_{s+1}, \ldots, A_n$ are the factors of prime order.

Since G is cyclic there is a character χ of G such that χ is an isomorphism from G to the cyclic group of roots of unity of order $|G|$. Clearly χ is not the principal character of G, hence

$$0 = \chi(G) = \chi(A_1) \cdots \chi(A_n).$$

It follows that $\chi(A_i) = 0$ for some i, $1 \le i \le n$. In the $s + 1 \le i \le n$ case A_i is a cyclic subset of prime order, say

$$A_i = \{e, a, a^2, \ldots, a^{p-1}\}.$$

If $\chi(a) = 1$, then we end up with the $0 = \chi(A_i) = p$ contradiction. Therefore $\chi(a) \ne 1$ and consequently

$$0 = \chi(A_i) = \frac{1 - \chi(a^p)}{1 - \chi(a)}.$$

Hence $\chi(a^p) = 1$. Since χ is a bijection it follows that $a^p = e$, that is, A_i is a subgroup of G. This is not the case. Thus $1 \le i \le s$, say $i = 1$, holds.

Let $|A_1| = pq$, where p, q are distinct primes. Let $|G| = p^\alpha q^\beta m$, where $p \nmid m$, $q \nmid m$. There are basis elements u, v, w of G such that

$$|u| = p^\alpha, \ |v| = q^\beta, \ |w| = m.$$

Let

$$\chi(u) = \rho, \ \chi(v) = \sigma, \ \chi(w) = \tau$$

be roots of unity of orders p^α, q^β, m respectively. G is a direct product of two cyclic subgroups H, K of orders $p^\alpha q^\beta$, m respectively. Suppose that the H-part of each element of A_1 is equal to e. From $\chi(A_1) = 0$ it follows that $\chi'(A_1) = 0$ for each character χ' of G for which

$$\chi'(u) = \rho, \ \chi'(v) = \sigma.$$

By the partition lemma (Lemma 3.2.7), there are subsets U, V of G such that

$$A_1 = U \langle u^{p^{\alpha-1}} \rangle \cup V \langle v^{q^{\beta-1}} \rangle,$$

where the products are direct and the union is disjoint. The cardinalities give that $pq = |U|p + |V|q$. It follows that either $|U| = 0$ or $|V| = 0$. If $U = \emptyset$, then A_1 is periodic with period $v^{q^{\beta-1}}$. If $V = \emptyset$, then A_1 is periodic with period $u^{p^{\alpha-1}}$. This is a contradiction.

The above consideration can also be presented in the following way. Let $|A_i| = p(i)q(i)$, where $p(i), q(i)$ are distinct primes $1 \le i \le s$. Let $|G| = p(i)^{\alpha(i)}q(i)^{\beta(i)}m(i)$, where $p(i) \nmid m(i)$, $q(i) \nmid m(i)$. G is a direct product of two cyclic subgroups $H(i)$, $K(i)$ of order $p(i)^{\alpha(i)}q(i)^{\beta(i)}$, $m(i)$ respectively. If for each i, $1 \le i \le s$ the $H(i)$-parts of elements of A_i are all e, then one of the factors $A_1, \ldots, A_s$ is periodic. Call the quantity

$$h(A_1, \ldots, A_n) = \prod_{i=1}^{s} \prod_{a \in A_i} |a_{|K(i)}|$$

the height of the factorization $G = A_1 \cdots A_n$. Here $a_{|K(i)}$ is the $K(i)$-part of a. In the $h(A_1 \ldots, A_n) = 1$ case one of the factors $A_1, \ldots, A_s$ is periodic. For the remaining part of the proof consider a factorization $G = A_1 \cdots A_n$ in which none of the factors is periodic and the height of the factorization is minimal.

By Lemma 6.1.1, A_1 can be replaced by B_1, where B_1 consists of the K-parts of elements of A_1. The height of the factorization $G = B_1 A_2 \cdots A_n$ is smaller than the height of the factorization $G = A_1 A_2 \cdots A_n$ so by the minimality of the height one of the factors $B_1, A_2, \ldots, A_n$ is periodic. Only B_1 can be periodic. We may assume that B_1 is periodic with period $u^{p^{\alpha-1}}$. Plainly B_1 can be represented in the form

$$B_1 = \left\{ u^{ip^{\alpha-1}+a(j)} v^{b(j)} : 0 \le i \le p-1,\ 1 \le j \le q \right\}$$

and therefore A_1 can be represented in the form

$$A_1 = \left\{ u^{ip^{\alpha-1}+a(j)} v^{b(j)} w^{c(i,j)} : 0 \le i \le p-1,\ 1 \le j \le q \right\}.$$

Here $0 \le a(j) \le p^{\alpha-1} - 1$, $0 \le b(j) \le q^{\beta} - 1$, $0 \le c(i,j) \le m - 1$. The sum

$$\chi(A_1) = \sum_{i=0}^{p-1} \sum_{j=1}^{q} \rho^{ip^{\alpha-1}+a(j)} \sigma^{b(j)} \tau^{c(i,j)}$$

is equal to 0. Replacing ρ by x we get the polynomial

$$P(x) = \sum_{i=0}^{p-1} \sum_{j=1}^{q} x^{ip^{\alpha-1}+a(j)} \sigma^{b(j)} \tau^{c(i,j)}.$$

The coefficients of $P(x)$ are in the $(q^{\beta}m)$th cyclotomic field. The (p^{α})th cyclotomic polynomial

$$F_{p^{\alpha}}(x) = \sum_{i=0}^{p-1} x^{ip^{\alpha-1}}$$

is irreducible over the $(q^\beta m)$th cyclotomic field as p^α and $(q^\beta m)$ are relatively prime. Since ρ is a common root of $F_{p^\alpha}(x)$ and $P(x)$ it follows that $F_{p^\alpha}(x)$ divides $P(x)$ over the $(q^\beta m)$th cyclotomic field.

We sort the exponents $a(1), \ldots, a(q)$ into classes. Namely, we put two exponents into the same class if they are equal. Let $C_1, C_2, \ldots$ be these classes. We may assume that C_1 consists of $a(1), \ldots, a(t)$. Note that the exponents $b(1), \ldots, b(t)$ are distinct as the H-parts of the elements of A_1 are distinct. It follows that the sums

$$\sum_{j=1}^{t} \sigma^{b(j)} \tau^{c(i,j)}$$

are equal for each i, $0 \le i \le p-1$. Consequently

$$\sum_{j=1}^{t} \sigma^{b(j)} \tau^{c(0,j)} - \sum_{j=1}^{t} \sigma^{b(j)} \tau^{c(i,j)}$$

is equal to 0 for each i, $1 \le i \le p-1$. Replacing σ by y we get the polynomials

$$Q_i(y) = \sum_{j=1}^{t} y^{b(j)} \tau^{c(0,j)} - \sum_{j=1}^{t} y^{b(j)} \tau^{c(i,j)}$$

for $1 \le i \le p-1$. The coefficients of $Q_i(y)$ are in the mth cyclotomic field. The (q^β)th cyclotomic polynomial

$$F_{q^\beta}(y) = \sum_{i=0}^{q-1} y^{iq^{\beta-1}}$$

is irreducible over the mth cyclotomic field as q^β and m are relatively prime. Since σ is a common root of $F_{q^\beta}(y)$ and $Q_i(y)$ it follows that $F_{q^\beta}(y)$ divides $Q_i(y)$ over the mth cyclotomic field.

We distinguish two cases depending on whether there is only one class of the exponents $a(1), \ldots, a(q)$, that is, $t = q$ or there are more than one classes, that is, $t < q$. If $t < q$, then $b(1), \ldots, b(t)$ cannot be a permutation of

$$0q^{\beta-1}, 1q^{\beta-1}, \ldots, (q-1)q^{\beta-1}. \tag{1}$$

It follows that

$$\tau^{c(0,1)} - \tau^{c(i,1)} = \tau^{c(0,2)} - \tau^{c(i,2)} = \cdots = \tau^{c(0,t)} - \tau^{c(i,t)} = 0$$

for each i, $1 \le i \le p-1$. This in turn implies that

$$c(0, i) = c(1, i) = \cdots = c(p-1, i)$$

for each i, $1 \leq i \leq t$. A similar argument using the other classes $C_2, C_3, \ldots$ gives

$$c(0, i) = c(1, i) = \cdots = c(p - 1, i)$$

for each i, $t + 1 \leq i \leq q$. Thus A_1 is periodic with period $u^{p^{\alpha-1}}$.

Turn to the case when $t = q$. If $b(1), \ldots, b(q)$ is not a permutation of (1), then as before we get that A_1 is periodic with period $u^{p^{\alpha-1}}$. We may assume that $b(1), \ldots, b(q)$ is a permutation of (1). It follows that

$$\tau^{c(0,1)} - \tau^{c(i,1)} = \tau^{c(0,2)} - \tau^{c(i,2)} = \cdots = \tau^{c(0,q)} - \tau^{c(i,q)} \tag{2}$$

for each i, $1 \leq i \leq p - 1$. If none of the differences in (2) is equal to 0, then the four roots of unity lemma gives that

$$c(i, 1) = c(i, 2) = \cdots = c(i, q)$$

for each i, $0 \leq i \leq p - 1$. Now A_1 is periodic with period $v^{q^{\beta-1}}$. In the remaining part assume that the differences in (2) are equal to zero for $1 \leq i \leq r$ and are not equal to zero for $r + 1 \leq i \leq p - 1$. The contradiction that A_1 is periodic follows seamlessly.

This completes the proof.

6.3 Nonperiodic factorizations

We construct factorizations for finite cyclic groups such that none of the factors is periodic. At the end of the section we present an application of these constructions. The next lemma is a variant of Lemma 2.3.1.

Lemma 6.3.1. *Let H be a subgroup of the finite cyclic group G, where $|G : H| = t \geq 2$. If $(r_1, \ldots, r_n)$ is not a periodicity forcing factorization type for H, then $(r_1, \ldots, r_n, t)$ and*

$$(r_1, \ldots, r_{i-1}, r_i t, r_{i+1}, \ldots, r_n)$$

are not periodicity forcing factorization types for G for each i, $1 \leq i \leq n$.

Proof. The factor group G/H is cyclic and so there is an element $b \in G$ such that Hb is a generator element of G/H. Let $B = \{e, b, b^2, \ldots, b^{t-1}\}$ and note that $G = HB$ is a normalized factorization of G. As $(r_1, \ldots, r_n)$ is not a periodicity forcing factorization type for H, there is a normalized factorization $H = A_1 \cdots A_n$ with $|A_i| = r_i$, where none of the factors is periodic. Set $C = \{e, b, b^2, \ldots, b^{t-2}, b^{t-1}d\}$, where $d \in G$. If B is a subgroup of G, then we choose d from $H \setminus \{e\}$. In the $t \geq 3$ case C is a simulated subset and C is not periodic. In the $t = 2$ case C is a cyclic subset. Now $|b| = 2$ and $|d| \neq 2$ as a cyclic group cannot have two elements of order 2. It follows that C is not periodic. If B is not a subgroup of G, then we set $d = e$. In this case C is a cyclic subset and C is not periodic. The

partition	group type
5	(p^5)
$4+1$	(p^4, q)
$3+2$	(p^3, q^2)
$3+1+1$	(p^3, q, r)
$2+2+1$	(p^2, q^2, r)
$2+1+1+1$	(p^2, q, r, s)
$1+1+1+1+1$	(p, q, r, s, t)

Table 1

factorization $G = HC = A_1 \cdots A_n C$ shows that $(r_1, \ldots, r_n, t)$ is not a periodicity forcing factorization type for G. In the remaining part of the proof we would like to establish that $A_n C$ is not periodic and so the factorization $G = A_1 \cdots A_{n-1}(A_n C)$ gives that $(r_1, \ldots, r_{n-1}, r_n t)$ is not a periodicity forcing factorization type for G. This proves the lemma in the $i = n$ case. The other cases can be handled by relabelling. As $A_n \subset H$, the members of this partition

$$A_n C = A_n \cup A_n b \cup \cdots \cup A_n b^{t-2} \cup A_n b^{t-1} d.$$

are in distinct cosets of G modulo H. By Observation 2.3.1, it follows that $A_n C$ is not periodic.

This completes the proof.

For an integer $n \geq 2$ let $\nu(n)$ be denote the number of the not necessarily distinct prime factors of n. In other words if $p_1^{\alpha(1)} \cdots p_s^{\alpha(s)}$ is the canonical prime factorization of n, then $\nu(n)$ is equal to $\alpha(1) + \cdots + \alpha(s)$.

Lemma 6.3.2. *Let G be a cyclic group with $\nu(|G|) = 5$. If G is not of type (p^5), (p^4, q), then there is a factorization $G = BA_1 A_2$, such that $\nu(|B|) = 3$, $\nu(|A_1|) = \nu(|A_2|) = 1$ and none of the factors is periodic.*

Proof. The 5 can be partitioned into a sum of smaller positive integers in seven different ways, as listed in Table 1. Here p, q, r, s, t are distinct primes. Thus there are seven nonisomorphic cyclic groups of order n, where $\nu(n) = 5$.

There are five nonisomorphic cyclic groups satisfying the condition of the lemma. In each case we will exhibit a suitable factorization. The constructions follow a common plan described in Section 2.3. We look for B in the form

$$B = k_0 K_0 L_0 \cup k_1 K_1 L_1 \cup \cdots \cup k_n K_n L_n,$$

type of G	type of H	a	b	c	d
(p^3, q^2)	(p^2, q^2)	p	p	q	q
(p^3, q, r)	(p^2, q, r)	p	p	r	q
(p^2, q^2, r)	(p, q^2, r)	p	r	q	q
(p^2, q, r, s)	(p, q, r, s)	r	p	s	q
(p, q, r, s, t)	(q, r, s, t)	s	q	t	r

Table 2

$$A_1 = [x, a], \ A_2 = [y, c],$$

i	K_i	L_i
0	$\langle x^a \rangle$	$[\langle y^c \rangle, x]$
1	$[\langle x^a \rangle, y]$	$\langle y^c \rangle$
2	$\langle x^a \rangle$	$\langle y^c \rangle$
$\vdots$	$\vdots$	$\vdots$
$p - 1$	$\langle x^a \rangle$	$\langle y^c \rangle$

Table 3

group type	factorization type
(p^3, q^2)	$(p^2 q, p, q)$
(p^3, q, r)	$(p^2 q, p, r)$
(p^2, q^2, r)	(pqr, p, q)
(p^2, q, r, s)	$(p^2 q, r, s)$
(p, q, r, s, t)	(pqr, s, t)

Table 4

where k_i is an element and K_i, L_i are subsets of G. If all the products $K_i L_i A_1 A_2$ are equal to the subgroup H of G and the elements k_i form a complete set of representatives of G modulo H, then $G = BA_1 A_2$ is a factorization of G. This idea goes back to G. Hajós, improved and extended by N. G. de Bruijn.

For a given G we choose a subgroup H of type (ab, cd) shown in Table 2. As G/H is cyclic and $|G : H| = p$, it follows that there is a $z \in G$ such that

$$G = \{e, z, z^2, \ldots, z^{p-1}\}H = [z, p]H$$

is a factorization of G. We choose basis elements x, y for H with $|x| = ab$, $|y| = cd$. Set $k_i = z^i$, $0 \le i \le p - 1$, K_i, L_i by Table 3. We claim that $K_i L_i A_1 A_2 = \langle x, y \rangle$ for each i, $0 \le i \le p - 1$. Indeed,

$$
\begin{aligned}
(K_0 A_1) L_0 A_2 &= \big(\langle x \rangle L_0\big) A_2 = \langle x, y^c \rangle A_2 = \langle x, y \rangle, \\
(L_1 A_2) K_1 A_1 &= \big(\langle y \rangle K_1\big) A_1 = \langle x^a, y \rangle A_1 = \langle x, y \rangle, \\
(K_2 A_1)(L_2 A_2) &= \cdots = (K_{p-1} A_1)(L_{p-1} A_2) = \langle x, y \rangle.
\end{aligned}
$$

Thus $G = BA_1 A_2$ is a factorization of G. As $x^a \ne e$, $y^c \ne e$, it follows that A_1 and A_2 are not periodic.

This completes the proof.

Corollary 6.3.1. *Let G be a cyclic group with $\nu(|G|) = 5$. If G is not of type (p^5), (p^4, q), then there is a factorization $G = BC$ of G such that $\nu(|B|) = 3$, $\nu(|C|) = 2$ and none of the factors is periodic.*

Proof. By Lemma 6.3.2 there is a factorization $G = BA_1 A_2$ of G for which $\nu(|B|) = 3$, $\nu(|A_1|) = \nu(|A_2|) = 1$ and none of the factors is periodic. The group types and the corresponding factorization types appearing in the constructions are summarized in Table 4. If $C = A_1 A_2$ is not periodic, then we get a factorization $G = BC$ of G, where $\nu(|B|) = 3$, $\nu(|C|) = 2$ and none of the factors is periodic. Assume that $C = A_1 A_2$ is periodic. Now Theorem 5.4.2 is applicable and this gives the contradiction that either A_1 or A_2 is periodic.

This completes the proof.

Theorem 6.3.1. *Let G be a finite cyclic group not of type (p^α), (p^α, q). Let $t_1, \ldots, t_n$ be integers such that*

    (1) $n \ge 2$,
    (2) $3 \le t_1, 1 \le t_2, \ldots, 1 \le t_n$,
    (3) $t_1 + \cdots + t_n = \nu(|G|) \ge 5$.

Then there is a factorization $G = A_1 \cdots A_n$ such that $\nu(|A_1|) = t_1, \ldots, \nu(|A_n|) = t_1$ and none of the factors is periodic.

Proof. Let $\nu(|G|) = m$. Since $\nu(m) \ge 5$ and G is not of type (p^α), (p^α, q), it follows that there is a subgroup H_5 of G whose type is not (p^5), (p^4, q) and $\nu(|H_5|) = 5$. Also there is a chain of subgroups

$$H_5 \subset H_6 \subset \cdots \subset H_m = G,$$

where $\nu(|H_i|) = i$. By Lemma 6.3.4 and Corollary 6.3.1 there are factoring types (s_1, s_2, s_3), $(s_1, s_2 s_3)$ of H_5 that do not force periodicity and $\nu(s_1) = 3$, $\nu(s_2) = \nu(s_3) = 1$. Using Lemma 6.3.3 going up on the chain of subgroups we can see that $(t_1, \ldots, t_n)$ is not a periodicity forcing factorization type for G.

This completes the proof.

We present an application of the constructions we described earlier. Let Z be the set of integers and let A be a finite subset of Z. Suppose there is a subset B of Z such that each element z of Z is uniquely expressible in the form

$$z = a + b, \quad a \in A, \ b \in B.$$

It means that the sets $A + b$, $b \in B$ form a partition of Z or in other words translated copies of A *tile* Z. We also can say that $Z = A + B$ is an additive factorization of Z. The following result is due to G. Hajós.

Theorem 6.3.2. *If $Z = A + B$ is a factorization of Z and A is finite, then B is periodic.*

Proof. For $x, y \in Z$ we define

$$(-\infty, x] = \{z \in Z : z \le x\},$$

$$[x, \infty) = \{z \in Z : x \le z\},$$

$$[x, y] = \{z \in Z : x \le z \le y\}.$$

Consider a factorization $Z = A + B$. We may assume that A has no negative elements, 0 is the smallest element of A and d is the largest element of A. For $x, y \in B$ we define

$$B_{(-\infty, x]} = \{b \in B : b \le x\},$$

$$B_{[x, \infty)} = \{b \in B : x \le b\},$$

$$B_{[x, y]} = \{b \in B : x \le b \le y\}.$$

Let x be an element of B and consider the set $A + B_{(-\infty, x]}$. Note that $(-\infty, x] \subset A + B_{(-\infty, x]}$ and $x + d + 1 \notin A + B_{(-\infty, x]}$. There is a least integer u_x between $x + 1$ and $x + d + 1$ such that $u_x \notin A + B_{(-\infty, x]}$. We claim that $u_x \in B$. Indeed as $u_x \in Z$ it can be represented in the form

$$u_x = a + b, \quad a \in A, \ b \in B.$$

Here $b \notin B_{(-\infty, x]}$. If $a = 0$, then $u_x \in B$ as claimed. Suppose $a \ne 0$ and consider $u_x - a$. Since $u_x - a \in Z$ it can be represented in the form

$$u_x - a = a' + b', \quad a' \in A, \ b' \in B.$$

Here $b' \in B_{(-\infty, x]}$ because of the minimality of u_x. Therefore

$$a' + b' = u_x - a = 0 + b$$

and it follows that $a' = 0$, $b' = b$. But this last is not possible as $b' \in B_{(-\infty, x]}$ and $b \notin B_{(-\infty, x]}$. Thus $u_x \in B$.

Let x be an element of B and suppose that we know $B_{(-\infty, x]}$. By the above argument we can find the next element of B. This element is u_x. Clearly u_x depends on the pattern of the elements of $B_{[x-d, x]}$. There are only finitely many possible patterns of the elements of $B_{[x-d, x]}$. The number of the possible patterns is at most 2^d. Checking these patterns for $B_{[y-d, y]}$, where $y \in B$ and $y < x$ we find that there is a $y \in B$ such that $B_{[y-d, y]}$ has the same pattern as $B_{[x-d, x]}$. We can conclude that $B_{[y, \infty)}$ is a union of translated copies of $B_{[y, x]}$.

We will show that $B_{(-\infty, x]}$ is also a union of translates of $B_{[y, x]}$. From the factorization $Z = A + B$ we get that

$$Z = Z - d = (A - d) + B = A' + B$$

is a factorization of Z. Consider $A' + B_{[x, \infty)}$. Note that $[x, \infty) \subset A' + B_{[x, \infty)}$ and $x - d - 1 \notin A' + B_{[x, \infty)}$. There is a largest integer v_x between $x - d - 1$ and x_1 such that $v_x \notin A' + B_{[x, \infty)}$. We claim that $v_x \in B$. We then can go on to show that $B_{(-\infty, x]}$ is a union of translates of $B_{[y, x]}$.

This completes the proof.

Consider a factorization $Z = A + B$, where A is finite. As B is periodic, B can be written in the form $B = C + L$, where L is a subgroup of Z and C is not periodic. The subgroup L is an infinite cyclic group generated by, say l. Clearly $-l$ is also a generator of L and L has no other generators. We call the absolute value of l the *length of the period* of the factorization $Z = A + B$ or the length of the period of the tiling consisting of the translates of A. From the proof of Theorem 6.3.2 we can read off that if the diameter of the tile A is d, then the length of the period is at most 2^d. Let the diameter of A be d. We will show that there are tilings with translates of A such that the length of the period is large relative to the diameter.

Theorem 6.3.3. *There are infinitely many factorizations $Z = A + B$ of Z and a fixed positive constant c such that $p > cd^2$, where p is the length of the period of B and d is the diameter of A.*

Proof. Write B in the form $B = C + L$, where L is a subgroup of Z and C is a nonperiodic subset of Z. From $Z = A + (C + L)$ we get the factorization $Z/L = (A + L)/L + (C + L)/L$ of the factor group Z/L, where none of the factors is periodic. Conversely if there is a factorization $G = U + V$ of a finite cyclic group, then one can construct a factorization $Z = A + (C + L)$, where the length of the period of $(C + L)$ is $|G|$.

Let p, q be primes $q > p > 5$. We can choose q such that $p \leq q \leq 2p$ holds. Let G be a group of type $(3p, 5q, 2)$ with basis elements x, y, z, $|x| = 3p$, $|y| = 5q$, $|z| = 2$. Clearly G is a cyclic group. By the proof of Lemma 6.3.2, there is a factorization $G = U + V$, where

$$U = \{0, x, 2x\} + \{0, y, 2y, 3y, 4y\}$$

and none of the factors is periodic. (Select the last row in Table 2.) As $|x| = 3p$ and $|y| = 5q$, it follows that $x = 10q$, $y = 6p$ and the diameter of A is

$$d = 2x + 4y = 20p + 24q.$$

Hence

$$d \le 20p + 24 \cdot 2p = 68p.$$

Using $|G| = 30pq$ we get that

$$|G| \ge 30p^2 \ge 30\Big(\frac{d}{68}\Big)^2 = cd^2.$$

This completes the proof.

6.4   More on factoring by simulated subsets

This section continues the theme of Section 5.3 using results from Section 6.1.

Theorem 6.4.1. *Let $G = AB$ be a normalized factorization of the finite abelian group G, where A is simulated and $|B|$ is a prime. Then either A or B is a subgroup of G.*

Proof. We may assume that $e \in A$, $e \in B$, that is, the factorization $G = AB$ is normalized. Let $|B| = q$, where q is a prime. If there is a $b \in B \setminus \{e\}$ with $|b| = q$, then B differs from the subgroup $\langle b \rangle$ in at most $q - 2$ elements. Now by Theorem 5.3.1, A or B is periodic. Then by Lemma 5.3.4, it follows that A or B is a subgroup of G. We may assume that there is a $b \in B \setminus \{e\}$ such that $|b| \ne q$. In the factorization $G = AB$ replace B by $[b, q]$. By Lemma 1.4.3, this can be done. Now A is periodic with period b^q. Then by Lemma 5.3.4, we get that A is a subgroup of G.

   This completes the proof.

Theorem 6.4.2. *Let $G = A_1 \cdots A_n$ be a normalized factorization of the finite abelian group G, where for each i, $1 \le i \le n$ either A_i is simulated or $|A_i|$ is a prime. Then at least one of the factors $A_1, \ldots, A_n$ must be a subgroup of G.*

Proof. We proceed by induction on n. If $|A_i|$ is a prime for each i, $1 \le i \le n$, then by Rédei's theorem we are done. If A_i is simulated for each i, $1 \le i \le n$, then by Theorem 5.3.2, we are done. Thus we may assume that among the factors $A_1, \ldots, A_n$ there is at least one which is not simulated and another one whose cardinality is not a prime. The $n = 2$ case is settled in Theorem 6.4.1 and hence we may assume that $n \ge 3$.

Let A_i be a simulated factor and let H_i be the associated subgroup of G. Replace A_i by H_i in the factorization $G = A_1 \cdots A_n$ which yields the factorization

$$G/H_i = (A_1 H_i)/H_i \cdots (A_{i-1} H_i)/H_i \cdot (A_{i+1} H_i)/H_i \cdots (A_n H_i)/H_i$$

of the factor group G/H_i. By the inductive assumption at least one of the factors, say $(A_j H_i)/H_i$, is a subgroup of G/H_i, that is, $H_i A_j$ is a subgroup of G. Continuing in this way we get that for each simulated A_i there is a permutation $B_{i1}, \ldots, B_{in}$ of the factors

$$A_1, \ldots, A_{i-1}, H_i, A_{i+1}, \ldots, A_n$$

such that $B_{i1} = H_i$ and

$$B_{i1}, \ B_{i1} B_{i2}, \ldots, B_{i1} B_{i2} \cdots B_{in}$$

is a chain of subgroups of G. It is clear that there may be more than one permutation with this property.

We claim that there is an A_i such that in an associated permutation $B_{i1}, \ldots, B_{in}$ the last member B_{in} is not simulated.

To prove the claim start with a simulated A_i. We may assume that A_1 is a simulated factor since this is only a matter of indexing the factors. If A_1 is the only simulated factor among $A_1, \ldots, A_n$, then we are done. So we assume that there is a further simulated factor. Thus in the chain

$$B_{11}, \ B_{11} B_{12}, \ldots, B_{11} B_{12} \cdots B_{1n}$$

there is a subgroup

$$K_1 = H_1 A_{1,1} \cdots A_{1,r(1)} A_i,$$

where $A_{1,1}, \ldots, A_{1,r(1)}$ are not simulated subsets and A_i is a simulated factor. We may assume that $i = 2$. Starting with A_2 there is a subgroup

$$K_2 = H_2 A_{2,1} \cdots A_{2,r(2)} A_3,$$

where $A_{2,1}, \ldots, A_{2,r(2)}$ are not simulated subsets and A_3 is a simulated subset. Finally there is a subgroup

$$K_t = H_t A_{t,1} \cdots A_{t,r(t)} A_{t+1},$$

where $A_{t,1}, \ldots, A_{t,r(t)}$ are not simulated subsets and A_{t+1} is a simulated subset; further A_{t+1} coincides with some of $A_1, \ldots, A_t$. For the sake of simplicity we assume that $A_{t+1} = A_1$. Consider the subgroup $K = K_1 \cdots K_t$.

If two factors with double indices are equal on list (1), then one factor is equal to $\{e\}$. Thus we may assume that these factors on the list are distinct.

$$
\begin{array}{llll}
A_1, & A_{1,1}, & \ldots, & A_{1,r(1)}, \\
A_2, & A_{2,1}, & \ldots, & A_{2,r(2)}, \\
\vdots & \vdots & \ddots & \vdots \\
A_t, & A_{t,1}, & \ldots, & A_{t,r(t)}.
\end{array}
\tag{1}
$$

To prove the factors with double indices are distinct assume the contrary that, say $A_{1,i} = A_{2,j}$. Here we may suppose that i and j are the least indices with this property. Consider the subgroups

$$L_1 = H_1 A_{1,1} \cdots A_{1,i},$$

$$L_2 = H_2 A_{2,1} \cdots A_{2,j},$$

$$M = L_1 L_2 = H_1 A_{1,1} \cdots A_{1,i} H_2 A_{2,1} \cdots A_{2,j-1},$$

$$L = L_1 \cap L_2.$$

The right sides of the first three equations are partial products of a factorization of G. Because of these factorizations each $l \in L$ is uniquely expressible in the forms

$$l = h_1 a_{1,1} \cdots a_{1,i}, \tag{2}$$

$$l = h_2 a_{2,1} \cdots a_{2,j}, \tag{3}$$

$$l = h_1' a_{1,1}' \cdots a_{1,i}' h_2' a_{2,1}' \cdots a_{2,j-1}', \tag{4}$$

where $h_1, h_1' \in H_1$, $h_2, h_2' \in H_2$, $a_{1,s}, a_{1,s}' \in A_{1,s}$, $a_{2,s}, a_{2,s}' \in A_{2,s}$. From (2) and (4) it follows that

$$h_2' = a_{2,1}' = \cdots = a_{2,j-1}' = e.$$

From (3) and (4) it follows that

$$h_1' = a_{1,1}' = \cdots = a_{1,i}' = e.$$

Hence $l = e$ and so $L = \{e\}$. On the other hand, by Lemma 5.1.3, $L = A_{1,i} = A_{2,j}$.

The product of the distinct factors occurring on the list (1) forms a factorization of K. If K is a proper subgroup of G, then by the inductive assumption we are done. Hence we may assume that $K = G$ and so the factors on the above list are distinct. As there is a nonsimulated factor among $A_1, \ldots, A_n$, there is an i, $1 \le i \le t$ with $r(i) \ge 1$, say $r(t) \ge 1$. Note that in the permutation $B_{11}, \ldots, B_{1n}$ belonging to A_1 $B_{1n} = A_{t,r(t)}$, that is, B_n is not simulated. This proves the claim.

Thus in the factorization $G = A_1 \cdots A_n$ we can rearrange the factors such that A_1 is simulated and A_n is of prime cardinality and

$$H_1, \ H_1 A_2, \ldots, H_1 A_2 \cdots A_n$$

are subgroups of G. Let $a_1, \ldots, a_s$ be all the elements of $A_1 \setminus H_1$. Here $s \le p - 2$, where p is the least prime divisor of $|A_1|$. Set $K = H_1 A_2 \cdots A_{n-1}$.

If $a_1, \ldots, a_s \in K$, then $K = A_1 A_2 \cdots A_{n-1}$ is a factorization of K and so by the inductive assumption one of the factors $A_1, \ldots, A_{n-1}$ is a subgroup of K and hence of G. We may assume that one of $a_1, \ldots, a_s$, say a_1, belongs to $G \setminus K$. Multiplying the factorization $G = K A_n$ by b^{-1}, where $b \in A_n$ we get the factorization $G = K(A_n b^{-1})$. Equivalently $A_n b^{-1}$ is a complete set of representatives modulo K. Hence there is an element $x \in A_n b^{-1}$ such that the coset $x^{-1}K$ contains a_i. Let us write x in the form $x = d_{i,b}b^{-1}$, where $d_{i,b} \in A_n$. In other words we may say that for each i, $1 \le i \le s$ and $b \in A_n$ there is an element $d_{i,b} \in A_n$ such that $a_i d_{i,b}b^{-1} \in K$.

First we show that for each fixed i, $1 \le i \le s$ the map $b \to d_{i,b}$, $b \in A_n$ is a permutation of the elements of A_n. To prove this claim consider $b, c \in A_n$ and assume that $d_{i,b} = d_{i,c}$. Now $a_i d_{i,b}b^{-1}, a_i d_{i,c}c^{-1} \in K$ imply

$$bc^{-1} = \left(a_i d_{i,b}b^{-1}\right)^{-1}\left(a_i d_{i,c}c^{-1}\right) \in K$$

and so b and c are congruent modulo K.

Secondly we show that we may focus our attention on the case when not only $a_i d_{i,b}b^{-1} \in K$ but in fact $a_i d_{i,b}b^{-1} \in H_1$. To prove this consider

$$C = \left(A_1 \setminus \{a_1, \ldots, a_s\}\right) \cup \{a_1 d_{1,b}b^{-1}, \ldots, a_s d_{s,b}b^{-1}\}.$$

Note that $K = CA_2 \cdots A_{n-1}$ is a factorization of K. Indeed, $C \subset K$ and products coming from $CA_2 \cdots A_{n-1}$ occur among products coming from $A_1 A_2 \cdots A_{n-1}(A_n b^{-1})$. This latter is a factorization of G and so the products are distinct.

By the inductive assumption from the factorization

$$K = CA_2 \cdots A_{n-1}$$

it follows that one of the factors $C, A_2, \ldots, A_{n-1}$ is a subgroup of K. If this is not C we are done. Thus we may assume that C is a subgroup of K. As $|A_1|$ is not a prime, $|A_1 \setminus H_1| > |H_1|/2$. This implies $C = H_1$ and so $a_i d_{i,b}b^{-1} \in H_1$.

Let the prime $|A_1|$ be q. As $a_1 d_{1,b}b^{-1} \in H_1$ for each $b \in A_n$ we have that

$$\prod_{b \in A_n} a_1 d_{1,b}b^{-1} = a^q \left(\prod_{b \in A_n} d_{1,b}\right)\left(\prod_{b \in A_n} b^{-1}\right) = a^q \in H_1.$$

Thus the order of the subgroup $L = H_1\langle a_1 \rangle$ of G is $|H_1|q$. Let t be the length of the shortest cycle of the permutation $b \to d_{1,b}, b \in A_n$. A similar argument gives that $a^t \in H_1$. If $t < q$, then since q is a prime $a^t, a^q \in H_1$ imply $a_1 \in H_1$ which is not the case. Consequently $t = q$, that is, the permutation consists of one cycle.

Our last claim is that $L = A_1 A_n$ is a factorization of L. To prove this claim first we verify that $A_n \subset L$. As $e \in A_n$, there is a $d_{1,e} \in A_n$ with $a_1 d_{1,e}e^{-1} \in H_1$. From this it follows that $d_{1,e} \in L$. Let $d_{1,e} = x$. As $x \in A_n$, there is a $d_{1,x} \in A_n$ with $a_1 d_{1,x}x^{-1} \in H_1$. This gives $d_{1,x} \in L$. Continuing in this way we get that $A_n \subset L$.

Elements of A_n are incongruent modulo K and so they are incongruent modulo H_1. Therefore $L = H_1\langle a_1 \rangle = H_1 A_n$. For each i, $1 \le i \le s$, there is a $d_{i,e} \in A_n$ with $a_i d_{i,e} \in H_1$. Hence $a_i \in H_1(d_{i,e})^{-1} \subset H_1 A_n = L$. Thus $A_1 \subset L$ and so $L = A_1 A_n$ is a factorization of L. Therefore Theorem 6.4.1 completes the proof.

The next theorem is a variant of Theorem 6.4.2.

Theorem 6.4.3. *Let $G = A_1 \cdots A_n$ be a normalized factorization of the finite abelian group G and let p_i be the least prime factor of $|A_i|$. If, for each i, there exists a subgroup H_i such that $|A_i| = |H_i| \leq |A_i \cap H_i| + p_i - 1$ and if $|A_1|, \ldots, |A_n|$ are pair-wise relatively prime, then $A_i = H_i$ for some i.*

Proof. Introduce the notation

$$|A_i| = |H_i| = l_i, \quad |A_i \setminus H_i| = m_i, \quad H_i = \{h_{ij} : 1 \leq j \leq l_j\},$$

$$A_i = \{h_{ij} : 1 \leq j \leq l_i - m_i\} \cup \{h_{ij}d_{ij} : l_i - m_i + 1 \leq j \leq l_j\},$$

where $d_{ij} \neq e$.

Suppose first that $m_1 = \cdots = m_n = 1$. If each $|A_i| \geq 3$, then the result follows by Theorem 1.2.1. If for some i, $|A_i| = 2$, then the result follows by Theorem 1.3.2. So we may suppose that $m_i > 1$ for some i. We proceed by induction on $|G|$ and for a fixed G we proceed by induction on the quantity $m_1 + \cdots + m_n$.

If $|A_i| \geq 2p_i$, then the proof of Lemma 5.3.2 shows that $\chi(A_i) = 0$ implies $\chi(H_i)$ for each character χ of G and so that A_i may be replaced by H_i.

If $|A_i| = l_i \leq 2p_i - 1$, then since p_i is the least prime factor of l_i it follows that $p_i = l_i$. Let

$$B_i = \prod_{\substack{j=1 \\ j \neq i}}^{n} A_j.$$

Then $A_i B_i = G$. Now $|A_i| = p_i$ and since $|A_i|$ and $|B_i|$ are relatively prime it follows that H_i is the p_i-component of G and $|H_i| = |A_i| = p_i$ implies that this p_i-component is cyclic. We can now apply Theorem 6.1.1 to deduce that either A_i or B_i is periodic. Since $|A_i|$ is a prime, A_i periodic and normalized implies $A_i = H_i$. If $A_i \neq H_i$, then let M be the subgroup of periods of B_i, and let $B_i = MC$. Then $G = A_iMC$ gives $G/M = (A_iM)/M \cdot (CM)/M$. Once again Theorem 6.1.1 applies, and as $(CM)/M$ is not periodic we have $(A_iM)/M$ is periodic and so equal to $(H_iM)/M$. It follows that $A_iM = H_iM$ and so that $G = A_iMC = H_iMC = H_iB_i$. Thus, in all cases, we may replace A_i by H_i in this particular factorization.

So we may replace A_i by H_i and obtain a factorization

$$G/H_i = \prod_{\substack{j=1 \\ j \neq i}}^{n} (A_j H_i)/H_i.$$

We may apply the inductive assumption on $|G|$ here to obtain $s = f(i) \neq i$ with $A_s H_i = H_s H_i$. It follows that $d_{sj} \in H_i$ for each j, $l_s - m_s + 1 \leq j \leq l_s$. Thus there is a cycle, say $(1, 2, \ldots, r)$ with

$$f(1) = 2, f(2) = 3, \ldots, f(r-1) = r, f(r) = 1.$$

If $r < n$, then as before we have $A_1 \cdots A_r = H_1 \cdots H_r$ and the inductive assumption on $|G|$ implies that $A_i = H_i$ for some i, $1 \le i \le r$. So we may suppose that $r = n$.

If for each $\chi \in \mathrm{Ann}(A_i)$ and some j we have $\chi(d_{ij}) = 1$, then we may replace $h_{ij}d_{ij} \in A_i$ by h_{ij} to obtain A_i' and then replace A_i by A_i' in the factorization of G. This reduces m_i by 1 and so by the inductive assumption on $m_1 + \cdots + m_n$, we have either $A_t = H_t$ for some $t \ne i$ as required, or else $A_i' = H_i$. This second gives $m_i = 1$. Thus we may suppose that there exist i, j and $\chi \in \mathrm{Ann}(A_i)$ with $\chi(d_{ij}) \ne 1$. Now from $0 = \chi(A_i) = \chi(H_i)$ we have, after cancellation of terms, a sum of at most $p_i - 1$ $|G|$th roots of unity equal to another such sum. This contradicts Lemma 3.1.5 unless both sums are empty. So there must exist t with $\chi(h_{ij}d_{ij}) = \chi(h_{it})$. In the cycle above we have $r = n$ and so there exists $s \ne i$ with $d_{ij} \in H_s$, where $f(s) = i$. Now $\chi(d_{ij})$ is an $|H_s|$th root of unity and $\chi(h_{ij}), \chi(h_{it})$ are $|H_i|$th roots of unity. Now $|H_i|, |H_s|$ are relatively prime which contradicts $1 \ne \chi(d_{ij}) = \chi(h_{ij})^{-1}\chi(h_{it})$.

This completes the proof.

The following exercise deals with an example which shows that the conditions of Theorem 6.4.3 cannot be removed.

Exercise 6.4.1. Let G be a group of type (p, q, r), where p, q, r are primes such that $3 \le p < q < r$. Let x, y, z be a basis of G, that is, $G = \langle x \rangle \times \langle y \rangle \times \langle z \rangle$, where $|x| = p$, $|y| = q$, $|z| = r$. Set

$$A_1 = \langle x \rangle \cup y\langle x \rangle \cup \cdots \cup y^{q-2}\langle x \rangle \cup y^{q-1}z\langle x \rangle,$$

$$A_2 = \{e, z, z^2, \ldots, z^{r-2}, z^{r-1}x\}.$$

(1) Verify that $G = A_1 A_2$ is a factorization of G.
(2) Check that A_1 differs from the subgroup $\langle x \rangle \times \langle y \rangle$ in p elements.
(3) Check that A_2 differs from the subgroup $\langle z \rangle$ in one element.
    (A_1 is periodic with period x but A_1 is not a subgroup.)

We can now make a similar improvement to Theorem 6.4.2 for cyclic groups.

Theorem 6.4.4. *Let G be a cyclic group and let $G = A_1 \cdots A_n$ be a normalized factorization. Let p_i be the least prime factor of $|A_i|$ and suppose that for each i there is a subgroup H_i such that $|A_i| = |H_i| \le |A_i \cap H_i| + p_i - 1$. Then $A_i = H_i$ for some i.*

Proof. The replacement argument proceeds as in Theorem 6.4.3 but now G cyclic implies that H_i cyclic is already known. Then $G = H_1 \cdots H_n$ and G is cyclic implies that $|A_i| = |H_i|$, $1 \le i \le n$, are pair-wise relatively prime and the result follows.

Use the ideas of construction (1) in the proof of Lemma 2.3.3 in the following exercise.

Exercise 6.4.2. Let G be the direct product of the elementary groups $K_1, \ldots, K_n$ of orders $p_1^4, \ldots, p_n^4$ respectively, where $p_1, \ldots, p_n$ are distinct primes. Show that there are subsets $A_1, \ldots, A_{2n}$ and subgroups $H_1, \ldots, H_{2n}$ of G such that $|H_i| = |A_i| = |A_i \cap H_i| + p_i - 1$, where p_i is the least prime factor of $|A_i|$ and $G = A_1 \cdots A_{2n}$ is a normalized factorization.

This example shows that Theorem 6.4.2 does not hold when $p_i - 2$ is replaced by $p_i - 1$. Turn to our second example. Let G be generated by the elements x, y, u, v, such that $|x| = q$, $|y| = r$, $|u| = s$, $|v| = t$, where $q < r$, $s < t$ are distinct odd primes. Consider the following subsets of G.

$$A = \langle x \rangle \cup y\langle x \rangle \cup \cdots \cup y^{r-2}\langle x \rangle \cup y^{r-1}u\langle x \rangle,$$

$$B = \langle u \rangle \cup v\langle u \rangle \cup \cdots \cup v^{t-2}\langle u \rangle \cup v^{t-1}x\langle u \rangle.$$

Clearly, A differs from $\langle x, y \rangle$ in q elements and B differs from $\langle u, v \rangle$ in s elements. This construction can be extended for more than two factors.

Exercise 6.4.3. Let G be the direct product of the cyclic groups $K_1, \ldots, K_n$ of orders $q_1 r_1 s_1 t_1, \ldots, q_n r_n s_n t_n$ respectively, where $q_i < r_i, s_i < t_i$ are distinct odd primes. Show that there are subsets $A_1, \ldots, A_{2n}$ and subgroups $H_1, \ldots, H_{2n}$ of G such that $|H_i| = |A_i| = |A_i \cap H_i| + p_i$, where p_i is the least prime factor of $|A_i|$ and $G = A_1 \cdots A_{2n}$ is a normalized factorization.

This example shows that Theorems 6.4.3 and 6.4.4 do not hold when $p_i - 1$ is replaced by p_i.

Chapter 7

p-groups

7.1 Groups of order p^4

Let $G = AB_1 \cdots B_n$ be a factorization of the finite abelian group G, where each B_i is either cyclic or simulated and $|A|$ is a product of two (not necessarily distinct) primes. Does it follow that at least one of the factors $A, B_1, \ldots, B_n$ is periodic?

This section gives an answer in the affirmative in a particular case. Namely, let G be a group of order p^4, where p is a prime. If $G = ABC$ is a factorization of G, where B and C may be cyclic or simulated, then at least one of the factors A, B, C must be periodic. The emphasis is on the fact that $|A|$ is not a prime and nothing is assumed about the structure of A.

At some instances it will be convenient to work in the group ring $Z(G)$. To the subset A of G we assign the element

$$\overline{A} = \sum_{a \in A} a$$

of $Z(G)$. We will use the next argument several times. Let $G = ABC$ be a factorization of G, where

$$B = \{e, b, b^2, \ldots, b^{r-1}\} = [b, r],$$
$$C = \{e, c, c^2, \ldots, c^{s-2}, c^{s-1}v\} = [\langle c \rangle, v].$$

Replace C by $\langle c \rangle$ in the factorization $G = ABC$ to get the factorization $G = AB\langle c \rangle$. This can be done by Lemma 2.2.3. The factorizations $G = ABC$ and $G = AB\langle c \rangle$ correspond to the equations $\overline{G} = \overline{ABC}$ and $\overline{G} = \overline{AB\langle c \rangle}$ respectively in the group ring $Z(G)$. Subtracting the first from the second we get $0 = \overline{AB}(c^{s-1} - c^{s-1}v)$ and so $0 = \overline{AB}(e - v)$. From this by multiplying by $e - b$ we get the equation $0 = \overline{A}(e - b^r)(e - v)$. By the partition lemma, there are subsets U, V of G such that $A = U\langle b^r \rangle \cup V\langle v \rangle$, where the union is disjoint and the products are direct. Analogous results hold when both B and C are cyclic or both are simulated. For easier reference we state them as a lemma.

Lemma 7.1.1. *Let $G = ABC$ be a factorization of the finite abelian group G, where the factors B and C are one of the following*

$$B = [b, r], \qquad C = [c, s],$$
$$B = [\langle b \rangle, u], \qquad C = [\langle c \rangle, v],$$
$$B = [b, r], \qquad C = [\langle c \rangle, v].$$

Then there are subsets U, V of G such that A can be represented in the following forms respectively,

$$A = U \langle b^r \rangle \cup V \langle c^s \rangle,$$
$$A = U \langle u \rangle \cup V \langle v \rangle,$$
$$A = U \langle b^r \rangle \cup V \langle v \rangle,$$

where the unions are disjoint and the products are direct.

Lemma 7.1.2. *Let p be a prime and let G be a group of type (p^λ, p^μ) with basis elements x, y. If $G = A[x, p][y, p]$ is a normalized factorization of G, then $A \subset \langle x^p, y \rangle$ or $A \subset \langle x, y^p \rangle$.*

Proof. Choose $a', a \in A$ and let $a' a^{-1} = x^\alpha y^\beta$ be the representation of $a' a^{-1}$ in the basis x, y. We claim that $p \mid \alpha$ or $p \mid \beta$.

In order to prove the claim, assume the contrary, that $p \nmid \alpha$ and $p \nmid \beta$. Then by Lemma 1.3.1, in the factorization $G = A[x, p][y, p]$ the factor $[x, p]$ can be replaced by $[x^\alpha, p]$ and the factor $[y, p]$ can be replaced by $[y^\beta, p]$ to get the factorization $G = A[x^\alpha, p][y^\beta, p]$. Now the equation $a' = a x^\alpha y^\beta$ contradicts this factorization. This completes the proof of the claim.

Choose $a \in A$ and let $a = x^\alpha y^\beta$ be the representation of a in the basis x, y. The claim with the $a' = e$ choice implies that $p \mid \alpha$ or $p \mid \beta$. If $p \mid \alpha$ for each $a \in A$, then $A \subset \langle x^p, y \rangle$. If $p \mid \beta$ for each $a \in A$, then $A \subset \langle x, y^p \rangle$. Thus we may assume that there are $a, a' \in A$ such that $a = x^\alpha y^\beta$, $a' = x^\gamma y^\delta$, $p \mid \alpha$, $p \nmid \beta$, $p \nmid \gamma$, $p \mid \delta$. It follows that $p \nmid (\gamma - \alpha)$ and $p \nmid (\delta - \beta)$. On the other hand from $a' a^{-1} = x^{\gamma - \alpha} y^{\delta - \beta}$ we get that $p \mid (\gamma - \alpha)$ or $p \mid (\delta - \beta)$.

This completes the proof.

Theorem 7.1.1. *Let p be a prime and G be an abelian group of order p^4. Let $G = ABC$ be a factorization of G, where $|A| = p^2$, $|B| = |C| = p$. Further the factors B and C are cyclic or simulated. Then one of A, B, C is periodic.*

Proof. The type of G can be

$$(p^4), \ (p^3, p), \ (p^2, p^2), \ (p^2, p, p), \ (p, p, p, p).$$

We distinguish five cases depending on the type of G. Then we distinguish three subcases depending on whether: both B and C are cyclic; both B and C are simulated; B is cyclic and C is simulated.

Case 1. G is of type (p^4). This case is settled by Theorem 6.2.1.

Case 2. G is of type (p^3, p). Let $G = \langle x \rangle \times \langle y \rangle$, where $|x| = p^3$, $|y| = p$.
Subcase 2(a) Both B and C are cyclic, that is,

$$B = [b, p], \quad C = [c, p].$$

If $b^p = e$ or $c^p = e$, then we are done and so we assume that $|b| \geq p^2$, $|c| \geq p^2$. By Lemma 7.1.1,

$$A = U \langle b^p \rangle \cup V \langle c^p \rangle.$$

Let $b = x^\alpha y^\beta$ and $c = x^\gamma y^\delta$ be the basis representations of b and c. Now $b^p = x^{p\alpha}$ and $c^p = x^{p\gamma}$. The subgroups of $\langle x^p \rangle$ form a chain. Hence

$$\langle x^{p^2} \rangle \subset \langle x^{p\alpha} \rangle \cap \langle x^{p\gamma} \rangle = \langle b^p \rangle \cap \langle c^p \rangle.$$

Thus x^{p^2} is a period of A.

Subcase 2(b) Both B and C are simulated, that is,

$$B = [\langle b \rangle, u], \quad C = [\langle c \rangle, v].$$

Here $|b| = |c| = p$. If $u = e$ or $v = e$, then we are done and so we assume that $|u| \geq p$, $|v| \geq p$. By Lemma 1.2.3, we may assume that $|u| = |v| = p$. The elements of G of order p generate the subgroup $K = \langle x^{p^2}, y \rangle$ of order p^2. Now $K = BC$ is a factorization of K. By Rédei's theorem one of the factors B and C is a subgroup of K.

Subcase 2(c) B is cyclic and C is simulated, that is,

$$B = [b, p], \quad C = [\langle c \rangle, v].$$

Here we may assume that $|b| \geq p^2$ and $|c| = |v| = p$. By Lemma 7.1.1,

$$A = U \langle b^p \rangle \cup V \langle v \rangle.$$

If $U = \emptyset$ or $V = \emptyset$, then A is periodic and so we assume that $U \neq \emptyset$ and $V \neq \emptyset$. If $|b| = p^3$, then $|U||\langle b^p \rangle| \geq p^2$ and hence $V = \emptyset$. We assume that $|b| = p^2$. Therefore $\langle b^p \rangle = \langle x^{p^2} \rangle$. If $v \in \langle x^{p^2} \rangle$, then $\langle v \rangle = \langle x^{p^2} \rangle$ and so A is periodic. Thus we assume that $v \notin \langle x^{p^2} \rangle$.

Replace C by $\langle c \rangle$ in the factorization $G = ABC$ to get the factorization $G = AB\langle c \rangle$. Let us compute $A\langle c \rangle$.

$$\begin{aligned} A\langle c \rangle &= \left(U \langle x^{p^2} \rangle \cup V \langle v \rangle \right) \langle c \rangle \\ &= U \langle x^{p^2} \rangle \langle c \rangle \cup V \langle v \rangle \langle c \rangle. \end{aligned}$$

The product $AB\langle c \rangle$ is direct so the products $\langle x^{p^2} \rangle \langle c \rangle$ and $\langle v \rangle \langle c \rangle$ must be direct. They both must be equal to K. Note that x^{p^2}, v form a basis for K. By Lemma

1.2.3, in C, c can be replaced by c^i for each i, $1 \leq i \leq p - 1$, so we have $p - 1$ choices for c. Namely, c may be $x^{p^2} v^i$, $1 \leq i \leq p - 1$. Now

$$C = \{e, x^{p^2} v^i, x^{2p^2} v^{2i}, \ldots, x^{(p-2)p^2} v^{(p-2)i}, x^{(p-1)p^2} v^{(p-1)i+1}\}.$$

There is a j such that $1 \leq j \leq p - 2$ and

$$ji \equiv (p - 1)i + 1 \quad (\mathrm{mod}\ p)$$

since

$$(j + 1)i \equiv 1 \quad (\mathrm{mod}\ p)$$

is solvable. Let $t \in U$. The product $t\langle x^{p^2}\rangle C$ is direct since it is part of ABC. Compute $\langle x^{p^2}\rangle C$. Note that the sets

$$\langle x^{p^2}\rangle x^{jp^2} v^{ji} = \langle x^{p^2}\rangle v^{ij},$$

$$\langle x^{p^2}\rangle x^{(p-1)p^2} v^{(p-1)i+1} = \langle x^{p^2}\rangle v^{(p-1)i+1}$$

are parts of $\langle x^{p^2}\rangle C$ and they have the same elements. This is a contradiction.

Case 3. G is of type (p^2, p^2). Let $G = \langle x\rangle \times \langle y\rangle$, where $|x| = p^2$, $|y| = p^2$.

Subcase 3(a) Both B and C are cyclic, that is,

$$B = [b, p], \quad C = [c, p].$$

If $b^p = e$ or $c^p = e$, then we are done and so we assume that $|b| = p^2$, $|c| = p^2$. Let $L = \langle b\rangle$. If $c \in L$, then $L = BC$ is a factorization of L and so by Rédei's theorem one of the factors is a subgroup of L. Thus we may assume that $c \notin L$. We may choose x, y to be b, c respectively. Now

$$B = [x, p], \quad C = [y, p].$$

From the factorization $G = ABC$ by Lemma 7.1.2 it follows that

$$A \subset \langle x^p, y\rangle \qquad \text{or} \qquad A \subset \langle x, y^p\rangle.$$

Let $M = \langle x^p, y\rangle$ and $N = \langle x, y^p\rangle$. If $A \subset \langle x^p, y\rangle = M$, then $M = AC$ is a factorization and so $\overline{A}(e - y^p) = 0$ shows that y^p is a period of A. Similarly if $A \subset \langle x, y^p\rangle = N$, then $N = AB$ is a factorization and so $\overline{A}(e - x^p) = 0$ shows that x^p is a period of A.

Subcase 3(b) Both B and C are simulated, that is,

$$B = [\langle b\rangle, u], \quad C = [\langle c\rangle, v].$$

Here $|b| = |c| = p$ and we may assume that $|u| = p$, $|v| = p$. The elements of G of order p generate the subgroup $K = \langle x^p, y^p\rangle$ of order p^2. Now $K = BC$ is a factorization of K. By Rédei's theorem one of the factors B and C is a subgroup of K.

Subcase 3(c) B is cyclic and C is simulated, that is,

$$B = [b, p], \quad C = [\langle c \rangle, v].$$

Here we may assume that $|b| \geq p^2$ and $|c| = |v| = p$. By Lemma 7.1.1,

$$A = U \langle b^p \rangle \cup V \langle v \rangle.$$

If $U = \emptyset$ or $V = \emptyset$, then A is periodic and so we assume that $U \neq \emptyset$ and $V \neq \emptyset$. If $v \in \langle b^p \rangle$, then $\langle v \rangle = \langle b^p \rangle$ and so A is periodic. Thus we assume that $v \notin \langle b^p \rangle$. Now b^p, v form a basis for $K = \langle x^p, y^p \rangle$.

In the factorization $G = ABC$ replace C by $\langle c \rangle$ to obtain the factorization $G = AB \langle c \rangle$. Compute $A \langle c \rangle$.

$$
\begin{aligned}
A \langle c \rangle &= \left(U \langle b^p \rangle \cup V \langle v \rangle \right) \langle c \rangle \\
&= U \langle b^p \rangle \langle c \rangle \cup V \langle v \rangle \langle c \rangle.
\end{aligned}
$$

Both $\langle b^p \rangle \langle c \rangle$ and $\langle v \rangle \langle c \rangle$ must be direct and equal to K. Note that b^p, v form a basis for K. By Lemma 1.2.3, in C, c can be replaced by c^i for each i, $1 \leq i \leq p - 1$, so we have $p - 1$ choices for c. Namely, c may be $b^p v^i$, $1 \leq i \leq p - 1$. Now

$$C = \{ e, b^p v^i, b^{2p} v^{2i}, \ldots, b^{(p-2)p} v^{(p-2)i}, b^{(p-1)p} v^{(p-1)i+1} \}.$$

There is a j such that $1 \leq j \leq p - 2$ and

$$ji \equiv (p - 1)i + 1 \pmod{p}$$

since

$$(j + 1)i \equiv 1 \pmod{p}$$

is solvable. Let $t \in U$. The product $t \langle b^p \rangle C$ is direct since it is part of ABC. Compute $\langle b^p \rangle C$. Note that the sets

$$\langle b^p \rangle b^{jp} v^{ji} = \langle b^p \rangle v^{jp},$$

$$\langle b^p \rangle b^{(p-1)p} v^{(p-1)i+1} = \langle b^p \rangle v^{(p-1)i+1}$$

are parts of $\langle b^p \rangle C$ and they have the same elements. This is a contradiction.

Case 4. G is of type (p^2, p, p). Let $G = \langle x \rangle \times \langle y \rangle \times \langle z \rangle$, where $|x| = p^2$, $|y| = |z| = p$.

Subcase 4(a) Both B and C are cyclic, that is,

$$B = [b, p], \quad C = [c, p].$$

Here we may assume that $|b| = p^2$, $|c| = p^2$. By Lemma 7.1.1

$$A = U \langle b^p \rangle \cup V \langle c^p \rangle.$$

Let $b = x^\alpha y^\beta z^\gamma$ and $c = x^\delta y^\epsilon z^\mu$ be the basis representations of b and c. Now $b^p = x^{p\alpha}$ and $c^p = x^{p\delta}$. The subgroups of $\langle x^p \rangle$ form a chain. So $\langle x^p \rangle = \langle b^p \rangle = \langle c^p \rangle$. Thus x^p is a period of A.

Subcase 4(b) Both B and C are simulated, that is,

$$B = [\langle b \rangle, u], \quad C = [\langle c \rangle, v].$$

Here $|b| = |c| = p$ and by Lemma 1.2.3, we may assume that $|u| = p$, $|v| = p$. Replace B, C by $\langle b \rangle, \langle c \rangle$ in the factorization $G = ABC$ to get the factorization $G = A\langle b \rangle \langle c \rangle$. This gives that the product $\langle b \rangle \langle c \rangle$ is direct.

If $u, v \in \langle b, c \rangle$, then $BC = \langle b, c \rangle$ is a factorization and so by Rédei's theorem we are done. We assume that $u \notin \langle b, c \rangle$ and $v \in \langle b, c, u \rangle$, that is, $v = b^\alpha c^\beta u^\gamma$ Now

$$B = [\langle b \rangle, u], \quad C = [\langle c \rangle, b^\alpha c^\beta u^\gamma],$$

$$A = U\langle u \rangle \cup V \langle b^\alpha c^\beta u^\gamma \rangle.$$

Here we assume that $U \neq \emptyset$ and $V \neq \emptyset$ since otherwise A is periodic.

Assume first that $\beta = 0$. If $\gamma = 0$, then $\alpha \neq 0$. By Lemma 1.2.3, we may assume that $\alpha = 1$. Let $t \in V$. The product $t\langle b \rangle B$ is direct since the product ABC is direct. On the other hand $b \in \langle b \rangle$ and $b \in B$. This is a contradiction.

If $\gamma \neq 0$, then by Lemma 1.2.3, we may assume that $\gamma = 1$. If $\alpha = 0$, then u is a period of A. Thus we assume that $\alpha \neq 0$. Let $t \in V$. The product $t\langle b^\alpha u \rangle B$ is direct. The elements $b^{(p-1)\alpha+p-1}, e, b, b^2, \ldots, p^{p-2}$ belong to $\langle b^\alpha u \rangle B$. So

$$(p-1)\alpha + p - 1 \not\equiv 0, 1, 2, \ldots, p-2 \quad (\text{mod } p)$$

and hence

$$(p-1)\alpha + p - 1 \equiv p - 1 \quad (\text{mod } p).$$

Thus $\alpha = 0$. But we know this is not the case.

Secondly assume that $\beta \neq 0$. By Lemma 1.2.3, we may assume that $\beta = 1$. Now

$$C = [\langle c \rangle, b^\alpha c u^\gamma] = \{e, c, c^2, \ldots, c^{p-2}, b^\alpha u^\gamma\}.$$

Let $t \in U$. The product $t\langle u \rangle BC$ is direct. Clearly $\langle u \rangle BC = \langle b, u \rangle C$. But this a contradiction since $b^\alpha u^\gamma \in \langle b, u \rangle$ and $b^\alpha u^\gamma \in C$.

Subcase 4(c) B is cyclic and C is simulated, that is,

$$B = [b, p], \quad C = [\langle c \rangle, v].$$

We may assume that $|b| = p^2$ and $|c| = |v| = p$. By Lemma 7.1.1,

$$A = U\langle b^p \rangle \cup V \langle v \rangle.$$

We assume that $U \neq \emptyset$ and $V \neq \emptyset$ since otherwise A is periodic.

Let $t \in U$. The product $t\langle b^p \rangle C$ is direct since it is part of the product ABC. If $c \in \langle b^p \rangle$, then $c = b^{pi}$ for some i, $1 \leq i \leq p-1$. This leads to the contradiction

$c \in \langle b^p \rangle$ and $c \in C$. Thus we assume that $c \notin \langle b^p \rangle$. We distinguish two cases depending on whether $v \in \langle b^p, c \rangle$ or $v \notin \langle b^p, c \rangle$.

If $v \in \langle b^p, c \rangle$, then $v = b^{p\alpha} c^\beta$. If $\beta = 0$, then A is periodic by b^p. If $\beta \neq 0$, then by Lemma 1.2.3, we may assume that $\beta = 1$. Now

$$C = [\langle c \rangle, b^{p\alpha}] = \{e, c, c^2, \ldots, c^{p-2}, b^{p\alpha}\}.$$

Let $t \in U$. The product $t\langle b^p \rangle C$ is direct since it is part of the product ABC. But $b^{p\alpha} \in \langle b^p \rangle$ and $b^{p\alpha} \in C$ is a contradiction.

Turn to the case when $\langle b^p, c, v \rangle$ is of type (p, p, p). From the factorization $G = ABC$ it follows that $0 = \overline{AC}(e - b^p)$ and so AC is periodic by b^p. Consequently $V\langle v \rangle C$ is periodic by b^p. Note that $\langle v \rangle C = \langle c, v \rangle$. If $t \in V$, then $e \in t^{-1}V\langle v \rangle C = t^{-1}V\langle c, v \rangle$. Now $\langle b^p \rangle \subset t^{-1}V\langle c, v \rangle$. As $t^{-1}V\langle c, v \rangle$ is a union of cosets modulo $\langle c, v \rangle$ and the elements of $\langle b^p \rangle$ are incongruent modulo $\langle c, v \rangle$, it follows that

$$p^3 = \left| \langle b^p \rangle \langle c, v \rangle \right| \leq \left| t^{-1}V\langle c, v \rangle \right| = |V|p^2.$$

This gives that $|V| \geq p$ and so we get the contradiction that $U = \emptyset$.

Case 5. G is of type (p, p, p, p). Each element of $G \setminus \{e\}$ is of order p and so a cyclic subset of G is a subgroup of G. Thus the only case we should consider is when both B and C are simulated, that is,

$$B = [\langle b \rangle, u], \quad C = [\langle c \rangle, v].$$

Here $|b| = |c| = p$ and we may assume that $|u| = p$, $|v| = p$. (This case has already been settled in Section 2.2. If you have read the proof of Theorem 2.2.1, then you can skip the remaining part of the argument.) Replace B, C by $\langle b \rangle, \langle c \rangle$ in the factorization $G = ABC$ to get the factorization $G = A\langle b \rangle \langle c \rangle$. This gives that the product $\langle b \rangle \langle c \rangle$ is direct. So the group $\langle b, c, u, v \rangle$ is one of the types (p, p), (p, p, p), (p, p, p, p).

Turn first to the case when $\langle b, c, u, v \rangle$ is of type (p, p). Now $u, v \in \langle b, c \rangle$. Further $BC = \langle b, c \rangle$ is a factorization and so by Rédei's theorem we are done.

Secondly consider the case when $\langle b, c, u, v \rangle$ is of type (p, p, p). We assume that $u \notin \langle b, c \rangle$ and $v \in \langle b, c, u \rangle$, that is, $v = b^\alpha c^\beta u^\gamma$. Now

$$B = [\langle b \rangle, u], \quad C = [\langle c \rangle, b^\alpha c^\beta u^\gamma],$$

$$A = U\langle u \rangle \cup V\langle b^\alpha c^\beta u^\gamma \rangle.$$

Here we assume that $U \neq \emptyset$ and $V \neq \emptyset$ otherwise A is periodic.

Assume first that $\beta = 0$. If $\gamma = 0$, then $\alpha \neq 0$. By Lemma 1.2.3, we may assume that $\alpha = 1$. Let $t \in V$. The product $t\langle b \rangle B$ is direct since the product ABC is direct. On the other hand $b \in \langle b \rangle$ and $b \in B$. This is a contradiction.

If $\gamma \neq 0$, then by Lemma 1.2.3, we may assume that $\gamma = 1$. If $\alpha = 0$, then u is a period of A. Thus we assume that $\alpha \neq 0$. Let $t \in V$. The product $t\langle b^\alpha u \rangle B$ is direct. The elements $b^{(p-1)\alpha + p - 1}, e, b, b^2, \ldots, p^{p-2}$ belong to $\langle b^\alpha u \rangle B$. So

$$(p - 1)\alpha + p - 1 \not\equiv 0, 1, 2, \ldots, p - 2 \pmod{p}$$

and hence

$$(p-1)\alpha + p - 1 \equiv p - 1 \pmod{p}.$$

Thus $\alpha = 0$. But we know this is not the case.

Secondly assume that $\beta \neq 0$. By Lemma 1.2.3, we may assume that $\beta = 1$. Now

$$C = [\langle c \rangle, b^\alpha c u^\gamma] = \{e, c, c^2, \ldots, c^{p-2}, b^\alpha u^\gamma\}.$$

Let $t \in U$. The product $t\langle u \rangle BC$ is direct. Clearly $\langle u \rangle BC = \langle b, u \rangle C$. But this a contradiction since $b^\alpha u^\gamma \in \langle b, u \rangle$ and $b^\alpha u^\gamma \in C$.

Finally turn to the case when $\langle b, c, u, v \rangle$ is of type (p, p, p, p). From the factorization $G = A\langle b \rangle \langle c \rangle$ it follows that A is a complete set of representatives modulo $\langle b, c \rangle$ and so

$$A = \{u^i v^j a_{ij} : 0 \leq i, j \leq p - 1, a_{ij} \in \langle b, c \rangle\}.$$

We may assume that $e \in A$, that is, $a_{00} = e$. We also know that

$$A = U\langle u \rangle \cup V\langle v \rangle.$$

Here we assume that $U \neq \emptyset$ and $V \neq \emptyset$ since otherwise A is periodic. As the roles of u and v are symmetric, we may assume that $e \in U$. Then $\langle u \rangle \subset A$. This means that $a_{i0} = e$ for each i, $0 \leq i \leq p - 1$. Let $u^i v^j a_{ij} \in V\langle v \rangle$. Now $u^i v^j a_{ij} \langle v \rangle \subset A$. This gives that a_{ij}'s are equal for each j, $0 \leq j \leq p - 1$. As $a_{i0} = e$, we have $a_{ij} = e$ for each j, $0 \leq j \leq p - 1$. Then $u^i v^j a_{ij} \langle v \rangle = u^i \langle v \rangle$. Therefore $u^i \in V\langle v \rangle$. On the other hand $u^i \in U\langle u \rangle$. This is a contradiction since $V\langle v \rangle$ and $U\langle u \rangle$ are disjoint.

This completes the proof.

7.2   Factoring 2-groups

This is another instance when we are looking for possible extensions of Rédei's theorem. We have seen examples that among 2-groups, results more general than Rédei's theorem hold. Here we will be more systematic.

Lemma 7.2.1. *Let A be a subset and χ_1, χ_2 be characters of a finite abelian group G such that $\mathrm{Ker}\chi_1 = \mathrm{Ker}\chi_2$ and $\chi_1(A) = 0$. Then also $\chi_2(A) = 0$.*

Proof. Let $K = \mathrm{Ker}\chi_1 = \mathrm{Ker}\chi_2$ and note that the factor group G/K is cyclic, say of order n. There is an $h \in G$ such that each element of G/K is in the form $h^i K$ for some i, $0 \leq i \leq n - 1$. Consequently, each $g \in G$ can be written in the form $g = h^{i(g)} k(g)$, where $k(g) \in K$. Clearly $\chi_1(h) = \rho_1$ and $\chi_2(h) = \rho_2$ are primitive (n)th roots of unity. To the sum

$$0 = \chi_1(A) = \sum_{a \in A} \left(\chi_1(h) \right)^{i(a)} \chi_1\left(k(a) \right) = \sum_{a \in A} \rho_1^{i(a)}$$

we assign the polynomial

$$A(x) = \sum_{a \in A} x^{i(a)}.$$

The (n)th cyclotomic polynomial $F(x)$ is irreducible over the rationals. Note that ρ_1 is a common root of $A(x)$ and $F(x)$. This implies that $F(x)$ divides $A(x)$ over the rationals. Hence $A(\rho_2) = 0$. However

$$A(\rho_2) = \sum_{a \in A} \rho_2^{i(a)} = \chi_2(A).$$

This completes the proof.

Theorem 7.2.1. *Let G be a group of type $(2^\alpha, 2)$. If $G = A_1 \cdots A_n$ is a factorization of G, then at least one of the factors $A_1, \ldots, A_n$ must be periodic.*

Proof. Let $G = \langle u, v \rangle$, where $|u| = 2^\alpha$ and $|v| = 2$. Let the characters χ_1 and χ_2 of G be defined by

$$\chi_1(u) = \rho, \qquad \chi_1(v) = 1,$$
$$\chi_2(u) = \rho, \qquad \chi_2(v) = -1,$$

where ρ is a primitive (2^α)th root of unity.

As χ_1 is not the principal character of G, it follows that $\chi_1(A_i) = 0$ for some i, $1 \le i \le n$. We may assume that $\chi_1(A_1) = 0$ since this is only a matter of relabelling the factors $A_1, \ldots, A_n$. Similarly, as χ_2 is not the principal character of G, it follows that $\chi_2(A_i) = 0$ for some i, $1 \le i \le n$. So either $\chi_2(A_1) = 0$ or $\chi_2(A_i) = 0$ for some i, $2 \le i \le n$. By relabelling the factors $A_2, \ldots, A_n$ we may assume that $\chi_2(A_1) = 0$ or $\chi_2(A_2) = 0$. We distinguish two cases.

Case 1: $\chi_1(A_1) = 0$ and $\chi_2(A_1) = 0$.

Case 2: $\chi_1(A_1) = 0$ and $\chi_2(A_2) = 0$.

In the definitions of χ_1 and χ_2 there are several choices for ρ. By Lemma 7.2.1, if Case 1 holds for one choice of ρ, then it holds for each choice of ρ. Similarly, if Case 2 holds for one choice of ρ, then it holds for all choices of ρ.

In Case 1 $\chi_1(A_1) = 0$ and $\chi_2(A_1) = 0$ for each choice of ρ. Note that from this it follows that

$$\operatorname{Ann}\left(\langle u^{2^{\alpha-1}} \rangle\right) \subset \operatorname{Ann}(A_1). \tag{1}$$

From (1) by Lemma 3.2.6, A_1 is periodic with period $u^{2^{\alpha-1}}$.

In Case 2 $\chi_1(A_1) = 0$ and $\chi_2(A_2) = 0$ for each choice of ρ. Note that from $\chi_1(A_1) = 0$ it follows that

$$\operatorname{Ann}\left(\langle u^{2^{\alpha-1}} \rangle\right) \cap \operatorname{Ann}\left(\langle u^{2^{\alpha-1}} v \rangle\right) \subset \operatorname{Ann}(A_1). \tag{2}$$

Further note that from $\chi_2(A_2) = 0$ it follows that

$$\operatorname{Ann}\left(\langle u^{2^{\alpha-1}} \rangle\right) \cap \operatorname{Ann}\left(\langle v \rangle\right) \subset \operatorname{Ann}(A_2). \tag{3}$$

From (2) and (3), by Lemma 3.2.7, there are subsets X, Y, U, V of G such that

$$A_1 = X\langle u^{2^{\alpha-1}}\rangle \cup Y\langle u^{2^{\alpha-1}}v\rangle,$$
$$A_2 = U\langle u^{2^{\alpha-1}}\rangle \cup V\langle v\rangle,$$

where the unions are disjoint and the products are direct.

If $X = \emptyset$, then A_1 is periodic with period $u^{2^{\alpha-1}}v$. If $U = \emptyset$, then A_2 is periodic with period v. So we assume that $X \neq \emptyset$ and $U \neq \emptyset$. There are elements $x \in X$ and $z \in U$. Multiplying the factorization $G = A_1 \cdots A_n$ by $g = x^{-1}z^{-1}$ we get the factorization

$$G = Gg = (x^{-1}A_1)(z^{-1}A_2)A_3 \cdots A_n.$$

Here $\langle u^{2^{\alpha-1}}\rangle \subset x^{-1}A_1$ and $\langle u^{2^{\alpha-1}}\rangle \subset z^{-1}A_2$ contradict the definition of the factorization.

This completes the proof.

Let G be a finite abelian group and let $A = \{e, a, b, c\}$ be a subset of G. We define a subset A' by $A' = \{e, a\}\{e, b\}$. Since the equation $c = abd$ is solvable for d, A can be written in the form $A = \{e, a, b, abd\}$. We need the next lemma.

Lemma 7.2.2. *If $|a| = 2$, then*
    (a) $\mathrm{Ann}(A) \subset \mathrm{Ann}(A')$,
    (b) A is periodic if and only if $d = e$,
    (c) $\chi(A) = 0$ *implies* $\chi(d) = 1$.

Proof. (a) Let χ be a character of G for which

$$0 = \chi(A) = 1 + \chi(a) + \chi(b) + \chi(c).$$

As $|a| = 2$, it follows that $\chi(a) = -1$ or $\chi(a) = 1$. If $\chi(a) = 1$, then $\chi(A) = 0$ gives that $\chi(b) = \chi(c) = -1$. Using this we have

$$\begin{aligned}
\chi(A') &= 1 + \chi(a) + \chi(b) + \chi(a)\chi(b) \\
&= 1 + 1 - 1 - 1 \\
&= 0.
\end{aligned}$$

If $\chi(a) = -1$, then $\chi(A) = 0$ gives that $\chi(b) = \rho$ and $\chi(c) = -\rho$, where ρ is a root of unity. Using this we have

$$\begin{aligned}
\chi(A') &= 1 + \chi(a) + \chi(b) + \chi(a)\chi(b) \\
&= 1 - 1 + \rho - \rho \\
&= 0.
\end{aligned}$$

(b) If $d = e$, then $A = A'$ and so A is periodic with period a. Conversely, assume that A is periodic with period g. Note that g^2 is also a period of A if $g^2 \neq e$. Using this observation we may assume that $|g| = 2$. From $e \in A$ it follows that $g \in A$.

If $g = a$, then

$$
\begin{aligned}
Aa &= \{a, e, ab, bd\} \\
&= \{e, a, b, abd\} \\
&= A.
\end{aligned}
$$

gives that $\{ab, bd\} = \{b, abd\}$. Here either $ab = b$ or $bd = b$. The first one leads to the contradiction $a = e$. The second one gives $d = e$.

If $g = b$, then

$$
\begin{aligned}
Ab &= \{b, ab, e, ad\} \\
&= \{e, a, b, abd\} \\
&= A.
\end{aligned}
$$

gives that $\{ab, ad\} = \{a, abd\}$. Hence either $ab = a$ or $ad = a$. The first one leads to the contradiction $b = e$. The second one gives $d = e$.

If $g = abd$, then

$$
\begin{aligned}
Aabd &= \{abd, bd, ab^2d, e\} \\
&= \{e, a, b, abd\} \\
&= A.
\end{aligned}
$$

gives that $\{ab^2d, bd\} = \{a, b\}$. Now either $ab^2d = b$ or $bd = b$. The first equality gives the contradiction $abd = e$, the second one provides $d = e$.

(c) If $\chi(A) = 0$, then by part (a), $\chi(A') = 0$ and so

$$
\begin{aligned}
0 &= \chi(A) - \chi(A') \\
&= \chi(ab)\chi(d) - \chi(ab) \\
&= \chi(ab)\big[\chi(d) - 1\big].
\end{aligned}
$$

This completes the proof.

Theorem 7.2.2. *Let G be a finite group of type $(2^\lambda, 2, \ldots, 2)$. If $G = A_1 \cdots A_n$ is a normalized factorization of G, where $|A_i|$ is either 2 or 4 for each i, $1 \le i \le n$, then at least one of the factors $A_1, \ldots, A_n$ is periodic.*

Proof. The $|G| = 2$ case is trivial. So we assume that $|G| \ge 4$ and proceed by induction on $|G|$. Clearly G is a direct product of its subgroups H and K of types (2^λ) and $(2, \ldots, 2)$ respectively. If $\lambda = 1$, then G is of type $(2, \ldots, 2)$. This special case is covered by Exercise 1.2.8. So for the remaining part of the proof we may assume that $\lambda \ge 2$. Let $H = \langle x \rangle$ and $K = \langle y_1, \ldots, y_s \rangle$, where $|x| = 2^\lambda$ and $|y_1| = \cdots = |y_s| = 2$. Consider a character χ of G that is faithful on H or equivalently for which $\chi(x) = \rho$, where ρ is a primitive (2^λ)th root of unity.

Let $A_i = \{e, a_i\}$ be a factor of order 2. If $0 = \chi(A_i) = 1 + \chi(a_i)$, then $\chi(a_i) = -1$ or $\chi(a_i^2) = 1$. Note that $a_i^2 \in H$. As χ is faithful on H it follows that $a_i^2 = e$. Therefore A_i is periodic with period a_i. So in the remaining part of the proof we may assume that $\chi(A_i) \neq 0$ when χ is faithful on H and $|A_i| = 2$. As χ is not the principal character of G, it follows that $0 = \chi(G) = \chi(A_1) \cdots \chi(A_n)$ and so $\chi(A_i) = 0$ for some i, $1 \leq i \leq n$. Thus we may assume that $|A_i| = 4$ for some i.

Let $A_i = \{e, a_i, b_i, c_i\}$ be a factor of order 4. If

$$0 = \chi(A_i) = 1 + \chi(a_i) + \chi(b_i) + \chi(c_i),$$

then one of $\chi(a_i)$, $\chi(b_i)$, $\chi(c_i)$ must be -1 and so one of a_i^2, b_i^2, c_i^2 must be e. Thus there is at least one factor of order 4 that contains at least one second order element. We choose the notation such that $A_1, \ldots, A_m$ are all the factors of order 4 containing at least one second order element. If $m = 1$, then $\chi(A_1) = 0$ for each χ that is faithful on H. Now, by Lemma 3.2.6, A_1 is periodic. So we may assume that $m \geq 2$.

Let us consider an $A_i = \{e, a_i, b_i, c_i\}$ with $1 \leq i \leq m$. We choose the notation such that $|a_i| = 2$. Further c_i can be written in the form $c_i = a_i b_i d_i$ with a suitable $d_i \in G$. By Lemma 7.2.2, A_i is periodic if and only if $d_i = e$. Thus we may assume that $d_i \neq e$. Also by Lemma 7.2.2, $\chi(A_i) = 0$ implies $\chi(d_i) = 1$. From this it follows that A_i can be replaced by

$$\{e, a_i, b_i, a_i b_i d_i^k\}$$

for each integer k. In particular, we may assume that $|d_i| = 2$ for each i, $1 \leq i \leq m$. Also A_i can be replaced by

$$\{e, a_i, b_i, a_i b_i\} = \{e, a_i\}\{e, b_i\}.$$

If $b_i^2 = e$, then each element of $A_i \setminus \{e\}$ is of order 2. We will say that A_i is a type 1 factor. Now A_i can be replaced by $H_i B_i$, where $H_i = \langle a_i, b_i \rangle$ and $B_i = \{e\}$. If $b_i^2 \neq e$, then a_i is the only second order element in A_i. We will say that A_i is a type 2 factor. In this case A_i can be replaced by $H_i B_i$, where $H_i = \{e, a_i\} = \langle a_i \rangle$ and $B_i = \{e, b_i\}$.

The subgroup H has a unique subgroup $L = \langle x^{2^{\lambda-1}} \rangle$ of order 2. From the factorization

$$G = H_1 B_1 \cdots H_m B_m A_{m+1} \cdots A_n$$

it follows that the product $H_1 \cdots H_m$ is direct. So there can be only one subgroup H_i for which $L \subset H_i$. Such an H_i does not necessarily exist. But if it does, then we choose the notation such that $L \subset H_1$. We claim that $L \not\subset H_1$ may be assumed.

In order to prove this claim let us consider $A_1 = \{e, a_1, b_1, c_1\}$ and distinguish two cases depending on whether A_1 is of type 1 or type 2.

Suppose first that A_1 is of type 1 and define the elements $d_{1,c}, d_{1,a}, d_{1,b}$ by the equations

$$c_1 = a_1 b_1 d_{1,c}, \quad a_1 = b_1 c_1 d_{1,a}, \quad b_1 = a_1 c_1 d_{1,b}.$$

Now A_1 can be written in the following forms and can be replaced by the following subgroups:

$$
\begin{aligned}
A_1 &= \{e, a_1, b_1, a_1 b_1 d_{1,c}\}, & H_{1,c} &= \langle a_1, b_1 \rangle, \\
A_1 &= \{e, b_1, c_1, b_1 c_1 d_{1,a}\}, & H_{1,a} &= \langle b_1, c_1 \rangle, \\
A_1 &= \{e, a_1, c_1, a_1 c_1 d_{1,b}\}, & H_{1,b} &= \langle a_1, c_1 \rangle
\end{aligned}
$$

respectively. If $L \subset H_{1,c}$, then one of a_1, b_1, $a_1 b_1$ is equal to $x^{2^{\lambda-1}}$. In the $a_1 = x^{2^{\lambda-1}}$ case $a_1 \notin H_{1,a}$, since obviously $a_1 \neq e$, $a_1 \neq b_1$, $a_1 \neq c_1$ and $a_1 = b_1 c_1$ combined with $c_1 = a_1 b_1 d_{1,c}$ leads to the $d_{1,c} = e$ contradiction. In the $b_1 = x^{2^{\lambda-1}}$ case $b_1 \notin H_{1,b}$, since clearly $b_1 \neq e$, $b_1 \neq a_1$, $b_1 \neq c_1$ and $b_1 = a_1 c_1$ leads to the $d_{1,c} = e$ contradiction. In the $a_1 b_1 = x^{2^{\lambda-1}}$ case $a_1 b_1 \notin H_{1,c}$, since $a_1 b_1 = e$, $a_1 b_1 = b_1$, $a_1 b_1 = c_1$, $a_1 b_1 = b_1 c_1$ leads in order to the $a_1 = b_1$, $a_1 = e$, $d_{1,c} = e$, $a_1 = c_1$ contradictions.

Suppose next that A_1 is of type 2. Now $a_1^2 = e$, $b_1^2 \neq e$ and A_1 can be written in the form $A_1 = \{e, a_1, b_1, a_1 b_1 d_1\}$ and can be replaced by $H_1 B_1$, where $H_1 = \{e, a_1\}$, $B_1 = \{e, b_1\}$. If $L \subset H_1$, then $a_1 = x^{2^{\lambda-1}}$. Now replace A_1 by

$$
\begin{aligned}
A_1' &= b_1^{-1} A_1 \\
&= \{b_1^{-1}, b_1^{-1} a_1, e, a_1 d_1\} \\
&= \{e, a_1', b_1', a_1' b_1' d_1'\},
\end{aligned}
$$

where $a_1' = a_1 d_1$, $b_1' = b_1^{-1} a_1$, $d_1' = d_1$. The only second order element in A_1' is $a_1' = a_1 d_1$ which is not equal to $x^{2^{\lambda-1}}$. Here A_1' is replaceable by $H_1' B_1'$, where

$$
H_1' = \langle a_1' \rangle = \langle a_1 d_1 \rangle, \qquad B_1' = \{e, b_1'\}.
$$

Thus in each case we may assume that $L \not\subset H_1$.

Replace A_i by $H_i B_i$ in the factorization $G = A_1 \cdots A_n$ to get the factorization

$$
G = A_1 \cdots A_{i-1}(H_i B_i) A_{i+1} \cdots A_n,
$$

where $1 \leq i \leq m$. This leads to the factorization

$$
G/H_i = (A_1 H_i)/H_i \cdots (A_{i-1} H_i)/H_i \cdot (B_i H_i)/H_i \cdot
$$

$$
(A_{i+1} H_i)/H_i \cdots (A_n H_i)/H_i
$$

of the factor group G/H_i. Here $(A_j H_i)/H_i$ is equal to

$$
\{H_i, a_j H_i, b_j H_i, c_j H_i\} \quad \text{or} \quad \{H_i, a_j H_i\}
$$

and $(B_i H_i)/H_i$ is equal to

$$
\{H_i, b_i H_i\} \quad \text{or} \quad \{H_i\}.
$$

As $|G/H_i| < |G|$, by the inductive assumption it follows that $(B_i H_i)/H_i$ or $(A_j H_i)/H_i$ is periodic for some j, $1 \leq j \leq n$, $j \neq i$.

If $(B_i H_i)/H_i$ is periodic, then $\left|(B_i H_i)/H_i\right| = |B_i|$ must be 2 and consequently A_i must be of type 2. Since $(B_i H_i)/H_i$ is periodic, it follows that $(b_i H_i)^2 = b_i^2 H_i = H_i$ and so $b_i^2 \in H_i = \{e, a_i\}$. We know that $b_i^2 \neq e$ and hence $b_i^2 = a_i$. Let

$$a_i = x^\alpha y_1^{\alpha(1)} \cdots y_s^{\alpha(s)}, \quad b_i = x^\beta y_1^{\beta(1)} \cdots y_s^{\beta(s)},$$

where

$$\alpha = 2^{\lambda-1}, \quad 0 \leq \beta \leq 2^\lambda - 1, \quad 0 \leq \alpha(1), \beta(1), \ldots, \alpha(s), \beta(s) \leq 1.$$

Now

$$b_i^2 = (x^\beta y_1^{\beta(1)} \cdots y_s^{\beta(s)})^2 = x^{2\beta} = x^\alpha y_1^{\alpha(1)} \cdots y_s^{\alpha(s)} = a_i$$

gives that $\alpha(1) = \cdots = \alpha(s) = 0$ and so $L \subset H_i$ and this is a contradiction.

If $(A_j H_i)/H_i$ is periodic and $\left|(A_j H_i)/H_i\right| = |A_j| = 2$, then in a similar way $a_j^2 \in H_i$ and $a_j^2 \neq e$ lead to the contradiction $L \subset H_i$. Therefore if $(A_j H_i)/H_i$ is periodic, then $\left|(A_j H_i)/H_i\right| = |A_j| = 4$. The periodicity of $(A_j H_i)/H_i$ implies that $(A_j H_i)/H_i$ contains a second order element, say $(a_j H_i)^2 = a_j^2 H_i = H_i$. Hence $a_j^2 \in H_i$. As $a_j^2 \neq e$ in the known way leads to the contradiction $L \subset H_i$, it follows that $a_j^2 = e$. Thus A_j contains a second order element, that is $1 \leq j \leq m$. By Lemma 7.2.2, the periodicity of $(A_j H_i)/H_i$ implies that $d_j \in H_i$.

The summary of the above argument is that for each i, $1 \leq i \leq m$ there is a j, $1 \leq j \leq m$ such that $d_j \in H_i$ and $i \neq j$. We define a bipartite graph Γ whose nodes are $H_1, \ldots, H_m$ and $d_1, \ldots, d_m$ and if $d_j \in H_i$, then (H_i, d_j) is a directed edge of Γ. If $(H_i, d_j), (H_k, d_j)$ are edges of Γ with $i \neq k$, then $d_j \in H_i \cap H_k = \{e\}$ which is a contradiction. Thus for each d_j there is at most one H_i such that (H_i, d_j) is an edge of Γ. Further, for each H_i there is at least one d_j for which (H_i, d_j) is an edge of Γ. Therefore there is a one-to-one map f from $\{H_1, \ldots, H_m\}$ into $\{d_1, \ldots, d_m\}$ such that $\big(H_i, f(H_i)\big)$, $1 \leq i \leq m$ are all the edges of Γ.

Let us consider

$$A_m = \{e, a_m, b_m, c_m\}.$$

(Remember $m \geq 2$.) Assume first that A_m is of type 1 and define the elements $d_{m,c}$, $d_{m,b}$, $d_{m,a}$ by the equations

$$c = a_m b_m d_{m,c}, \quad b = a_m c_m d_{m,b}, \quad a = b_m c_m d_{m,a}.$$

Then A_m can be written in the following forms and can be replaced by the following subgroups:

$$\begin{aligned}
A_m &= \{e, a_m, b_m, a_m b_m d_{m,c}\}, \quad & H_{m,c} &= \langle a_m, b_m \rangle, \\
A_m &= \{e, a_m, c_m, a_m c_m d_{m,b}\}, \quad & H_{m,b} &= \langle a_m, c_m \rangle, \\
A_m &= \{e, b_m, c_m, b_m c_m d_{m,a}\}, \quad & H_{m,a} &= \langle b_m, c_m \rangle
\end{aligned}$$

respectively. These replacements give rise to the graphs Γ_c, Γ_b, Γ_a and the maps f_c, f_b, f_a respectively. The nodes $H_1, \ldots, H_{m-1}$ and $d_1, \ldots, d_{m-1}$ are common in these graphs. After removing the edges joining to $H_{m,c}, H_{m,b}, H_{m,a}$ and $d_{m,c}, d_{m,b}, d_{m,a}$

the remaining parts of the graphs are identical. From this it follows that $f_c(H_{m,c}) = f_b(H_{m,b}) = f_a(H_{m,a})$. Let d_j be this common value. This leads to the contradiction $d_j \in H_{m,c} \cap H_{m,b} \cap H_{m,a} = \{e\}$.

Next assume that A_m is of type 2. Then $a_m^2 = e$, $b_m^2 \neq e$ and A_m can be written in the form

$$A_m = \{e, a_m, b_m, a_m b_m d_m\}$$

and can be replaced by $H_m B_m$, where

$$H_m = \{e, a_m\}, \qquad B_m = \{e, b_m\}.$$

The factor A_m can be replaced by

$$\begin{aligned} A'_m &= b_m^{-1} A_m \\ &= \{b_m^{-1}, b_m^{-1} a_m, e, a_m d_m\} \\ &= \{e, a'_m, b'_m, a'_m b'_m d'_m\}, \end{aligned}$$

where $a'_m = a_m d_m$, $b'_m = b_m^{-1} a_m$, $d'_m = d_m$. Then A'_m can be replaced by $H'_m B'_m$, where

$$H'_m = \{e, a'_m\}, \qquad B'_m = \{e, b'_m\}.$$

The $A_m \to H_m B_m$ and $A'_m \to H'_m B'_m$ replacements give rise to the graphs Γ, Γ' and the maps f, f' respectively. The nodes $H_1, \ldots, H_{m-1}$ and $d_1, \ldots, d_{m-1}$ are common in these graphs. After removing the edges joining to H_m, H'_m the remaining parts of the graphs are identical. From this it follows that $f(H_m) = f'(H'_m)$. Let d_j be this common value. This gives the contradiction $d_j \in H_m \cap H'_m = \{e\}$.

This completes the proof.

Theorem 7.2.3. *Let G be a finite group of type $(2^\lambda, 2, \ldots, 2)$. If $G = AB$ is a normalized factorization and $|B| = 4$, then A or B is periodic.*

Proof. Obviously G is a direct product of its subgroups H and K of types (2^λ) and $(2, \ldots, 2)$ respectively. Let $H = \langle x \rangle$ and $K = \langle y_1, \ldots, y_s \rangle$, where $|x| = 2^\lambda$ and $|y_1| = \cdots = |y_s| = 2$. Further let $B = \{e, b_1, b_2, b_3\}$.

Consider a character χ of G which is faithful on H. Plainly, it means that $\chi(x) = \rho$, where ρ is a primitive (2^λ)th root of unity. If $\chi(A) = 0$ for each character χ of G that is faithful on H, then by Lemma 3.2.6, it follows that A is periodic with period $x^{2^{\lambda-1}}$. Thus we may assume that $\chi(B) = 0$ for some character χ of G which is faithful on H. From

$$0 = \chi(B) = 1 + \chi(b_1) + \chi(b_2) + \chi(b_3)$$

it follows that one of the terms is -1, say $\chi(b_1) = -1$. Hence $\chi(b_1^2) = 1$. Let $b_1 = x^\beta y_1^{\beta(1)} \cdots y_s^{\beta(s)}$. This shows that $b_1^2 = x^{2\beta} \in H$. As χ is faithful on H we get that $b_1^2 = e$. Therefore B contains a second order element. We can write b_3 in the form $b_3 = b_1 b_2 d$, where d is a suitable element of G. By Lemma 7.2.2, B is periodic if and only if $d = e$. So we may assume that $d \neq e$. Also by Lemma 7.2.2, $\chi(B) = 0$ implies $\chi(d) = 1$ for each character χ of G. In other words, if $\chi(d) \neq 1$, then $\chi(B) \neq 0$ and consequently $\chi(A) = 0$. From this by Lemma 3.2.6, it follows that A is periodic with period d.

This completes the proof.

In the proof of the next theorem we will watch the size of annihilators.

Theorem 7.2.4. *If $G = A_1 \cdots A_n$ is a normalized factorization, where G is a group of type $(2^\lambda, 2^\mu)$ and $|A_i| = 2$ or $|A_i| = 4$ for each i, $1 \leq i \leq n$, then at least one of the factors $A_1, \ldots, A_n$ is periodic.*

Proof. Let G be a group of type $(2^\lambda, 2^\mu)$ and consider a normalized factorization $G = A_1 \cdots A_n$ of G such that $|A_i| = 2$ or $|A_i| = 4$ for each i, $1 \leq i \leq n$. We may assume that

$$|A_1| = \cdots = |A_s| = 4, \qquad |A_{s+1}| = \cdots = |A_n| = 2$$

since this is only a matter of reindexing the factors. We proceed by induction on s.

If $s = 0$, then by Rédei's theorem one of the factors is periodic. For the remaining part we assume that $s \geq 1$. Let $A_s = \{e, a, b, c\}$ and introduce the elements d_a, d_b, d_c defined by the equations

$$a = bcd_a, \qquad b = acd_b, \qquad c = abd_c.$$

Note that if $d_a = e$, then $A_s = \{e, a, b, ab\} = \{e, a\}\{e, b\}$. Now s decreases and by the inductive assumption one of the factors is periodic. Thus we may assume that $d_a \neq e$. Similarly we may assume that $d_b \neq e$, $d_c \neq e$.

To the factor A_i of G we assign the subgroup K_i of G defined by

$$K_i = \bigcap_{\chi \in \mathrm{Ann}(A_i)} \mathrm{Ker}\chi.$$

We claim that it may be assumed that $K_s \neq \{e\}$. In order to prove this claim we distinguish two cases depending on whether A_s contains a second order element or not.

In the first case suppose that $a^2 = e$ and denote the subgroup $\langle d_b, d_c \rangle$ by L. Note that $L \neq \{e\}$. We will show that $L \subset \mathrm{Ker}\chi$ for each $\chi \in \mathrm{Ann}(A_s)$. Let us consider

$$0 = \chi(A) = 1 + \chi(a) + \chi(b) + \chi(c). \tag{4}$$

If $\chi(a) = 1$, then from (4) it follows that $\chi(b) = \chi(c) = -1$. Hence

$$\begin{aligned} \chi(d_b) &= \chi(b)\chi(a^{-1})\chi(c^{-1}) \\ &= \chi(b)\left[\chi(a)\right]^{-1}\left[\chi(c)\right]^{-1} \\ &= 1, \end{aligned}$$

$$\begin{aligned}
\chi(d_c) &= \chi(c)\chi(a^{-1})\chi(b^{-1}) \\
&= \chi(c)\left[\chi(a)\right]^{-1}\left[\chi(b)\right]^{-1} \\
&= 1.
\end{aligned}$$

If $\chi(a) = -1$, then from (4) it follows that $\chi(b) = -\chi(c)$ and so

$$\begin{aligned}
\chi(d_b) &= \chi(b)\chi(a^{-1})\chi(c^{-1}) \\
&= \chi(b)\left[\chi(a)\right]^{-1}\left[\chi(c)\right]^{-1} \\
&= 1,
\end{aligned}$$

$$\begin{aligned}
\chi(d_c) &= \chi(c)\chi(a^{-1})\chi(b^{-1}) \\
&= \chi(c)\left[\chi(a)\right]^{-1}\left[\chi(b)\right]^{-1} \\
&= 1.
\end{aligned}$$

Let us turn to the second case when A does not contain any element of order 2. Since the group G is of type $(2^\lambda, 2^\mu)$ it follows that the product of the subgroups $\langle d_a \rangle$, $\langle d_b \rangle$, $\langle d_c \rangle$ cannot be direct. Thus we may assume that after a suitable reordering of the elements d_a, d_b, d_c the inequality

$$\langle d_a \rangle \cap \langle d_b, d_c \rangle \neq \{e\}$$

holds. Let L denote the subgroup $\langle d_a \rangle \cap \langle d_b, d_c \rangle$. Since $L \neq \{e\}$ it will be enough to show that $L \subset \mathrm{Ker}\chi$ for each $\chi \in \mathrm{Ann}(A_s)$. Note that

$$\begin{aligned}
\chi(A) &= 1 + \chi(a) + \chi(b) + \chi(c) \\
&= 1 + \chi(a) + \chi(b) + \chi(abd_c) \\
&= 1 + \chi(a) + \chi(b) + \chi(ab) - \chi(ab) + \chi(abd_c) \\
&= \left[1 + \chi(a)\right]\left[1 + \chi(b)\right] - \left[1 - \chi(d_c)\right]\chi(ab).
\end{aligned}$$

From (4) it follows that

$$\left[1 + \chi(a)\right]\left[1 + \chi(b)\right] = \left[1 - \chi(d_c)\right]\chi(ab). \tag{5}$$

Similarly

$$\left[1 + \chi(a)\right]\left[1 + \chi(c)\right] = \left[1 - \chi(d_b)\right]\chi(ac), \tag{6}$$

$$\left[1 + \chi(b)\right]\left[1 + \chi(c)\right] = \left[1 - \chi(d_a)\right]\chi(bc). \tag{7}$$

Also from (4) it follows that at least one of $\chi(a)$, $\chi(b)$, $\chi(c)$ is equal to -1. If $\chi(a) = -1$, then by (5) and (6) we get $\chi(d_c) = \chi(d_b) = 1$. If $\chi(a) \neq -1$, then by (7) we get $\chi(d_a) = 1$.

Summing up our argument we may assume that $K_s \neq \{e\}$. Similarly, we may assume that $K_i \neq \{e\}$ for each i, $1 \leq i \leq s$.

Let $A_n = \{e, a\}$. Note that in the $a^2 = e$ case A_n is periodic so we assume that $a^2 \neq e$. This in turn implies that

$$\{e\} \neq \langle a^2 \rangle \subset K_n.$$

Similarly, we may assume that $K_i \neq \{e\}$ for each i, $s+1 \leq i \leq n$. Therefore, $K_i \neq \{e\}$ may be assumed for each factor. By Theorem 5.1.1, this is not possible. This contradiction completes the proof.

Theorem 7.2.5. *Let p be a prime and let G be an abelian group of type $(p^\alpha, p, \ldots, p)$. If $G = BA_1 \cdots A_n$ is a factorization of G, where $A_1, \ldots, A_n$ are cyclic subsets of cardinality p, then at least one of the factors must be periodic.*

Proof. Clearly, G is the direct product of its subgroups H, K of types (p^α), $(p, \ldots, p)$ respectively. Let a be a basis element of H. Choose a character χ of G such that $\chi(a) = \rho$, where ρ is a primitive (p^α)th root of unity. Obviously χ is not the principal character of G. From $0 = \chi(G) = \chi(B)\chi(A_1) \cdots \chi(A_n)$ it follows that $\chi(B) = 0$ or $\chi(A_i) = 0$ for some i, $1 \leq i \leq n$. If $\chi(A_i) = 0$ and $A_i = \{e, a_i, a_i^2, \ldots, a_i^{r(i)-1}\}$, where $a_i = a^\beta k$, $0 \leq \beta \leq p^\alpha - 1$, $k \in K$, then $1 = \chi(a_i^p) = \chi(a^{p\beta}) = \rho^{p\beta}$ gives that $p\beta \mid p^\alpha$. Hence $\beta \mid p^{\alpha-1}$ and so $|a_i| = p$. Therefore A_i is a subgroup of G. We may assume that $\chi(B) = 0$ for each χ with $\chi(a) = \rho$. Now by Lemma 3.2.6, B is periodic with period $a^{p^{\alpha-1}}$.

This completes the proof.

In the $p = 2$ special case we get the following result.

Corollary 7.2.1. *Let G be a group of type $(2^\alpha, 2, \ldots, 2)$ and let $G = BA_1 \cdots A_n$ be a factorization of G such that $|A_1| = \cdots = |A_n| = 2$. Then at least one of the factors must be periodic.*

Theorem 7.2.6. *Let G be a finite abelian 2-group and let $G = BA_1 \cdots A_n$ be a factorization of G such that $|B| = 4$, $|A_1| = \cdots = |A_n| = 2$. Then at least one of the factors is periodic.*

Proof. Set $C = A_1 \cdots A_n$. Let $B = \{e, a, b, c\}$. Assume first that one of $|a|$, $|b|$, $|c|$ is 2, say $|a| = 2$. There is a $d \in G$ such that $c = abd$. Consider a nonprincipal character χ of G. By Lemma 7.2.1, $\chi(B) = 0$ implies $\chi(d) = 1$. Thus if $\chi(d) \neq 1$, then $\chi(B) \neq 0$ and so $\chi(C) = 0$. By Lemma 3.2.6, $C = A_1 \cdots A_n$ is periodic. Then by Theorem 5.4.1, one of $A_1, \ldots, A_n$ is periodic. We may assume that none of $|a|$, $|b|$, $|c|$ is 2. Again there is a $d \in G$ such that $c = abd$. If $d = e$, then $B = \{e, a\}\{e, b\}$ and so by Rédei's theorem, one of the factors $\{e, a\}$, $\{e, b\}$, $A_1, \ldots, A_n$ is periodic. As $|a| \neq 2$, $|b| \neq 2$, we get that one of $A_1, \ldots, A_n$ is periodic. So we may assume that $d \neq e$. Consider a nonprincipal character χ of G. If $\chi(B) = 0$, then

$$0 = \chi(e) + \chi(a) + \chi(b) + \chi(abd).$$

By the four roots of unity lemma (Lemma 3.1.7), one of the following holds:

$$\chi(e) + \chi(a) = 0, \quad \chi(b) + \chi(abd) = 0,$$

$$\chi(e) + \chi(b) = 0, \quad \chi(a) + \chi(abd) = 0,$$

$$\chi(e) + \chi(abd) = 0, \quad \chi(a) + \chi(b) = 0.$$

From the first two it follows that $\chi(d) = 1$. Hence $\chi(d) \neq 1$, $\chi(B) = 0$ implies $\chi(abd) = \chi(c) = -1$.

As $\{e, c\} \subset B$, the product $\{e, a\}C$ is direct. We claim that $\{e, c\}C$ is periodic with period d. It is enough to verify that $\chi(\{e, a\}C) = \chi(\{e, a\}Cd)$ for each character χ of G. It is clear for the principal character. It is also clear for characters χ with $\chi(C) = 0$ or $\chi(d) = 1$ or $\chi(c) = -1$. So assume that $\chi(d) \neq 1$, $\chi(c) \neq -1$, $\chi(C) \neq 0$. As $\chi(C) \neq 0$, from the factorization $G = BC$ it follows that $\chi(B) = 0$. Now $\chi(B) = 0$ and $\chi(d) \neq 1$ implies $\chi(c) = -1$. This contradiction proves the claim. Therefore $\{e, c\}A_1 \cdots, A_n$ is periodic. By Theorem 5.4.1, one of $\{e, c\}$, $A_1, \ldots, A_n$ is periodic. As $|c| \neq 2$, one of $A_1, \ldots, A_n$ is periodic.

This completes the proof.

Theorem 7.2.7. *Let G be a group of type $(2^\lambda, 2^\lambda)$, with $\lambda \geq 2$. If $G = AB$ is a normalized factorization of G and $|A| = 4$, $\langle A \rangle = G$, then B is periodic.*

Proof. We divide the proof into four steps.

(1) We claim that there are elements x, y in A such that x, y form a basis for G. In order to prove the claim note, that because of its type G has a basis u, v with $|u| = |v| = 2^\lambda$. The elements of G whose order is less than or equal to $2^{\lambda-1}$ span the subgroup $\langle u^{2^{\lambda-1}}, v^{2^{\lambda-1}} \rangle$ of G. If A does not contain any element of order 2^λ, then $\langle A \rangle \subset \langle u^{2^{\lambda-1}}, v^{2^{\lambda-1}} \rangle \neq G$ which is a contradiction. Thus there is an element x of A with $|x| = 2^\lambda$. In the basis representation $x = u^\alpha v^\beta$, $0 \leq \alpha, \beta \leq 2^\lambda - 1$ of x at least one of α and β is odd. For the sake of definiteness we assume that α is odd. Now $\langle x \rangle \cap \langle v \rangle = \{e\}$ and consequently x, v is a basis for G. If $A \setminus \langle x \rangle = \emptyset$, then $\langle A \rangle \subset \langle x \rangle \neq G$. So $A \setminus \langle x \rangle \neq \emptyset$. Let $y = x^\gamma v^\delta$, $0 \leq \gamma, \delta \leq 2^\lambda - 1$ be an element of $A \setminus \langle x \rangle$. If for each choice of y the exponent δ is even, then $\langle A \rangle \subset \langle x, v^2 \rangle \neq G$. Therefore there is a y with odd δ. In this case $\langle x \rangle \cap \langle y \rangle = \{e\}$ and so x, y is a basis for G.

(2) By step (1) A can be written in the form $A = \{e, x, y, a\}$, where

$$a = x^\alpha y^\beta, \quad 0 \leq \alpha, \beta \leq 2^\lambda - 1.$$

Set $g = (xya)^{2^{\lambda-1}}$. We claim that $B = Bg$. To prove this claim it is enough to show that

$$\chi(B) = \chi(B)\chi(g) \tag{8}$$

holds for each character χ of G. If $\chi(B) = 0$ or $\chi(g) = 1$, then (8) is clearly satisfied. Assume that $\chi(B) \neq 0$ and $\chi(g) \neq 1$ for some character χ of G. Now χ is not the principal character of G. Hence $0 = \chi(G) = \chi(A)\chi(B)$ gives that $\chi(A) = 0$, that is,

$$0 = 1 + \chi(x) + \chi(y) + \chi(a).$$

The four roots of unity lemma gives that one of $\chi(x)$, $\chi(y)$, $\chi(a)$ is -1, say $\chi(x) = -1$ and then $\chi(y) = -\chi(a)$. Using this we get the contradiction

$$\chi(g) = [\chi(x)]^{2^{\lambda-1}}[\chi(y)]^{2^{\lambda-1}}[\chi(a)]^{2^{\lambda-1}} = [\chi(y)]^{2^\lambda} = 1.$$

If $g \neq e$, then B is periodic with period g. We may assume that $g = [x^{\alpha+1}y^{\beta+1}]^{2^{\lambda-1}}$ $= e$, that is, $2 \mid (\alpha+1)$, $2 \mid (\beta+1)$. In other words we may assume that α and β are odd.

(3) If B is periodic with period g, then we may assume that $|g| = 2$. The possible choices for g are $x^{2^{\lambda-1}}$, $y^{2^{\lambda-1}}$, $x^{2^{\lambda-1}}y^{2^{\lambda-1}}$.

Let $g = x^{2^{\lambda-1}}$ and let ρ be a primitive (2^{λ})th root of unity. Note that $\chi(g) \neq 1$ holds for each character χ of G for which

$$\chi(x) = \rho, \quad \chi(y) = \rho^m, \quad 0 \leq m \leq 2^{\lambda} - 1.$$

Now $\chi(A) = 0$ must hold for some of these characters otherwise B is periodic with period g. From

$$0 = \chi(A) = \rho^0 + \rho^1 + \rho^m + \rho^{\alpha+m\beta},$$

by the four roots of unity lemma, it follows that $0, 1, m, \alpha+m\beta$ is a rearrangement of $0, 1, 2^{\lambda-1}, 2^{\lambda-1}+1$. This leads to the following two possibilities.

$$\begin{aligned}
m &\equiv 2^{\lambda-1} &&\pmod{2^{\lambda}}, \\
\alpha + m\beta &\equiv 2^{\lambda-1}+1 &&\pmod{2^{\lambda}},
\end{aligned}$$

or

$$\begin{aligned}
m &\equiv 2^{\lambda-1}+1 &&\pmod{2^{\lambda}}, \\
\alpha + m\beta &\equiv 2^{\lambda-1} &&\pmod{2^{\lambda}}.
\end{aligned}$$

From the first two congruences we get

$$\alpha \equiv 2^{\lambda-1}\beta + 2^{\lambda-1} + 1 \pmod{2^{\lambda}}.$$

Replacing β by $2v + 1$ it follows that $\alpha \equiv 1 \pmod{2^{\lambda}}$ and so $\alpha = 1$. From the second two congruences we get

$$\alpha \equiv (2^{\lambda-1} - 1)\beta + 2^{\lambda-1} \pmod{2^{\lambda}}.$$

Replacing β by $2v + 1$ it follows that $\alpha + \beta \equiv 0 \pmod{2^{\lambda}}$.

Let $g = y^{2^{\lambda-1}}$. In this case we can argue as above to get that $\beta = 1$ or $\alpha + \beta \equiv 0 \pmod{2^{\lambda}}$ or otherwise B is periodic with period g.

Let $g = x^{2^{\lambda-1}}y^{2^{\lambda-1}}$ and let ρ let a primitive (2^{λ})th root of unity. Note that $\chi(g) \neq 1$ holds for each character χ of G for which

$$\chi(x) = \rho^m, \quad \chi(xy) = \rho, \quad 0 \leq m \leq 2^{\lambda} - 1.$$

Obviously, $\chi(y) = \rho^{1-m}$. Now $\chi(A) = 0$ must hold for some such character χ of G, otherwise B is periodic with period g. From

$$0 = \chi(A) = \rho^0 + \rho^m + \rho^{1-m} + \rho^{m\alpha+(1-m)\beta}$$

it follows that there is an integer l such that $0, m, 1 - m, m\alpha + (1 - m)\beta$ is a rearrangement of $0, 2^{\lambda-1}, l, l + 2^{\lambda-1} + 1$. We face the following three possibilities.

$$m \equiv 2^{\lambda-1} \pmod{2^\lambda},$$
$$1 - m + 2^{\lambda-1} \equiv m\alpha + (1 - m)\beta \pmod{2^\lambda},$$

$$1 - m \equiv 2^{\lambda-1} \pmod{2^\lambda},$$
$$m + 2^{\lambda-1} \equiv m\alpha + (1 - m)\beta \pmod{2^\lambda},$$

$$m\alpha + (1 - m)\beta \equiv 2^{\lambda-1} \pmod{2^\lambda},$$
$$m + 2^{\lambda-1} \equiv 1 - m \pmod{2^\lambda}.$$

From the last congruence it follows that

$$2m + 2^{\lambda-1} \equiv 1 \pmod{2^\lambda}$$

and so $0 \equiv 1 \pmod 2$. This contradiction sorts out the third case. From the first two congruences we get

$$\beta \equiv 2^{\lambda-1}\alpha + 2^{\lambda-1}\beta + 1 \pmod{2^\lambda}.$$

Replacing α by $2u + 1$ and replacing β by $2v + 1$ it follows that $\beta = 1$. A similar argument gives that from the second two congruences it follows that $\alpha = 1$. Summing up our considerations we can say that if (α, β) is not one of

$$(1, 1), \quad (1, 2^{\lambda-1}), \quad (2^{\lambda-1}, 1), \tag{9}$$

then B is periodic.

(4) Let $A = \{e, x, y, x^\alpha y^\beta\}$, where (α, β) is one of (9). We claim that $\langle B \rangle \neq G$.

In the $(\alpha, \beta) = (1, 1)$ case $A = \{e, x\}\{e, y\}$. By Lemma 7.1.2, from the factorization $G = B\{e, x\}\{e, y\}$ it follows that either $B \subset \langle x^2, y \rangle$ or $B \subset \langle x, y^2 \rangle$, that is, $\langle B \rangle \neq G$.

In the $(\alpha, \beta) = (1, 2^{\lambda-1})$ case $A = \{e, x, y, xy^{-1}\}$. Note that y and xy^{-1} span G and their product is x. This reduces the problem to the previous case. The $(\alpha, \beta) = (2^{\lambda-1}, 1)$ case can be settled in a similar way. This proves the claim.

Using $\langle B \rangle \neq G$ we show that B is periodic. Let $H = \langle B \rangle$. Restricting the factorization $G = AB$ to H we get the factorization $H = (A \cap H)B$. The set $A \cap H$ has two elements, say $A \cap H = \{e, d\}$. Plainly $|d| \neq 2$. By Lemma 2.1.1, B is periodic with period d.

This completes the proof.

7.3 Nonperiodic factorizations of 2-groups

The next results enable us to extend a factorization from a subgroup to the whole group. They are variants of Lemma 2.3.1. In the first lemma we treat elementary 2-groups. The second lemma deals with nonelementary 2-groups.

Lemma 7.3.1. *Let G be a finite group of type $(2, \ldots, 2)$ and let H be a subgroup of G with $|G : H| = m \geq 4$. If $(q_1, \ldots, q_n)$ is not a periodicity forcing factorization type for H, then $(q_1, \ldots, q_n, m)$ and $(q_1, \ldots, q_{i-1}, q_i m, q_{i+1}, \ldots, q_n)$ are not periodicity forcing factorization types for G, for each i, $1 \leq i \leq n$.*

Proof. The factor group G/H is of type $(2, \ldots, 2)$ and so there is a subgroup $K = \{k_1, \ldots, k_m\}$ of G such that $Hk_1, \ldots, Hk_m$ are all the elements of G/H. In other words $Hk_1, \ldots, Hk_m$ form a partition of G and $G = HK = H\{k_1, \ldots, k_m\}$ is a factorization of G. Let $d \in H \setminus \{e\}$ and set $B = \{k_1, \ldots, k_{m-1}, k_m d\}$. We choose the notation such that $k_m \neq e$. Note that $Hk_m d = Hk_m$ and so $Hk_1, \ldots, Hk_{m-1}, Hk_m d$ form a partition of G. Thus $G = HB$ is a normalized factorization of G. As $(q_1, \ldots, q_n)$ is not a periodicity factorization type of H, there is a normalized factorization $H = A_1 \cdots A_n$ of H with $|A_i| = q_i$ and none of the factors is periodic. Since $d \neq e$, it follows that B is not periodic. The factorization $G = HB = A_1 \cdots A_n B$ shows that $(q_1, \ldots, q_n, m)$ is not a periodicity forcing factorization type for G.

Next we show that $A_n B$ is not periodic. So the factorization $G = A_1 \cdots A_{n-1}(A_n B)$ gives that $(q_1, \ldots, q_{n-1}, q_n m)$ is not a periodicity forcing factorization type for G. This proves the $i = n$ case of the lemma. The other cases can be settled in a similar manner. Since $A_n \subset H$, it follows that the subsets in the partition

$$A_n B = A_n k_1 \cup \cdots \cup A_n k_{m-1} \cup A_n k_m d$$

are in distinct cosets of G modulo H. By Observation 2.3.1, it follows that $A_n B$ is not periodic.

This completes the proof.

Lemma 7.3.2. *Let G be a finite abelian 2-group whose type is not $(2, \ldots, 2)$ and let H be a subgroup of G for which $|G : H| = 2$. If $(q_1, \ldots, q_n)$ is not a periodicity forcing factorization type for H, then $(q_1, \ldots, q_n, 2)$ and $(q_1, \ldots, q_{i-1}, q_i 2, q_{i+1}, \ldots, q_n)$ are not periodicity forcing factorization types for G, for each i, $1 \leq i \leq n$.*

Proof. The factor group G/H consists of two cosets H and Hb. Set $B = \{e, b\}$. The cosets H, Hb form a partition of G and so $G = HB = H\{e, b\}$ is a factorization of G. We claim that we can choose b such $|b| \geq 4$. Choose an a from $G \setminus H$. If $|a| \geq 4$, then a is a suitable choice for b. So we may assume that each element a of $G \setminus H$ has order 2. Choose an h from H. If $|h| \geq 4$, then any ah is a suitable choice for b. So we may assume that each element h of $H \setminus \{e\}$ has order 2. Thus each element of $G \setminus \{e\}$ has order 2. This means that G is of type $(2, \ldots, 2)$, which is not the case.

As $(q_1, \ldots, q_n)$ is not a periodicity forcing factorization type for H, there is a normalized factorization $H = A_1 \cdots A_n$ of H such that the factors are not periodic and $|A_1| = q_1, \ldots, |A_n| = q_n$. Since $|b| \geq 4$, it follows that B is not periodic and the factorization $G = HB = A_1 \cdots A_n B$ shows that $(q_1, \ldots, q_n, 2)$ is not a periodicity forcing factorization type for G.

group type	factorization type
$(2^3, 2^2)$	$(2^3, 2, 2)$
$(2^2, 2^2, 2)$	$(2^3, 2, 2)$
$(2^2, 2, 2, 2, 2)$	$(2^3, 2^2, 2)$
$(2^3, 2, 2, 2)$	$(2^3, 2^2, 2)$
$(2^4, 2, 2)$	$(2^3, 2^2, 2)$
$(2, 2, 2, 2, 2, 2, 2)$	$(2^3, 2^2, 2^2)$
$(2, 2, 2, 2, 2, 2, 2, 2)$	$(2^4, 2^2, 2^2)$
$(2, 2, 2, 2, 2, 2)$	$(2^3, 2^3)$

Table 1

i	K_i	L_i
0	$\langle x^2 u \rangle$	$\langle z \rangle$
1	$\langle x^2 \rangle$	$\{e, xz\}$

i	K_i	L_i
0	$\langle x^4 z \rangle$	$\langle y \rangle$
1	$\langle x^4 \rangle$	$\{e, x^2 y\}$

Table 2   Table 3

Next we show that $A_n B$ is not periodic and so the factorization $G = A_1 \cdots A_{n-1}(A_n B)$ gives that $(q_1, \ldots, q_{n-1}, q_n 2)$ is not a periodicity forcing factorization type for G. This proves the lemma in the $i = n$ case. The remaining cases can be proved in a similar way. The sets in the partition $A_n B = A_n \cup A_n b$ are in distinct cosets of G modulo H. By Observation 2.3.1, the subset $A_n B$ is not periodic.

This completes the proof.

We will show by means of explicit constructions that the factorization types listed in Table 1 do not force periodicity for the corresponding types of groups.

(1) Let G be of type $(2^3, 2^2)$. In construction (5) Lemma 2.3.4 the condition $p \geq 3$ was not used. So it can be applied when $p = 2$.

(2) Let G be of type $(2^2, 2^2, 2)$. We can use construction (3) in Lemma 2.3.4.

(3) Let G be of type $(2^2, 2, 2, 2, 2)$ with basis x, y, z, u, v where $|x| = 2^2$, $|y| = |z| = |u| = |v| = 2$. Set $k_i = y^i$, $0 \leq i \leq 1$, K_i, L_i as shown in Table 2 and

$$A_1 = \{e, u, v, uvz\}, \quad A_2 = \{e, x\},$$

(4) Let G be of type $(2^3, 2, 2, 2)$ and let x, y, z, u be a basis for G, where $|x| = 2^3$, $|y| = |z| = |u| = 2$. Set $k_i = x^i$, $0 \leq i \leq 1$, K_i, L_i as shown in Table 3 and

$$A_1 = \{e, z, u, zuy\}, \quad A_2 = \{e, x^2\}.$$

i	K_i	L_i
0	$\langle x^8 \rangle$	$\{e, x^4yz\}$
1	$\langle x^8 y \rangle$	$\{e, yz\}$

Table 4

i	K_i	L_i
0	$\langle v \rangle$	$\langle xw \rangle$
1	$\langle w \rangle$	$\langle zv \rangle$

Table 5

i	K_i	L_i
0	$\langle xw, yt \rangle$	$\langle v \rangle$
1	$\langle w, t \rangle$	$\langle zv \rangle$

Table 6

i	K_i	L_i
0	$\langle z, u \rangle$	$\langle zv, uw \rangle$
1	$\langle zw, uv \rangle$	$\langle v, w \rangle$

Table 7

(5) Let G be of type $(2^4, 2, 2)$ with basis elements x, y, z, where $|x| = 2^4$, $|y| = |z| = 2$. Set $k_i = x^i$, $0 \le i \le 1$, K_i, L_i as shown in Table 4 and

$$A_1 = \{e, x^4\}, \quad A_2 = \{e, y, x^2, x^2 z\}.$$

(6) Let G be of type $(2, 2, 2, 2, 2, 2, 2)$ with a basis s, x, y, z, u, v, w. Set $k_i = s^i$, $0 \le i \le 1$, K_i, L_i as shown in Table 5 and

$$A_1 = \{e, x, y, xyv\}, \quad A_2 = \{e, z, u, zuw\}.$$

(7) Let G be of type $(2, 2, 2, 2, 2, 2, 2, 2)$ with basis elements s, x, y, z, u, v, w, t. Set $k_i = s^i$, $0 \le i \le 1$, K_i, L_i as shown in Table 6 and

$$A_1 = \{e, x, y, xyv\}, \quad A_2 = \{e, z, u, zuw\}.$$

(8) Let G be of type $(2, 2, 2, 2, 2, 2)$ and let x, y, z, u, v, w be a basis for G. This construction is based on the same ideas as constructions in Lemma 2.3.3. Set K_i, L_i as shown in Table 7 and

$$A_1 = x^0 K_0 \cup x^1 K_1, \quad A_2 = y^0 L_0 \cup y^1 L_1.$$

By Theorem 7.2.1, if G is a finite group of type $(2^\lambda, 2)$, then each factorization type for G forces periodicity. Thus the classification of periodicity forcing factorization types is solved for groups of type $(2^\lambda, 2)$. We are now in position to characterize the periodicity forcing factorization types for elementary 2-groups.

Theorem 7.3.1. *Let G be a finite group of type $(2, \ldots, 2)$. The factorization type $(q_1, \ldots, q_n)$ of G with $q_1 \ge \cdots \ge q_n \ge 2$ is not periodicity forcing if*
      $n = 2$ and $q_2 \ge 8$,
      $n \ge 3$ and $q_1 \ge 8$, $q_2 \ge \cdots \ge q_n \ge 4$.

It is periodicity forcing if

$q_n = 2$ *independently of the value of* n,

$n = 1$ *independently of the value of* q_1,

$n = 2$ *and* $q_2 \mid 4$ *independently of the value of* q_1,

$n \geq 3$ *and* $q_1 \mid 4, \ldots, q_n \mid 4$.

Proof. Let G be a finite group of type $(2, \ldots, 2)$ and let $G = A_1 \cdots A_n$ be a normalized factorization of G with $|A_1| = q_1, \ldots, |A_n| = q_n$, such that $q_1 \geq \cdots \geq q_n \geq 2$. Note that $q_n \geq 2$ implies $|G| \geq 2$.

Let $n = 2$. Now $q_1 \geq q_2 \geq 8$ implies that $|G| \geq 2^6$. Construction (8) shows that $(2^3, 2^3)$ is not a periodicity forcing factorization type. Constructions (6), (7) with the grouping $G = B(A_1 A_2)$ give that $(2^3, 2^4)$, $(2^4, 2^4)$ are not periodicity forcing factorization types. Using Lemma 7.3.1 we can conclude that (q_1, q_2) is not a periodicity forcing factorization type for each $q_1 \geq q_2 \geq 8$.

Let $n \geq 3$. With the help of Lemma 7.3.1 (q_1, q_2), $q_1 \geq q_2 \geq 8$ can be extended to each factorization type $(q_1, \ldots, q_n)$, $q_1 \geq \cdots \geq q_n \geq 4$ and $q_1 \geq q_2 \geq 8$. Thus we may assume that $q_2 = \cdots = q_n = 4$. Constructions (6), (7) give that $(2^3, 2^2, 2^2)$, $(2^4, 2^2, 2^2)$ are not periodicity forcing factorization types for G. With the help of Lemma 7.3.1 we can extend these to each factorization type $(q_1, \ldots, q_n)$, $q_1 \geq 8$, $q_2 = \cdots = q_n = 4$.

On the other hand, if $q_n = 2$, then $A_n = \{e, a_n\}$ is periodic since $|a_n| = 2$. If $n = 1$, then $G = A_1$ and so A_1 is periodic. So for the remaining part of the proof we may assume that $n \geq 2$ and $q_1 \geq \cdots \geq q_n \geq 4$.

Let $n = 2$. If $q_2 = 4$, then as A_2 is simulated, A_1 or A_2 is periodic independently of the value of $|A_1| = q_1$.

Suppose that $n \geq 3$. In the $q_1 = \cdots = q_n = 4$ case by Exercise 1.2.8, one of the factors $A_1, \ldots, A_n$ is periodic.

This completes the proof.

Theorem 7.3.2. *Let G be a finite abelian group of type $(2^\lambda, 2, \ldots, 2)$, where $\lambda \geq 2$ and the rank of G is at least four. The factorization type $(q_1, \ldots, q_n)$ of G with $q_1 \geq \cdots \geq q_n \geq 2$ is not periodicity forcing if*

$n = 2$ *and* $q_2 \geq 8$,

$n \geq 3$ *and* $q_1 \geq 8$, $q_2 \geq 4$.

It is periodicity forcing if

$n = 1$ *independently of the value of* q_1,

$n = 2$ *and* $q_2 \mid 4$ *independently of the value of* q_1,

$n \geq 3$ *and* $q_1 \mid 4, \ldots, q_n \mid 4$.

Proof. Let $G = A_1 \cdots A_n$ be a normalized factorization of G with $|A_1| = q_1, \ldots,$ $|A_n| = q_n$, such that $q_1 \geq \cdots \geq q_n \geq 2$. Note that $q_n \geq 2$ implies $|G| \geq 2$.

Let $n = 2$. As $q_1 \geq q_2 \geq 8$, it follows that $|G| \geq 2^6$. Constructions (3), (4) with the grouping $G = B(A_1 A_2)$ give that $(2^3, 2^3)$ is not a periodicity forcing factorization type for G. Using Lemma 7.3.2 we can conclude that (q_1, q_2) is not a periodicity forcing factorization type if $q_1 \geq q_2 \geq 8$.

Let $n \geq 3$. From $q_1 \geq 8$, $q_2 \geq 4$, $q_3 \geq 2$ we get that $|G| \geq 2^6$. Constructions (3), (4) show that $(2^3, 2^2, 2)$ is not a periodicity forcing factorization type for G. Using Lemma 7.3.2 we can extend this to each factorization type $(q_1, \ldots, q_n)$, $q_1 \geq \cdots \geq q_n \geq 2$ with $q_1 \geq 8$, $q_2 \geq 4$.

On the other hand if $n = 1$, then $G = A_1$ gives that A_1 is periodic. So we may assume that $n \geq 2$.

Let $n = 2$. If $q_2 = 2$, then A_2 is cyclic. This gives that A_1 or A_2 is periodic. If $q_2 = 4$, then Theorem 7.2.3 gives that A_1 or A_2 is periodic.

Let $n \geq 3$ and $q_1 \mid 4, \ldots, q_n \mid 4$. Now, Theorem 7.2.2 provides that one of the factors is periodic.

This completes the proof.

Theorem 7.3.3. *Let G be a finite group of type $(2^\lambda, 2, 2)$, $\lambda \geq 2$. The factorization type $(q_1, \ldots, q_n)$ of G with $q_1 \geq \cdots \geq q_n \geq 2$ is not periodicity forcing if*

$n = 2$ *and* $q_2 \geq 8$,

$n \geq 3$ *and* $q_1 \geq 8$, $q_2 \mid 4$, $q_3 \mid 4, \ldots, q_n \mid 4$.

It is periodicity forcing if

$n = 1$ *independently of the value of* q_1,

$n = 2$ *and* $q_2 \mid 4$ *independently of the value of* q_1,

$n \geq 3$ *and* $q_1 \mid 4, \ldots, q_n \mid 4$.

Proof. Let $n \geq 2$. Now $q_1 \geq q_2 \geq 8$ forces that $|G| \geq 2^6$. Construction (5) with the $G = B(A_1 A_2)$ grouping gives that $(2^3, 2^3)$ is not a periodicity forcing factorization type for G. With the aid of Lemma 7.3.2 it can be extended to any factorization type $(q_1, \ldots, q_n)$, $q_1 \geq q_2 \geq 8$. The remaining part of the proof is similar to the proof of the previous theorem.

This completes the proof.

Theorem 7.3.4. *Let G be a finite abelian 2-group whose type is not $(2^\lambda, 2, \ldots, 2)$. The factorization type $(q_1, \ldots, q_n)$ of G with $q_1 \geq \cdots \geq q_n \geq 2$ is not periodicity forcing if*

$n = 2$ *and* $q_1 \geq 8$, $q_2 \geq 4$,

$n \geq 3$ *and* $q_1 \geq 8$.

It is periodicity forcing if

$n = 1$ *independently of the value of* q_1,

$n = 2$ *and* $q_2 = 2$ *independently of the value of* q_1.

The $n \geq 3$ and $q_1 \mid 4, \ldots, q_n \mid 4$ case is undecided.

Proof. As G is not of type $(2^\lambda, 2, \ldots, 2)$, it is of type $(t_1, \ldots, t_s)$, $t_1 \geq \cdots \geq t_s \geq 2$, with $s \geq 2$ and $t_2 \geq 4$.

Let $n = 2$. As $q_1 \geq 8$, $q_2 \geq 4$, it follows that $|G| \geq 2^5$. Constructions (1), (2) with grouping $G = B(A_1 A_2)$ give that $(2^3, 2^2)$ is not a periodicity forcing factorization type for G. Using Lemma 7.3.2 this can be extended to each factorization type (q_1, q_2), $q_1 \geq 8$, $q_2 \geq 4$.

Let $n \geq 3$. Constructions (1), (2) give that $(2^3, 2, 2)$ is not a periodicity forcing factorization type for G. Using Lemma 7.3.2 we can extend this to each factorization type $(q_1, \ldots, q_n)$, $q_1 \geq \cdots \geq q_n \geq 2$ with $q_1 \geq 8$.

On the other hand in the $n = 1$ case $G = A_1$ and so A_1 is periodic as $|G| \geq 2$. If $n = 2$ and $q_2 = 2$, then Lemma 7.2.1 gives that A_1 or A_2 is periodic.

This completes the proof.

7.4 Groups with elementary 2-component

By Theorem 7.2.2, if a finite 2-group G is factored in the form $G = BA_1 \cdots A_n$, where $|B| = 4$, $|A_1| = \cdots = |A_n| = 2$, then at least one of the factors is periodic. We extend this result showing that if $G = BA_1 \cdots A_n$ is a factorization of the finite abelian group G such that $|B| = 4$ and each $|A_i|$ is a prime, then one of the factors is periodic.

We have seen in Exercise 1.2.8 that in each normalized factorization of a finite elementary 2-group by subsets of cardinality four at least one of the factors is a subgroup. Looking for a possible common generalization of Rédei's theorem and the previous result the following question comes to mind. If a finite abelian group whose 2-component is elementary is factored into normalized subsets of cardinality of four or prime, does a subgroup always occur among the factors? The answer to this question is "no". To see this consider a group G of type $(2, 2, 3)$ with basis elements x, y, z such that $|x| = |y| = 2$ and $|z| = 3$. Let

$$A_1 = \{e, x, yz, xyz\}, \qquad A_2 = \{e, z, xz^2\}.$$

Now $G = A_1 A_2$ is a normalized factorization of G and none of the factors is a subgroup of G.

Rédei's theorem can be reformulated in the following way. In each factorization of a finite abelian group by subsets of prime cardinality at least one of the factors is periodic. Similarly the result of Exercise 1.2.8 can be stated in the following way. In each factorization of a finite elementary 2-group by subsets of cardinality four at least one of the factors is periodic.

As a common generalization of these results we prove the following. If $G = A_1 \cdots A_n$ is a factorization of the finite abelian group G whose 2-component is an elementary group and if $|A_i|$ is either a prime or 4, then at least one of the factors must be periodic.

Lemma 7.4.1. *If A is a subset of G such that $|A| = 4$, then A can be replaced by*

$$A' = \{a_{|2}(a_{|2'})^v : a \in A\}$$

for each exponent v.

Proof. By the character test for replacement it suffices to show that for each character χ of G for which $\chi(A) = 0$ it follows that $\chi(A') = 0$. To show this let χ be a character of G such that $\chi(A) = 0$ and let $A = \{e, ax, by, cz\}$, where $|a| = |b| = |c| = 2$ and x, y, z are $2'$-elements. Now

$$\chi(A) = \chi(e) + \chi(ax) + \chi(by) + \chi(cz)$$

is a vanishing sum of roots of unity. By the four roots of unity lemma, the sum of four roots of unity can be zero only if the four numbers can be arranged into two couples such that the members in any couple sum to zero.

We may assume that $\chi(e) + \chi(ax) = 0$ and $\chi(by) + \chi(cz) = 0$. It follows that

$$-\chi(a) = \chi(x^{-1}), \quad -\chi(bc^{-1}) = \chi(y^{-1}z).$$

The left sides have 2-power orders, the right side have odd orders, and so

$$-\chi(a) = 1, \quad \chi(x^{-1}) = 1, \quad \chi(bc^{-1}) = 1, \quad \chi(y^{-1}z) = 1.$$

From these we have $\chi(e) = -\chi(a)$, $\chi(e) = \chi(x)$ and $\chi(b) = -\chi(c)$, $\chi(y) = \chi(z)$. Using this information we can see that $\chi(A') = 0$.

This completes the proof.

Theorem 7.4.1. *Let $G = BA_1 \cdots A_n$ be a factorization of the finite abelian group G such that $|A_1|, \ldots, |A_n|$ are primes and $|B| = 4$. Then at least one of the factors $A_1, \ldots, A_n$ is periodic.*

Proof. We may assume that $|A_1| = \cdots = |A_s| = 2$ and $|A_{s+1}|, \ldots, |A_n|$ are odd primes. In order to prove the theorem we assume the contrary, that there is a counterexample. In the $n = 0$ case the theorem holds. So in a counterexample, $n \geq 1$.

Let us consider an A_i and assume that $|A_i| = p$ is a prime. If each $a \in A_i \setminus \{e\}$ has order p, then do nothing with A_i. If there is an element $a \in A_i \setminus \{e\}$ with $p^2 \mid |a|$, then replace A_i by $A_i' = \{e, a, a^2, \ldots, a^{p-1}\}$. Clearly A_i' is not periodic. If there is an element $a \in A_i$ whose order is not a prime power then there is an integer s such that $A_i^s = \{a^s : a \in A_i\}$ contains only (p, q)-elements, where q is a prime distinct from p. In this case replace A_i by A_i^s. By Lemma 1.4.3, this can be done. In short we may assume that there is a counterexample in which each A_i contains only (p, q)-elements.

If each of $B, A_1, \ldots, A_s$ contains only 2-elements, then the product $BA_1 \cdots A_s$ forms a factorization of the 2-component of G and so by Theorem 7.2.6, at least one of the factors $B, A_1, \ldots, A_s$ is periodic. We choose a counterexample with minimal n and among these we choose one for which the quantity

$$h(B, A_1, \ldots, A_n) = \left(\prod_{b \in B} |b_{|2'}| \right) \left(\prod_{i=1}^{s} \prod_{a \in A_i} |a_{|2'}| \right)$$

is minimal. We may assume that one of $B, A_1, \ldots, A_s$ contains not only 2-elements.

Let us consider B. Clearly $B = \{e, ax, by, cz\}$, where the orders of a, b, c are 2-powers and the orders of x, y, z are odd. Assume first that each of x, y, z is equal to e. This case can be settled by the method of the proof of Theorem 7.2.6. In this case $B = \{e, a, b, c\}$. If one of a, b, c has order 2, say $|a| = 2$, then the squares of e and a are equal. From the factorization $G = B(A_1 \cdots A_n)$ by Lemma 8.2.2, it follows that B or $A_1 \cdots A_n$ is periodic. As B is not periodic, $A_1 \cdots A_n$ is periodic. In this case the conditions of the Theorem 5.4.2 are satisfied and so by this theorem, one of the factors $A_1, \ldots, A_n$ is periodic. This is a contradiction. We may assume that none of a, b, c has order 2. Write c in the form $c = abd$, where d is an element of G. Now B is in form

$$B = \{e, a, b, abd\}.$$

If $d = e$, then $B = \{e, a\}\{e, b\}$. Here $\{e, a\}$, $\{e, b\}$ are not subgroups of G as $|a| \neq 2$, $|b| \neq 2$. From the factorization $G = \{e, a\}\{e, b\}A_1 \cdots A_n$, by Rédei's theorem, it follows that at least one of the factors is periodic. This is not possible. Thus we may assume that $d \neq e$. If there is no character χ of G for which $\chi(B) = 0$, then $0 = \chi(G) = \chi(A_1 \cdots A_n)$ for each nonprincipal character χ of G. By Exercise 3.2.2, we get the contradiction $G = A_1 \cdots A_n$. Consider a character χ of G for which $\chi(B) = 0$. Now

$$0 = \chi(B) = \chi(e) + \chi(a) + \chi(b) + \chi(abd).$$

The sum of four complex numbers whose length are all 1 is 0. This can happen only in the following three ways:

$$\chi(e) + \chi(a) = 0, \quad \chi(b) + \chi(abd) = 0,$$

$$\chi(e) + \chi(b) = 0, \quad \chi(a) + \chi(abd) = 0,$$

$$\chi(e) + \chi(abd) = 0, \quad \chi(a) + \chi(b) = 0.$$

From $\chi(a) = -1$, $\chi(b) = -\chi(abd)$ it follows that $\chi(d) = 1$. From $\chi(b) = -1$, $\chi(a) = -\chi(abd)$ it follows that $\chi(d) = 1$. In other words $\chi(B) = 0$, $\chi(d) \neq 1$ implies $\chi(abd) = -1$.

Set $C = \{e, abd\}A_1 \cdots A_n$. Notice that the product is direct. We claim that $Cd = C$. We verify the claim showing that $\chi(Cd) = \chi(C)$ for each character χ of G. It is clear when $\chi(d) = 1$ or $\chi(C) = 0$. So suppose that $\chi(d) \neq 1$, $\chi(C) \neq 0$. Now χ is not the principal character of G. From the factorization $G = BA_1 \cdots A_n$ it follows that $\chi(B) = 0$ or $\chi(A_i) = 0$ for some i, $1 \leq i \leq n$. As $\chi(C) \neq 0$ we get $\chi(B) = 0$. But $\chi(B) = 0$, $\chi(d) \neq 1$ implies $\chi(\{e, abd\}) = 0$ and $\chi(C) = 0$. Therefore $\chi(Cd) = \chi(C)$. Thus C is periodic with period d. Theorem 5.4.2 gives that one of the factors $A_1, \ldots, A_n$, $\{e, abd\}$ is periodic. This is a contradiction.

Turn to the cases when not all of x, y, z is equal to e. By Lemma 7.4.1, B can be replaced by $B' = \{e, a, b, c\}$. From the factorization $G = B'A_1 \cdots A_n$ by the minimality of the counterexample, it follows that one of the factors B', $A_1, \ldots, A_n$, is periodic. This leads to a contradiction unless B' is periodic. By relabelling a, b, c we may assume that $B' = \{e, a\}\{e, b\}$, where $|a| = 2$. Thus $B = \{e, ax, by, abz\}$, that is, $c = ab$. We distinguish five cases.

Case 1. Two of x, y, z are equal to e.

Case 2. One of x, y, z is equal to e.

Case 3. None of x, y, z is equal to e and $|x||y||z|$ is not a power of a prime.

Case 4. x, y, z are nonidentity p-elements and two of them generate the same subgroup.

Case 5. x, y, z are of order p and $z = x^k y$.

Turn to Case 1. Suppose that exactly one of x, y, z is not equal to e. Now factor B is in one of the following forms:

$$B = \{e, ax, b, ab\}, \quad x \neq e,$$

$$B = \{e, a, by, ab\}, \quad y \neq e,$$

$$B = \{e, a, b, abz\}, \quad z \neq e.$$

In the last two cases B contains an element of order 2. This leads to the contradiction that one of the factors $B, A_1, \ldots, A_n$ is periodic. We may assume that the first possibility occurs. Consider a character χ of G with $\chi(B) = 0$. Now

$$0 = \chi(B) = \chi(e) + \chi(ax) + \chi(b) + \chi(ab).$$

It can happen only in one of the following three ways:

$$1 + \chi(ax) = 0, \quad \chi(b) + \chi(bc) = 0,$$

$$1 + \chi(b) = 0, \quad \chi(ax) + \chi(bc) = 0,$$

$$1 + \chi(bc) = 0, \quad \chi(ax) + \chi(b) = 0.$$

From $\chi(ax) = -1$, $\chi(b) = -\chi(ab)$ it follows that $\chi(x) = 1$. From $\chi(b) = -1$, $\chi(ax) = -\chi(ab)$ it follows that $\chi(x) = 1$. From $\chi(ab) = -1$, $\chi(ax) = -\chi(b)$ it follows that $\chi(x) = 1$. In short $\chi(B) = 0$ implies $\chi(x) = 1$. Thus B can be replaced by

$$B' = \{e, ax, bx, abx^2\} = \{e, ax\}\{e, bx\}.$$

From the factorization $G = \{e, ax\}\{e, bx\}A_1 \cdots A_n$ by Rédei's theorem it follows that one of the factors is periodic. This is a contradiction.

Turn to Case 2 when exactly 2 of x, y, z are not equal to e. Now B is in one of the following forms.

$$B = \{e, ax, by, ab\}, \quad x \neq e, \; y \neq e,$$

$$B = \{e, ax, b, abz\}, \quad x \neq e, \; z \neq e,$$

$$B = \{e, a, by, abz\}, \quad y \neq e, \; z \neq e.$$

In the third case B contains an element of order 2. This leads to the contradiction that one of the factors is periodic. We may assume that the first two possibilities occur. Suppose

$$B = \{e, ax, by, ab\}.$$

Consider a character χ of G for which $\chi(B) = 0$. We can conclude that

$$0 = \chi(B) = \chi(e) + \chi(ax) + \chi(by) + \chi(ab)$$

can hold only in the following ways:

$$1 + \chi(ax) = 0, \quad \chi(by) + \chi(ab) = 0,$$

$$1 + \chi(by) = 0, \quad \chi(ax) + \chi(ab) = 0,$$

$$1 + \chi(ab) = 0, \quad \chi(ax) + \chi(by) = 0.$$

From $\chi(ax) = -1$, $\chi(by) = -\chi(ab)$ it follows that $\chi(x) = 1$, $\chi(y) = 1$. From $\chi(by) = -1$, $\chi(ax) = -\chi(ab)$ it follows that $\chi(x) = 1$, $\chi(y) = 1$. From $\chi(ab) = -1$, $\chi(ax) = -\chi(by)$ it follows that $\chi(x) = \chi(y)$. In short $\chi(B) = 0$ implies $\chi(x) = \chi(y)$.

If $x = y$, then

$$B = \{e, ax, by, ab\} = \{e, ax\}\{e, ab\}.$$

In the $|ab| = 2$ case B is periodic which is a contradiction. We may assume that $|ab| \neq 2$. From the factorization

$$G = \{e, ax\}\{e, ab\} A_1 \cdots A_n$$

by Rédei's theorem, it follows that one of the factors is periodic. So we may assume that $x \neq y$. Set $v = x^{-1}y$. Note that $v \neq e$ and $\chi(B) = 0$ implies $\chi(v) = 1$. This gives that B can be replaced by

$$B' = \{e, av, bv, abv^2\} = \{e, av\}\{e, bv\}.$$

The factorization $G = \{e, av\}\{e, bv\} A_1 \cdots A_n$ contradicts Rédei's theorem.

Suppose $B = \{e, ax, b, abz\}$. This case can be settled in a similar manner.

Turn to Case 3 when none of x, y, z is equal to e and there are distinct primes p, q such that both p and q divides $|x||y||z|$. Now $B = \{e, ax, by, abz\}$. Assume first that $|b| \neq 2$. Replace B by $B' = \{e, ax^p, by^p, abz^p\}$ in the factorization $G = BA_1 \cdots A_n$. In the factorization $G = B'A_1 \cdots A_n$ the value of h decreased. By the minimality of the counterexample, B' is periodic. This gives that $x^p = e$, that is, $|x| \mid p$. Replacing B by $\{e, ax^q, by^q, abz^q\}$ gives $x^q = e$, that is, $|x| \mid q$. Therefore $x = e$. This is a contradiction. So we may assume that $|b| = 2$. Then $|ab| = 2$. In this situation the roles of a, b, ab are symmetric and we write B in the form $B = \{e, ax, by, cz\}$. Replacing B by $B' = \{e, ax^p, by^p, cz^p\}$ gives that B' is periodic. By symmetry we may assume that $x^p = e$ and $y^p = z^p$. Replacing

B by $B'' = \{e, ax^q, by^q, cz^q\}$ gives that B'' is periodic. We face the following possibilities:

$$x^q = e, \quad y^q = z^q,$$

$$y^q = e, \quad x^q = z^q,$$

$$z^q = e, \quad x^q = y^q.$$

In the first case $x^p = e$, $x^q = e$ imply the contradiction $x = e$. In the second case from $x^p = e$, $y^q = e$ it follows that $|x| = p$, $|y| = q$. From $z^{pq} = (z^p)^q = (y^p)^q = (y^q)^p = e$ it follows that $|z| \mid pq$. Set $z = \alpha\beta$ where $|\alpha|$ is a p-power and $|\beta|$ is a q-power. From $y^p = z^p = (\alpha\beta)^p = \beta^p$ it follows that $y = \beta$. From $x^q = z^q = (\alpha\beta)^q = \alpha^q$ it follows that $x = \alpha$. Hence $z = xy$ and

$$B = \{e, ax, by, abz\} = \{e, ax\}\{e, by\}.$$

By Rédei's theorem, one of the factors in the factorization $G = \{e, ax\}\{e, by\}$ $A_1 \cdots A_n$ is periodic. This is a contradiction. The third possibility can be settled in a similar way.

Turn to Case 4 when x, y z are p-elements and two of x, y, z generate the same subgroup. This can happen in the following ways:

$$\langle x \rangle = \langle y \rangle, \quad y = x^s, \ p \nmid s,$$

$$\langle x \rangle = \langle z \rangle, \quad z = x^s, \ p \nmid s,$$

$$\langle y \rangle = \langle z \rangle, \quad z = y^s, \ p \nmid s.$$

In the first case $B = \{e, ax, bx^s, abz\}$. Consider a character χ of G with $\chi(B) = 0$. We conclude that

$$0 = \chi(B) = \chi(e) + \chi(ax) + \chi(bx^s) + \chi(abz)$$

can hold only in the following ways

$$1 + \chi(ax) = 0, \quad \chi(bx^s) + \chi(abz) = 0,$$

$$1 + \chi(bx^s) = 0, \quad \chi(ax) + \chi(abz) = 0,$$

$$1 + \chi(abz) = 0, \quad \chi(ax) + \chi(bx^s) = 0.$$

From $\chi(a) = -1$, $\chi(x) = 1$ we get $\chi(x^s) = \chi(z)$ then $\chi(z) = 1$. From $\chi(b) = -1$, $\chi(x^s) = 1$ we get $\chi(x) = \chi(z)$ then $\chi(z) = 1$. From $\chi(ab) = -1$, $\chi(z) = 1$ we get $\chi(z) = 1$. In short $\chi(B) = 0$ implies $\chi(z) = 1$. So B can be replaced by

$$B' = \{e, az, bz, abz^2\} = \{e, az\}\{e, bz\}.$$

From the factorization $G = \{e, az\}\{e, bz\}A_1 \cdots A_n$ by Rédei's theorem we get the contradiction that one of the factors is periodic. The second and third possibilities can be settled in similar ways.

Turn to Case 5 when $|x| = |y| = |z| = p$ and $z = x^k y$. Now $B = \{e, ax, by, abx^k y\}$. If $k = 0$, then $\langle y \rangle = \langle z \rangle$ and by Case 4, we are done. If $k = 1$, then

$$B = \{e, ax, by, abxy\} = \{e, ax\}\{e, by\}$$

and by Rédei's theorem we are done. If $k = -1$, then

$$B = \{e, ax, by, abx^{-1}y\} = \{e, ax\}\{a, abx^{-1}y\}$$

and by Rédei's theorem we are done. We may assume that $2 \leq k \leq p - 2$. Consider a character χ of G with $\chi(B) = 0$. Now

$$0 = \chi(B) = \chi(e) + \chi(ax) + \chi(by) + \chi(abx^k y).$$

This can hold only in the following ways:

$$1 + \chi(ax) = 0, \quad \chi(by) + \chi(abx^k y) = 0,$$

$$1 + \chi(by) = 0, \quad \chi(ax) + \chi(abx^k y) = 0,$$

$$1 + \chi(abx^k y) = 0, \quad \chi(ax) + \chi(by) = 0.$$

If $\chi(a) = -1$, $\chi(x) = 1$, then we get $\chi(x) = 1$. If $\chi(b) = -1$, $\chi(y) = 1$, then we get $\chi(x^{k-1}) = 1$. If $\chi(ab) = -1$, $\chi(x^k y) = 1$, then we get $\chi(x^{k+1}) = 1$. In short $\chi(B) = 0$ implies $\chi(x) = 1$ and so B can be replaced by

$$B' = \{e, ax, bx, abx^2\} = \{e, ax\}\{e, bx\}.$$

From the factorization $G = \{e, ax\}\{e, bx\}A_1 \cdots A_n$ by Rédei's theorem we get the contradiction that one of the factors is periodic.

In the factorization $G = BA_1 \cdots A_n$ replace B by

$$B' = \{e, a\}\{e, b\}.$$

Set $H_0 = \{e, a\} = \langle a \rangle$, $A_0 = \{e, b\}$. We get the factorization $G = H_0 A_0 A_1 \cdots A_n$ and then the factorization

$$G/H_0 = (A_0 H_0)/H_0 \cdots (A_n H_0)/H_0$$

of the factor group G/H_0. Here $(A_i H_0)/H_0$ stands for $\{aH_0 : a \in A_i\}$. By Rédei's theorem, one of the factors, say $(A_0 H_0)/H_0$, is a subgroup of G/H_0. It means that $A_0 H_0 = H_1$ is a subgroup of G. Then we get the factorization

$$G/H_1 = (A_1 H_1)/H_1 \cdots (A_n H_1)/H_1$$

of the factor group G/H_1. Continuing in this way we get that there is a maximal subgroup M of G such that M contains all but one of the factors $H_0, A_0, \ldots, A_n$. We may assume that

(i) $A_0 \not\subseteq M$ or

(ii) one of $A_1, \ldots, A_s$ is not contained by M, say $A_s \not\subseteq M$ or

(iii) one of $A_{s+1}, \ldots, A_n$ is not contained by M, say $A_n \not\subseteq M$.

If $A_0 \not\subseteq M$, then $|G : M| = 2$ and each $2'$-element of G is in M. In particular $x \in M$. So $\{e, ax\} \subset M$. Therefore $M = \{e, ax\}A_1 \cdots A_n$ is a factorization of M. By Rédei's theorem one of the factors is a subgroup of M. This is a contradiction.

If $A_s \not\subseteq M$, then $|G : M| = 2$ and each $2'$-element of G is in M. In particular $x, y, z \in M$. Hence $B = \{e, ax, by, cz\} \subset M$ and so

$$M = BA_1 \cdots A_{s-1}A_{s+1} \cdots A_n$$

is a factorization of M. By the minimality of the counterexample one of the factors is periodic. This is a contradiction. We may assume that $A_n \not\subseteq M$. This means $|G : M| = p$ is an odd prime. If $\{x, y, z\} \subset M$, then $B \subset M$ and so $M = BA_1 \cdots A_{n-1}$ is a factorization of M. By the minimality of the counterexample one of the factors is periodic. This is a contradiction. We may assume that $\{x, y, z\} \not\subseteq M$. By Lemma 1.4.3, A_n can be replaced by A_n' such that A_n' contains only p-elements. Choose an element $c \in A_n' \setminus \{e\}$ and replace A_n' by $C = \{e, c, c^2, \ldots, c^{p-1}\}$ to get the factorization $G = BA_1 \cdots A_{n-1}C$. As $G = M(Cd^{-1})$ is a factorization of G for each $d \in C$, Cd^{-1} is a complete set of representatives modulo M. There are elements $u_d, v_d, w_d \in Cd^{-1}$ such that

$$(ax)^{-1} \in u_d M, \quad (by)^{-1} \in v_d M, \quad (abz)^{-1} \in w_d M.$$

Hence $axu_d, byv_d, abw_d \in M$. Set

$$D_d = \{e, (ax)u_d, (by)v_d, (abz)w_d\}.$$

Note that $M = D_d A_1 \cdots A_{n-1}$ is a factorization of M. Indeed, products coming from $D_d A_1 \cdots A_{n-1}$ are among the products coming from $BA_1 \cdots A_{n-1}$ (Cd^{-1}). By the minimality of the counterexample D_d is periodic. If $|b| \neq 2$, then the periodicity of D_d gives that $xu_d = e$ and $yv_d = zw_d$ for each $d \in C$. We get

$$x^{-1} \in \bigcap_{d \in C} Cd^{-1}.$$

By Lemma 1.2.1, C is periodic. As $|C| = p$ we get $C = \langle c \rangle$, that is, $|c| = p$. Choose d to be e and consider $xu_e = e$, $yv_e = zw_e$. As $u_e, v_e, w_e \in \langle c \rangle$, we get that there is an integer k such that $z = x^k y$. Now $B = \{e, ax, by, abx^k y\}$ and by Case 5 we are done. We may assume that $|b| = 2$. In this situation $|ab| = 2$ and the roles of a, b, ab are symmetric. We write B back in the form $B = \{e, ax, by, cz\}$, where $c = ab$. By the periodicity of D_d, one of the following holds:

$$xu_d = e, \quad yv_d = zw_d,$$

$$yv_d = e, \quad xu_d = zw_d,$$

$$zw_d = e, \quad xu_d = yv_d.$$

Suppose there are $g, h \in C$ such that $g \neq h$ and

$$xu_g = e, \quad yv_g = zw_g,$$

$$yv_h = e, \quad xu_h = zw_h.$$

Then $x = c^i$, $y = c^j$ for some i, j, $-p + 1 \leq i, j \leq p - 1$. Here $i \neq 0$, $j \neq 0$ as $x \neq e$, $y \neq e$. If $i = j$, then $x = y$ and by Case 4, we are done. Therefore $i \neq j$ and $y = x^s$, where $p \nmid s$. Again by Case 4, we are done. We may assume that $xu_d = e$ for each $d \in C$. It means that

$$x^{-1} \in \bigcap_{d \in C} Cd^{-1}$$

and by Lemma 1.2.1, C is periodic. Since $|C| = p$, we get that C is a subgroup of G and so $|c| = p$. Using the fact that D_d is periodic for $d = e$ by symmetry we may assume that $xu_e = e$, $yv_e = zw_e$. As $u_e, v_e, w_e \in \langle c \rangle$, it follows that there is an integer k for which $z = x^k y$. Now by Case 5, we are done.

This completes the proof.

Note that construction (1) in the proof of Lemma 2.3.1 is applicable for groups of type (p, p, q), where $p \geq 4$, $q \geq 2$. (Even if p, q are not primes.) Let p, q be primes with $p \geq 5$ and let G be a group with basis elements x, y, z, $|x| = |y| = p$, $|z| = q$. Set

$$A_1 = \{e, xy, (xy)^2, \ldots, (xy)^{p-3}, x^{p-2}y^{p-1}, x^{p-1}y^{p-2}\},$$

$$B = \langle x \rangle \cup z \langle x \rangle \cup \cdots \cup z^{q-2} \langle x \rangle \cup z^{q-1} \langle y \rangle.$$

A routine computation shows that $G = BA_1$ is a factorization and none of the factors is periodic. Clearly $|A_1| = p$, $|B| = pq$.

Let $G = BA_1 \cdots A_n$ be a factorization of G such that $|A_1|, \ldots, |A_n|$ are primes and $|B|$ is a product of two primes, say $|B| = pq$. Does it follow that at least one of the factors is always periodic? The example above shows that the answer is "no" if $p \geq 5$. The $p = q = 2$ case is settled by Theorem 7.4.1. The answer is not known when $p = q = 3$ or $p = 3$, $q = 2$. We spell this out as an open problem.

Problem 7.4.1. Let $G = BA_1 \cdots A_n$ be a factorization of a finite abelian group G such that $|A_1|, \ldots, |A_n|$ are primes and $|B|$ is 6 or 9. Does it follow that one of the factors is periodic?

We turn to groups with elementary 2-component.

Theorem 7.4.2. *Let $G = A_1 \cdots A_n$ be a factorization of the finite abelian group G whose 2-component is elementary. If each $|A_i|$ is either a prime or 4, then at least one of the factors must be periodic.*

Proof. If $|A_i|$ is a prime for each i, $1 \leq i \leq n$, then by Rédei's theorem one of the factors is periodic. Let m be the number of factors among $A_1, \ldots, A_n$ for which $|A_i| = 4$. The case $m = 0$ is settled. We assume that $m \geq 1$ and we proceed by induction on m. We may suppose that $|A_1| = 4$ since this is only a matter of rearranging the factors. Suppose each A_i with $|A_i| = 4$ contains only 2-elements. Now A_i is simulated. Indeed, $A_i = \{e, a_i, b_i, c_i\}$, where $|a_i| = |b_i| = |c_i| = 2$. There is an element $d_i \in G$ such that $c_i = a_i b_i d_i$. Clearly $H_i = \{e, a_i, b_i, a_i b_i\} = \langle a_i, b_i \rangle$ is a subgroup of G. Now by Theorem 6.4.2, one of the factors is periodic.

Consider the quantity

$$h(A_1, \ldots, A_n) = \prod_{|A_i|=4} \prod_{a \in A_i} |a_{|2'}|.$$

The case $h(A_1, \ldots, A_n) = 1$ is settled. We assume that

$$h(A_1, \ldots, A_n) \geq 2$$

and we proceed by induction on $h(A_1, \ldots, A_n)$. We may assume that A_1 contains not only 2-elements. If $n = 1$, then $G = A_1$, that is A_1 is a subgroup of G. We assume that $n \geq 2$ and we proceed by induction on n.

Let $A_1 = \{e, ax, by, cz\}$, where $|a| = |b| = |c| = 2$ and x, y, z are $2'$-elements. Let H_1 be the subgroup $\{e, a, b, ab\}$ and let $c = dab$. Replace A_1 by $A_1' = \{e, a, b, dab\}$. (Let $v = 0$ in Lemma 7.4.1.) Now

$$h(A_1, A_2, \ldots, A_n)/h(A_1', A_2, \ldots, A_n) = |x||y||z| > 1.$$

By the inductive assumption one of the factors $A_1', A_2, \ldots, A_n$ is periodic. If this is not A_1', then we are done. Thus we may assume that A_1' is periodic. Then $A_1' = H_1$ and $c = ab$.

We distinguish 5 cases.

Case 1: $x = e$, $y = e$, $z \neq e$,

Case 2: $x = e$, $y \neq e$, $z \neq e$,

Case 3: $x \neq e$, $y \neq e$, $z \neq e$ are not p-elements for any prime p,

Case 4: $x \neq e$, $y \neq e$, $z \neq e$ are p-elements and $\langle x \rangle = \langle y \rangle$,

Case 5: $x \neq e$, $y \neq e$, $z \neq e$ are p-elements $|x| = p$ and $z = x^k y$.

In Case 1 for each character χ of G for which $\chi(A_1) = 0$ it follows that $\chi(H_1) = 0$ and $\chi(z) = 1$. Hence A_1 can be replaced by

$$\{e, az, bz, abz^2\} = \{e, az\}\{e, bz\}$$

which is a product of two nonsubgroup factors of prime cardinality. The number of the factors with $|A_i| = 4$ decreased and so by the inductive assumption one of the factors $A_2, \ldots, A_n$ is periodic.

If in Case 2 $y = z$, then

$$A_1 = \{e, a, by, aby\} = \{e, a\}\{e, by\}$$

is periodic. Assume that $y \neq z$ and let $|y| = r$ and $|z| = s$. Replace A_1 by $A'_1 = \{e, a, b, cz^r\}$. (Let $v = r$ in Lemma 7.4.1.) Now

$$h(A_1, A_2, \ldots, A_n)/h(A'_1, A_2, \ldots, A_n) = |y||z|/|z^r| > 1.$$

By the inductive assumption one of the factors $A'_1, A_2, \ldots, A_n$ is periodic. If this is not A'_1, then we are done. We may assume that A'_1 is periodic. Hence $z^r = e$ and so s divides r. A similar argument shows that we need to consider the case when r divides s and hence $r = s$.

Consider a character χ of G for which $\chi(A_1) = 0$. We get that $\chi(y^{-1}z) = 1$. Since $y \neq z$, $w = y^{-1}z \neq e$. Now A_1 can be replaced by

$$\{e, aw, bw, abw^2\} = \{e, aw\}\{e, bw\}$$

which is a product of nonsubgroup factors of prime cardinality. The number of the factors with $|A_i| = 4$ decreased. By the inductive assumption one of the factors $A_2, \ldots, A_n$ is periodic.

Turn to Case 3 and assume that x, y, z are not p-elements for any prime p. There are distinct primes p and q such that both divide $|x||y||z|$. Replace A_1 by $A'_1 = \{e, ax^p, by^p, cz^p\}$. Here

$$h(A_1, A_2, \ldots, A_n)/h(A'_1, A_2, \ldots, A_n) = |x||y||z|/|x^p||y^p||z^p| > 1.$$

By the inductive assumption one of the factors $A'_1, A_2, \ldots, A_n$ is periodic. If this is not A'_1, then we are done. We may assume that A'_1 is periodic. Further we may assume that $x^p = e$ and $y^p = z^p$ since this is only a matter of reordering the elements of A_1. After a similar argument we may assume that $y^q = e$ and $x^q = z^q$. We get that $z = xy$ and

$$A_1 = \{e, ax, by, abxy\} = \{e, ax\}\{e, by\}$$

is a product of nonsubgroup factors of prime cardinality. The number of factors of composite order decreased. By the inductive assumption one of the factors $A_2, \ldots, A_n$ is periodic.

Consider Case 4 and assume that $y = x^s$, where p does not divide s. Now $A_1 = \{e, ax, bx^s, cz\}$. Let χ be a character of G for which $\chi(A_1) = 0$. It follows that $\chi(z) = 1$. Therefore A_1 can be replaced by

$$\{e, az, bz, abz^2\} = \{e, az\}\{e, bz\}$$

which is a product of nonsubgroup factors of prime cardinality. The number of the factors of composite cardinality decreased. By the inductive assumption one of the factors $A_2, \ldots, A_n$ is periodic.

In Case 5, $A_1 = \{e, ax, by, abx^k y\}$. If $k = 0$, then by Case 4 we are done. If $k = 1$, then

$$A_1 = \{e, ax, by, abxy\} = \{e, ax\}\{e, by\},$$

that is A_1 is a product of two nonsubgroup normalized factors. The situation is similar when $k = -1$. Namely,

$$\{e, ax\}\{e, abx^{-1}y\} = \{e, ax, abx^{-1}y, by\}.$$

Thus we assume that $2 \leq k \leq p-2$. Let χ be a character of G for which $\chi(A_1) = 0$. We can conclude that $\chi(x) = 1$. Hence A_1 can be replaced by

$$\{e, ax, by, abxy\} = \{e, ax\}\{e, by\}$$

which is a product of nonsubgroup normalized factors of prime cardinality.

Thus by Case 4 we need to consider only the case when x, y, z are nonidentity p-elements for some prime p. Replace A_1 by the subgroup $H_1 = \{e, a, b, ab\}$. This leads to the factorizations $G = H_1 A_2 \cdots A_n$ and

$$G/H_1 = (A_2 H_1)/H_1 \cdots (A_n H_1)/H_1.$$

By the inductive assumption one of the factors, say $(A_2 H_1)/H_1$ is a periodic subset of G/H_1. There is a subgroup K and a subset C_2 of G such that $KC_2 = A_2 H_1$. Here $C_2 = \{e\}$ or $|C_2| = 2$ depending on $|A_2|$ is a prime or four. Then consider the factorization

$$G/K = (C_2 K)/K \cdot (A_3 K)/K \cdots (A_n K)/K.$$

Continuing in this way finally we have that there is a maximal subgroup M of G such that M contains all but one of the factors $H_1, A_2, \ldots, A_n$. We simply assume that $A_n \not\subset M$. Restricting the factorization $G = H_1 A_2 \cdots A_n$ to M we have the factorization

$$M = H_1 A_2 \cdots A_{n-1}(M \cap A_n).$$

If $M \cap A_n \neq \{e\}$, then $|A_n| = 4$ and $|G : M| = 2$. Then each $2'$-element of G is in M. In particular $x, y, z \in M$. Thus $M = A_1 A_2 \cdots A_{n-1}(M \cap A_n)$ is also a factorization. Here $|M \cap A_n| = 2$ and so the number of factors of composite cardinality decreased. By the inductive assumption one of the factors $A_1, A_2, \ldots, A_{n-1}, M \cap A_n$ is periodic. If this is not $M \cap A_n$, then we are done. If $M \cap A_n$ is periodic, then A_n contains a 2-element. Now by Cases 1 and 2 one of the factors is periodic. We assume that $M \cap A_n = \{e\}$ and so $|A_n| = p$ is a prime. If $p = 2$, then $|G : M| = 2$. As before now each $2'$-element of G is in M. In particular $x, y, z \in M$. Hence from factorization $M = H_1 A_2 \cdots A_{n-1}$ we have the factorization $M = A_1 A_2 \cdots A_{n-1}$. By the inductive assumption on the number of the factors it follows that one of the factors $A_1, \ldots, A_{n-1}$ is periodic. We need to consider only the case when $p \geq 3$.

By Lemma 1.4.2, A_n can be replaced by $A_n' = \{a_{|p} : a \in A_n\}$. Then choose $r \in A_n' \setminus \{e\}$. By Lemma 1.4.3, A_n' can be replaced by $C = \{e, r, r^2, \ldots, r^{p-1}\}$. Let $l \in C$. Since $G = M(l^{-1}C)$ is a factorization, $l^{-1}C$ is a complete set of representatives modulo M. There are elements $u_l, v_l, w_l \in l^{-1}C$ such that

$$(ax)^{-1} \in u_l M, \quad (by)^{-1} \in v_l M, \quad (cz)^{-1} \in w_l M.$$

Hence $axu_l, byv_l, czw_l \in M$. Let $D^{(l)} = \{e, axu_l, byv_l, czw_l\}$. Note that $M = D^{(l)} A_2 \cdots A_{n-1}$ is a factorization. By the inductive assumption one of the factors is periodic. If this is not $D^{(l)}$, then we are done. Assume that $D^{(l)}$ is periodic for each $l \in C$. At least one of the equations holds for each $l \in C$

$$xu_l = e, \quad yv_l = e, \quad zw_l = e.$$

If $xu_k = e$ and $yv_l = e$ for $k, l \in C$, then $x = r^i$ and $y = r^j$ for some i and j, $-p+1 \leq i, j \leq p-1$. If $i = 0$ or $j = 0$, then by Case 1, one of the factors $A_1, \ldots, A_n$ is periodic. If $i \neq 0$ and $j \neq 0$, then $y = x^s$, where p does not divide s. Now by Case 4 we are done. Thus we may assume that $xu_l = e$ for each $l \in C$. Therefore $u_l = x^{-1}$ is a fixed element of $l^{-1}C$ for each $l \in C$. Now

$$x^{-1} \in \bigcap_{l \in C} l^{-1}C.$$

By Lemma 1.2.1, C is periodic and so $C = \langle r \rangle$, that is $|r| = p$.

As $D^{(l)}$ is periodic for each $l \in C$, $D^{(e)}$ is periodic. We may assume that $xu_e = e$ and $yv_e = zw_e$ since this is only a matter of rearranging the elements of $D^{(e)}$. Now $v_e = r^i$ and $w_e = r^j$, where $1 \leq i, j \leq p-1$. If $i = j$, then $y = z$ and by Case 4 one of the factors $A_1, \ldots, A_n$ is periodic. If $i > j$, then let $s = i - j$. Then $yr^s = z$. As $xu_e = e$ and $u_e \in \langle r \rangle$, $A_1 = \{e, ax, by, cx^k y\}$ with a suitable integer k. Thus by Case 5 we are done.

This completes the proof.

Chapter 8

The Hajós property

8.1 Groups without Hajós 2-property

In Section 1.1 we mentioned that a finite abelian group G possesses the Hajós 2-property if and only if G is a subgroup of a group of type

$$
\begin{array}{cccc}
(p^{\alpha},q), & (p^2,q^2), & (p^2,q,r), & (p,q,r,s), \\
(p^3,2,2), & (p^2,2,2,2), & (p,2^2,2), & (p,2,2,2,2), \\
(p,q,2,2), & (p,3,3), & (3^2,3), & (2^{\alpha},2), \\
& (2^2,2^2), & (p,p). &
\end{array}
$$

Here p,q,r,s are distinct primes, the $p=2$ or $p=3$ cases are not excluded and $\alpha \geq 3$ is an integer.

In this section we prove one half of this statement. Namely, we show that if the finite abelian group G is a not a subgroup of a group of the above types, then G has a factorization $G = AB$, where neither A nor B is periodic. The argument we use will be constructive and elementary. There are overlaps with Sections 2.3, 6.3, and 7.3. This is deliberate. The constructions described in these sections serve multiple purposes and sometimes have to be more detailed than we need here.

Lemma 8.1.1. *Let H be a subgroup of the finite abelian group G such that $H \neq \{e\}$ and $H \neq G$. Assume that AB and AC are normalized factorizations of H such that A is not periodic, B and C are periodic but without common period. Then G does not possess the Hajós 2-property.*

Proof. Let L be a complete set of representatives modulo H in G. Choose L such that $e \in L$. Introduce the notation $L' = L \setminus \{e\}$. Now $G = HL$ is a normalized factorization of G. Using B, C and L' construct the subset D by $D = B \cup CL'$. We claim that $G = AD$ is a factorization of G. Indeed,

$$
AD = A(B \cup CL') = AB \cup ACL' = H \cup HL' = HL = G.
$$

We claim that D is not periodic. Assume now the contrary, that D is periodic and let $g \in G$ be a period of D. As $B \subset D$ and $B \subset H$, it follows that $gB \subset gD = D$ and $gB \subset gH = lH$ with a suitable $l \in L$. Thus $gB \subset lH \cap D$. Let us compute $lH \cap D$. If $l = e$, then $lH \cap D = H \cap D = B$. If $l \neq e$, that is, when $l \in L'$, then $lH \cap D = lC$. Hence $gB \subset B$ or $gB \subset lC$. From these and from $|B| = |gB| = |C| = |lC|$, it follows that $gB = B$ or $gB = lC$. In the second case the periods of B are also periods of C and so this case is not possible. Therefore g is a period of B.

We intend to show that g is a period of C as well. To do so consider the subset $l'C$, where $l' \in L'$. As $l'C \subset D$ and $l'C \subset l'H$, it follows that $gl'C \subset gD = D$ and $gl'C \subset gl'H$. Further $g \in B \subset H$ and so $gl'H = l'H$. Hence $gl'C \subset l'H \cap D = l'C$. Thus $gl'C = l'C$. Therefore g is a period of C.

This completes the proof.

Lemma 8.1.2. *Assume that the order of the finite abelian group G is composite and G is not of type $(2, 2)$. Then G admits a normalized factorization $G = AH$, where A is a nonperiodic subset of G and H is a proper subgroup of G.*

Proof. Since $|G|$ is composite, there is a subgroup K in G such that the index $|G : K| = p$ is a prime. If C is a complete set of representatives modulo K such that $e \in C$ and C is not periodic, then we are done with the choice $H = K$, $A = C$. Thus assume that in each time when C is a complete set of representatives modulo K, then C is periodic. In the $e \in C$ case C must be a subgroup of order p. Consequently each element of $C \setminus \{e\}$ is of order p. We show that each element of $G \setminus \{e\}$ is of order p. To do this let $k \in K$ and $c \in G \setminus K$. Clearly $kc \notin K$ and so

$$e = (kc)^p = k^p c^p = k^p.$$

Thus G is an elementary p-group. Therefore $G = K \times L$, where L is a subgroup of order p.

If $p \geq 3$, then the choice $H = K$ and

$$A = \{kl\} \cup \left(L \setminus \{l\}\right), \quad \text{where} \quad k \in K \setminus \{e\}, \quad l \in L \setminus \{e\} \tag{1}$$

is suitable.

If $p = 2$, then since G is not of type $(2, 2)$, G can be written in the form $G = K \times L$, where $|L| = 4$ and $|K| \geq 2$. Now again the choice (1) is satisfactory.

This completes the proof.

Lemma 8.1.3. *Assume that the finite abelian group G has a proper subgroup H of form $H = H_1 \times H_2$. In the following two cases G does not have the Hajós 2-property.*

    (i) $|H_1|$ and $|H_2|$ are composite numbers and none of H_1, H_2 is of type, $(2, 2)$.

    (ii) $|H_1| = |H_2| \geq 4$ and H_1, H_2 are cyclic groups.

Proof. By the hypotheses (i) and Lemma 8.1.2, there are normalized factorizations $H_1 = T_1 A_1$ and $H_2 = T_2 A_2$ such that T_1, T_2 are subgroups and A_1, A_2 are nonperiodic subsets of G. Let us introduce the $T_1' = T_1 \setminus \{e\}$ and $T_2' = T_2 \setminus \{e\}$ notations. Construct the subsets A, B, C in the following way

$$A = A_1 A_2, \qquad B = T_1 \cup (u_1 T_1) T_2', \qquad C = T_2 \cup (u_2 T_2) T_1',$$

where $u_1 \in H_1 \setminus T_1$ and $u_2 \in H_2 \setminus T_2$. It is clear that elements of T_1' are all the periods of B and similarly elements of T_2' are all the periods of C. Since $T_1' \cap T_2' = \emptyset$, by Lemma 8.1.1, it will be enough to verify that AB and AC are factorizations of H. But this latter holds.

$$
\begin{aligned}
AB &= A_1 A_2 (T_1 \cup u_1 T_1 T_2') \\
&= A_1 A_2 T_1 \cup A_1 A_2 u_1 T_1 T_2' \\
&= A_2 H_1 \cup A_2 H_1 u_1 T_2' \\
&= A_2 H_1 \cup A_2 H_1 T_2' \\
&= A_2 H_1 (\{e\} \cup T_2') \\
&= A_2 H_1 T_2 \\
&= H_1 H_2 \\
&= H.
\end{aligned}
$$

Similarly,

$$
\begin{aligned}
AC &= A_1 A_2 (T_2 \cup u_2 T_2 T_1') \\
&= A_1 A_2 T_2 \cup A_1 A_2 u_2 T_2 T_1' \\
&= A_1 H_2 \cup A_1 H_2 u_2 T_1' \\
&= A_1 H_2 \cup A_1 H_2 T_1' \\
&= A_1 H_2 (\{e\} \cup T_1') \\
&= A_1 H_2 T_1 \\
&= H_1 H_2 \\
&= H.
\end{aligned}
$$

In the case (ii) let $H_1 = \langle g \rangle$ and $H_2 = \langle h \rangle$, $|g| = |h| = u$. Construct the subsets A, B, C in the following way:

$$A = \{e, gh, (gh)^2, \ldots, (gh)^{u-3}, g^{u-2} h^{u-1}, g^{u-1} h^{u-2}\},$$

$$B = H_1, \quad C = H_2.$$

As A is not periodic it will be enough to show that AB and AC are factorizations of H. But this can be verified easily.

This completes the proof.

Lemma 8.1.4. *Assume that the finite abelian group G has a proper subgroup K which itself has a proper subgroup H. In the following two cases G does not possess the Hajós 2-property.*

(i) H is type $(2, 2, 2, 2)$ or (p, p), where p is an odd prime.

(ii) $H = H_1 \times H_2$, where $|H_1|$ is composite and H_2 is of type $(2, 2)$.

type of G	type of H_1	type of H_2	remark
(p^α, q, t)	$(p^{\alpha-1})$	(q, t)	$\alpha \geq 3$
(p^2, q^2, t)	(p^2)	(q^2)	
(p^2, q, r, t)	(p^2)	(q, r)	
(p, q, r, s, t)	(p, q)	(r, s)	
$(p, 2^2, 2, t)$	$(p, 2)$	(2^2)	$p \geq 3$
$(p, q, 2, 2, t)$	$(p, 2)$	$(q, 2)$	$p \geq 3$
$(2^\alpha, 2, t)$	$(2^{\alpha-1})$	$(2, t)$	$\alpha \geq 4, \quad t \geq 3$
$(2^2, 2^2, t)$	(2^2)	(2^2)	

Table 1

type of G	type of H_1	type of H_2	remark
(p, p, t)	(p)	(p)	$p \geq 3$

Table 2

type of G	type of H
$(p, 2, 2, 2, 2, t)$	$(2, 2, 2, 2)$
$(p, 3, 3, t)$	$(3, 3)$
$(3^2, 3, t)$	$(3, 3)$

Table 3

type of G	type of H_1	type of H_2	remark
$(p^3, 2, 2, t)$	(p^2)	$(2, 2)$	
$(p^2, 2, 2, 2, t)$	(p^2)	$(2, 2)$	
$(2^\alpha, 2, t)$	$(2^{\alpha-2})$	$(2, t)$	$\alpha \geq 4, \quad t = 2$

Table 4

Proof. In the case (i) there are normalized factorizations in the form $G = KB$ and $K = HA$. Hence $G = HAB$ is also a factorization. Let us introduce the $B' = B \setminus \{e\}$ and $A' = A \setminus \{e\}$ notations. Note that there are subgroups L_1, L_2, L_3, L_4 of H such that

$$H = L_i \times L_j, \quad \text{for each} \quad 1 \le i < j \le 4$$

and

$$|L_1| = |L_2| = |L_3| = |L_4|.$$

In the way we saw in the proof of Lemma 8.1.1 it can be shown that the subsets

$$D = L_1 \cup L_2 A', \qquad M = L_3 \cup L_4 B'$$

are not periodic. Next we show that $G = DM$ is a factorization of G. Indeed

$$\begin{aligned}
DM &= (L_1 \cup L_2 A')(L_3 \cup L_4 B') \\
&= L_1 L_3 \cup L_1 L_4 B' \cup L_2 L_3 A' \cup L_2 L_4 A' B' \\
&= H \cup HB' \cup HA' \cup HA'B' \\
&= H(\{e\} \cup B' \cup A' \cup A'B') \\
&= HAB \\
&= G.
\end{aligned}$$

Turn to the case (ii). The case when H_1 is of type $(2,2)$ is settled by the previous argument. We assume that H_1 is not of type $(2,2)$. By Lemma 8.1.2, there is a normalized factorization in the form $H_1 = TB_1$ such that T is a proper subgroup of H_1 and B_1 is a nonperiodic subset of G. There is a normalized factorization in form $K = HA_1$ as well. Let $H_2 = \langle g \rangle \times \langle h \rangle$. (Remember $|g| = |h| = 2$.) Let us introduce the

$$A_1' = A_1 \setminus \{e\}, \quad B_1' = B_1 \setminus \{e\}, \quad T' = T \setminus \{e\}$$

notations. Then define the subsets A, B, C by

$$A = \{e, b_1'gh\}T \cup \big(\{e, gh\} \cup \{g, h\}T'\big)A_1',$$

$$B = \{e, g\}B_1, \quad C = \{e, h\}B_1,$$

where $b_1' \in B_1'$. The only period of B is g and the only period of C is h. So B and C are without common period. The elements of T' are all the periods of $\{e, b_1'gh\}T$ and gh is the only period of $\{e, gh\} \cup \{g, h\}T'$. Thus these periodic subsets are also without common periods. In the way we saw in the proof of Lemma 8.1.1 one can prove that the subset A is not periodic. By Lemma 8.1.1, it will be enough to show that AB and AC are factorizations of K. Indeed,

$$\begin{aligned}
& AB \\
&= \{e, b_1'gh\}T\{e, g\}B_1 \cup \{e, gh\}A_1'\{e, g\}B_1 \cup \{g, h\}T'A_1'\{e, g\}B_1 \\
&= H_1\{e, g, b_1'gh, b_1'h\} \cup H_2 A_1' B_1 \cup H_2 T' A_1' B_1
\end{aligned}$$

$$
\begin{aligned}
&= & H_1 H_2 \cup H_2 A_1' B_1\big(\{e\} \cup T'\big) \\
&= & H_1 H_2 \cup H_2 A_1' B_1 T \\
&= & H_1 H_2 \cup H_1 H_2 A_1' \\
&= & H\big(\{e\} \cup A_1'\big) \\
&= & H A_1 \\
&= & K.
\end{aligned}
$$

Similarly,

$$
\begin{aligned}
& & AC \\
&= & \{e, b_1' gh\} T \{e, h\} B_1 \cup \{e, gh\} A_1' \{e, h\} B_1 \cup \{g, h\} T' A_1' \{e, h\} B_1 \\
&= & H_1 \{e, h, b_1' gh, b_1' g\} \cup H_1 A_1' B_1 \cup H_1 T' A_1' B_1 \\
&= & H_1 H_2 \cup H_1 A_1' B_1 \big(\{e\} \cup T'\big) \\
&= & H_1 H_2 \cup H_2 A_1' B_1 T \\
&= & H_1 H_2 \cup H_1 H_2 A_1' \\
&= & H\big(\{e\} \cup A_1'\big) \\
&= & H A_1 \\
&= & K.
\end{aligned}
$$

This completes the proof.

Lemma 8.1.5. *If H is a subgroup of the finite abelian group and H does not possess the Hajós 2-property, then neither does G.*

Proof. In the $H = G$ case there is nothing to prove. So we proceed by induction on the number of the prime factors of $|G : H|$.

First consider the case when $|G : H| = p$ is a prime. Since H does not possess the Hajós 2-property, there is a normalized factorization $H = AB$, where neither A nor B is periodic. Let C be a complete set of representatives of G modulo H. Now $G = HC$ is a factorization of G. We can choose C such that this factorization is normalized. Let $D = AC$. Then $DB = ACB = ABC = HC = G$. We show that if C is not periodic, then neither is D. To do this assume the contrary, that D is periodic with period g. Let $c \in C$. As $cA \subset D$ and $cA \subset cH$, it follows that $gcA \subset gD = D$ and $gcA \subset gcH = c'H$ with a suitable $c' \in C$. Hence $gcA \subset \big(c'H \cap D\big) = c'A$. Therefore $gcA = c'A$. The subset A is not periodic and so it follows that $gc = c'$. This means that $gC \subset C$ and consequently $gC = C$. The subset C is not periodic and so we get $g = e$ which is a contradiction.

As a next step we show that C can be chosen such that C is not periodic. If there is a complete set of representatives modulo H which is not a subgroup of G, then we are done. So we assume that each complete set of representatives modulo H that contains the identity element e is a subgroup of order p. In the way we saw in the proof of Lemma 8.1.2, it can be concluded that G is an elementary p-group.

If $p \geq 3$, then a subset defined in (1) can play the role of C. Thus assume that $p = 2$. Note that groups of type

$$(2), \ (2, 2), \ (2, 2, 2), \ (2, 2, 2, 2), \ (2, 2, 2, 2, 2)$$

have the Hajós 2-property. By Lemma 8.1.4 case (i) groups of type

$$(2,2,2,2,2,2), \quad (2,2,2,2,2,2,2), \ldots$$

do not possess the Hajós 2-property. By the hypotheses H is a group without Hajós 2-property and so $|H| \geq 2^6$. Hence $|G| \geq 2^7$ and therefore G is a group without Hajós 2-property.

Secondly assume that $|G : H|$ has n prime factors and consider a chain of subgroups

$$H = H_1 \subset H_2 \subset \cdots \subset H_n = G,$$

where the indices

$$|H_2 : H_1|, \ |H_3 : H_2|, \ldots, |H_n : H_{n-1}|$$

are primes. By the first part, H_2 does not possess the Hajós 2-property. Then H_3 is without Hajós 2-property. Continuing in this way finally we get that $H_n = G$ does not possess the Hajós 2-property.

This completes the proof.

Using Lemmas 8.1.1–8.1.5 we can prove the main result of the section in the following way.

Note that there are repetitions on the enclosed list of group types. In the case of groups of type $(p, 2^2, 2)$ and $(p, q, 2, 2)$ we may assume that $p \geq 3$ since otherwise these are subgroups of a group of type $(p^2, 2, 2)$.

In the case of groups of type $(2^\alpha, 2)$ we may assume that $\alpha \geq 4$ because of groups of type $(p^3, 2, 2)$.

Finally, in the case of groups of type (p, p) it may be assumed that $p \geq 5$ since the groups of type $(2, 2)$ and $(3, 3)$ are subgroups of groups of types $(p^3, 2, 2)$ and $(p, 3, 3)$ respectively.

Let $t \geq 3$ be an integer and apply Lemma 8.1.3 case (i) with the choice listed in Table 1. Apply Lemma 8.1.3 case (ii) with the choice in Table 2. Apply Lemma 8.1.4 case (i) with the choice indicated by Table 3. Finally apply Lemma 8.1.4 case (ii) with the choice of Table 4.

This completes the proof.

8.2 Groups with the Hajós 2-property

This section continues the theme of the previous section. We now show that groups on the list of Section 8.1 do indeed possess the Hajós 2-property. We deal with these groups in a case by case manner.

The cyclic groups on the list are of types (p^α, q), (p^2, q^2), (p^2, q, r), (p, q, r, s). If $G = AB$ is a factorization of G whose type is (p^α, q), then $|A|$ or $|B|$ must be a prime power. By Theorem 6.1.1, A or B is periodic. If $G = AB$ is a factorization of G, where $|G|$ is a product of four primes, then either one $|A|$ and $|B|$ is a prime or both $|A|$ and $|B|$ are product of two primes. By Theorem 6.2.1, A or B is periodic.

Lemma 8.2.1. *If $G = AB$ is a factorization of the finite abelian group G and $|A| = 2$ or 3, then A or B is periodic.*

Proof. We may assume that the factorization $G = AB$ is normalized. Let $A = \{e, a\}$. By Lemma 2.1.1, A or B is periodic. Let $A = \{e, a, b\}$. Now there is a $c \in G$ such that $b = a^2 c$, that is, $A = \{e, a, a^2 c\}$. By Lemma 2.1.1, A or B is periodic.

   This completes the proof.

   A group of type $(2^2, 2^2, 2)$ has factorizations $G = AB$ with $|A| = 4$ and $|B| = 8$ and neither A nor B is periodic. So Lemma 8.2.1 does not extend to the case $|A| > 3$. However in the case $|A| = 4$ one additional condition suffices.

Lemma 8.2.2. *Let $G = AB$ be a factorization of the finite abelian group G, where $|A| = 4$ and let two elements of A have a common square. Then either A or B is periodic.*

Proof. Let $A = \{u, v, w, x\}$ with $u^2 = v^2$. Then uAB and vAB are also factorizations of G, that is,

$$u^2 B \cup uvB \cup uwB \cup uxB,$$
$$vuB \cup v^2 B \cup vwB \cup vxB$$

are partitions of G. Since $uB \cap vB = \emptyset$ we have $uwB = vxB$ and $uxB = vwB$. Hence B is periodic or $uw = vx$ and $ux = vw$. This implies that $v^{-1}u$ is a period of A.

   This completes the proof.

   Lemmas 8.2.1 and 8.2.2 are sufficient to cover the cases of the groups of types $(2, 2, 2, 2, 2)$, $(2^2, 2, 2, 2)$, $(3^2, 3)$ and their subgroups.

Theorem 8.2.1. *The group of type $(2^2, 2^2)$ has the Hajós 2-property.*

Proof. Let $G = AB$ be a factorization of the group G of type $(2^2, 2^2)$. We may assume that the factorization is normalized. The cases $|A| = 2$ and $|B| = 2$ are covered by Lemma 8.2.1. So we may assume that $|A| = |B| = 4$. Set $H = \langle A \rangle$. As $A \subset H$ and $H \subset G$, it follows that $4 \leq |H| \leq 16$. We distinguish three cases depending on whether $|H| = 4$ or $|H| = 8$ or $|H| = 16$. In the $|H| = 4$ case $A = H$ and so A is clearly periodic. If $|H| = 8$, then restricting the factorization $G = AB$ to H gives the $H = G \cap H = A(B \cap H)$ normalized factorization of H. From $|H| = 8$ and $|A| = 4$, it follows that $|B \cap H| = 2$. Let $B \cap H = \{e, b\}$. If $b^2 \neq e$, then from the factorization $H = A\{e, b\}$ it follows that A is periodic with period b^2. If $b^2 = e$, then B has an element of order 2. Now Lemma 8.2.2 is applicable to factorization $G = AB$ and gives that A or B is periodic. In the $|H| = 16$ case $\langle A \rangle = G$ and Theorem 7.2.7 is applicable with the $\lambda = 2$ choice. This gives that B is periodic.

   The proof is complete.

Lemma 8.2.3. *Let $G = AB$ be a normalized factorization of the finite abelian group G, where $|A| = 4$. If G is of the following types:*

$$\text{(i)} \quad (p^\alpha, 2, \ldots, 2), \qquad \text{(ii)} \quad (p, q, 2, 2), \qquad \text{(iii)} \quad (p, 4, 2),$$

where p and q are distinct primes ($p = 2$ is included), then either A or B is periodic.

Proof. Clearly G is a direct product $H \times K$, where K is an elementary 2-group and H is of type (p^α), (p, q), $(p, 4)$ in the cases (i), (ii), (iii) respectively. In case (iii) we may assume that $p \geq 3$ since $p = 2$ is already covered by case (i). Thus we may assume that H is cyclic in each case. Consider a character χ of G which is faithful on H. If $\chi(A) = 0$, then by Lemma 3.1.7 it follows that $\chi(a) = -1$ for some $a \in A$. Now $|a| = 2$. Hence e and a have a common square and so either A or B is periodic. Thus we may assume that for each character χ of G which is faithful on H $\chi(A) \neq 0$, that is, $\chi(B) = 0$.

In Case (i) B is periodic.

In Case (ii) H has unique subgroups L and M of order p and q respectively. In Case (iii) H has unique subgroups L and M of order p and 2 respectively. In both cases, by Lemma 3.2.7 there are $X, Y \subset G$ such that $B = LX \cup MY$, where the products are direct and the union is disjoint. The cardinalities give that either $X = \emptyset$ or $Y = \emptyset$, that is, B is periodic.

This completes the proof.

By Theorem 6.1.1, if the p-component of the group is cyclic and one of the factors has cardinality a power of p, then one of the factors is periodic. This result covers many of the cases in the groups remaining. For instance groups of type $(p, 3, 3)$ have the Hajós 2-property. The cases $p = 2$ and $p = 3$ covered by Lemma 8.2.1. The case $p > 3$ is covered by Lemma 8.2.1 and the above remark.

Groups of type $(p, 2^2, 2)$ have the Hajós 2-property. To verify this claim let $G = AB$ be a factorization of a group G of type $(p, 2^2, 2)$. If $|A| = p$ or $|A| = 2$, then by the previous result, A or B is periodic. If $|A| = 4$, $|B| = 2p$, then by Lemma 8.2.3 either A or B is periodic.

Theorem 8.2.2. *Groups of type $(p, 2, 2, 2, 2)$, where p is a prime have the Hajós 2-property.*

Proof. Let $G = AB$ be a normalized factorization of the group G of type $(p, 2, 2, 2, 2)$. The cases $|A| = p$, $|A| = 2$, $|A| = 4$ are covered by previous results. So we need only to consider the case $|A| = 2p$, $|B| = 8$. Let a be an element of order p and let H be the subgroup of order 16. Let ρ be a primitive pth root of unity. Consider a character χ of G with $\chi(a) = \rho$ and $\chi(h) = 1$ for all $h \in H$. Then, as p does not divide 8, $\chi(B) = 0$ is not possible. Therefore $\chi(A) = 0$. Let

$$A = \{a^{\alpha(i)} h_i : 1 \leq i \leq 2p\},$$

where $0 \le \alpha(i) \le p - 1$ and $h_i \in H$. Then

$$0 = \chi(A) = \sum_{i=0}^{2p-1} \rho^{\alpha(i)}$$

implies that the integers $\alpha(i)$ are the numbers $0, 1, \ldots, p - 1$, each occurring with multiplicity 2. Thus we may represent A in the form

$$A = \{a^i h_i, a^i h'_i : 0 \le i \le p - 1\},$$

where $h_i, h'_i \in H$; further $h_0 = e$ since A is normalized and $h_i \ne h'_i$.

Let χ be a character G for which $\chi(a) = \rho$. Now

$$\chi(A) = \sum_{i=0}^{p-1} \rho^i \big(\chi(h_i) + \chi(h'_i)\big). \tag{1}$$

If $\chi(A) = 0$, then since the pth cyclotomic polynomial divides the polynomial

$$\sum_{i=0}^{p-1} x^i \big(\chi(h_i) + \chi(h'_i)\big)$$

associated with (1), it follows that

$$\chi(h_0) + \chi(h'_0) = \chi(h_1) + \chi(h'_1) = \cdots = \chi(h_{p-1}) + \chi(h'_{p-1}). \tag{2}$$

If

$$h_0^{-1} h'_0 = h_1^{-1} h'_1 = \cdots = h_{p-1}^{-1} h'_{p-1} = c,$$

then

$$h'_0 = h_0 c, \, h'_1 = h_1 c, \ldots, h'_{p-1} = h_{p-1} c$$

and so c is a period of A. Thus we may assume that $h_i^{-1} h'_i \ne h_j^{-1} h'_j$ for some i and j. Now $c = h_i^{-1} h'_i h_j (h'_j)^{-1}$ is not the identity element. We show that $\chi(c) = 1$. To do so we distinguish two cases. First assume that $\chi(h'_0) = 1$. Then from (2) we get that $\chi(h_i) = \chi(h'_i) = 1$, that is, $\chi(h_i^{-1} h'_i) = 1$ for each i, $0 \le i \le p - 1$. Therefore $\chi(c) = 1$. Secondly assume that $\chi(h'_0) = -1$. From (2) we get that $\chi(h_i) + \chi(h'_i) = 0$, that is, $\chi(h_i^{-1} h'_i) = -1$ for each i, $0 \le i \le p - 1$. Therefore again $\chi(c) = 1$.

Summing up our consideration we may say that $\chi(A) \ne 0$ and so $\chi(B) = 0$ for each character χ of G with $\chi(a) \ne 1$ and $\chi(c) \ne 1$. By Lemma 3.2.7, there are $X, Y \subset G$ such that $B = X\langle a \rangle \cup Y \langle c \rangle$, where the products are direct and the union is disjoint. For the cardinalities this gives $8 = xp + y2$, where $x = |X|$ and $y = |Y|$ are nonnegative integers. As x is an even number we get $4 = (x/2)p + y$. This implies $x = 0$ when $p \ge 5$. Therefore in this case c is a period of B.

In the remaining part we assume that $p = 3$, that is, G is of type $(3, 2, 2, 2, 2)$, $G = AB$ is a factorization of G, where $|A| = 2 \cdot 3 = 6$ and $|B| = 8$. We know that $B = X\langle a \rangle \cup Y\langle c \rangle$. The only case we need to consider is when $|X| = 2$ and $|Y| = 1$. We assume that $Y = \{e\}$ and $X = \{a^i k, a^j l\}$, $k, l \in H$, $0 \le i, j \le 2$. Now

$$B = \{ek, ak, a^2 k, el, al, al^2, e, c\}$$

and so $\langle B \rangle = \langle a \rangle \times \langle k, l, c \rangle$. Set $M = \langle B \rangle$. Let $d \in A$. Multiplying $G = AB$ by d^{-1} we get $G = (Ad^{-1})B$. Restricting this factorization to M we get the factorization $M = (Ad^{-1} \cap M)B$ of M. As $|B| = 8$ divides $|M|$, it follows that k, l, c are independent elements and so $|M| = 3 \cdot 8 = 24$. By Lemma 8.2.1, either $Ad^{-1} \cap M$ or B is periodic. We are done if B is periodic. We assume that $Ad^{-1} \cap M$ is periodic. This means that $Ad^{-1} \cap M = \langle a \rangle$ and so

$$\langle a \rangle \subset \bigcap_{d \in A} Ad^{-1}.$$

Therefore A is periodic.

This completes the proof.

Theorem 8.2.3. *Groups of type $(p, q, 2, 2)$, where p and q are odd distinct primes have the Hajós 2-property.*

Proof. Let $G = AB$ be a normalized factorization of the group G of type $(p, q, 2, 2)$. Here we assume that $p > q \ge 3$. If $|A|$ or $|B|$ is one of $2, 4, p, q$, then either A or B is periodic. So we need only consider the case $|A| = 2p$ and $|B| = 2q$. Let a and b be elements of order p and q respectively and let L be the subgroup of type $(2, 2)$. Clearly $G = \langle a \rangle \times \langle b \rangle \times L$. The factor A can be represented in the form

$$A = \{a^{\alpha(i)} h_i : 1 \le i \le 2p\},$$

where $0 \le \alpha(i) \le p - 1$, $h_i \in \langle b \rangle \times L$, $\alpha(0) = 0$, $h_0 = e$. Let ρ and σ be pth and qth primitive roots of unity.

Consider a character χ of G for which $\chi(a) = \rho$, $\chi(b) = 1$ and $\chi(l) = 1$ for each $l \in L$. If $\chi(B) = 0$, then it is a vanishing sum of pth roots of unity. By Lemma 3.1.1, this leads to the contradiction that p divides $2q$. Therefore $\chi(B) \ne 0$ and consequently $\chi(A) = 0$. Now

$$0 = \chi(A) = \sum_{i=0}^{2p-1} \rho^{\alpha(i)}$$

implies that the integers $\alpha(i)$ are $0, 1, \ldots, p - 1$ and each occurs with multiplicity 2. So A can be represented in the form

$$A = \{a^i h_i, a^i h_i' : 0 \le i \le p - 1\},$$

where $h_i, h_i' \in \langle b \rangle \times L$, $h_i \ne h_i'$ and $h_0 = e$.

Consider a character χ of G with $\chi(a) = \rho$ and $\chi(b) = \sigma$. If $\chi(B) = 0$ for each such character, then by Lemma 3.2.7, there are $X, Y \subset G$ such that $B = X\langle a \rangle \cup Y\langle b \rangle$, where the products are direct and the union is disjoint. For the cardinalities this gives $2q = xp + yq$, where $x = |X|$ and $y = |Y|$ are nonnegative integers. Clearly q divides x and so $2 = (x/q)p + y$. As $p \geq 5$ from this it follows that $x = 0$, that is, b is a period of B. Thus we may assume that $\chi(B) \neq 0$, consequently $\chi(A) = 0$ for at least one such character χ of G. Now

$$0 = \chi(A) = \sum_{i=0}^{p-1} \rho^i \big(\chi(h_i) + \chi(h_i') \big). \tag{3}$$

Since the pth cyclotomic polynomial divides the polynomial

$$0 = \sum_{i=0}^{p-1} x^i \big(\chi(h_i) + \chi(h_i') \big)$$

associated with (3) we get that

$$\chi(h_0) + \chi(h_0') = \chi(h_1) + \chi(h_1') = \cdots = \chi(h_{p-1}) + \chi(h_{p-1}'). \tag{4}$$

We distinguish three cases (i) $\chi(h_0') = 1$ (ii) $\chi(h_0') = -1$ (iii) $\chi(h_0')$ is either a $(2q)$th or a qth primitive root of unity. In case (i) $\chi(h_i) = \chi(h_i') = 1$ and so $h_i, h_i' \in L$ for each i, $0 \leq i \leq p-1$. In case (ii) $\chi(h_i) + \chi(h_i') = 0$ and so $h_i = d_i l_i$, $h_i' = d_i l_i'$, $d_i \in \langle b \rangle$, $l_i, l_i' \in L$ for each i, $0 \leq i \leq p-1$. In case (iii)

$$\chi(h_i) = 1 \qquad \text{and} \qquad \chi(h_i') = \chi(h_0') \tag{5}$$

or

$$\chi(h_i) = \chi(h_0') \qquad \text{and} \qquad \chi(h_i') = 1$$

for each i, $0 \leq i \leq p-1$. We may choose the notation such that always (5) holds. Hence $h_i = l_i$, $h_i' = dl_i'$, where $l_i, l_i' \in L$ and $d \in \langle b \rangle$. Summarizing our argument we may say that A can be represented in the forms

$$A = \{ a^i d_i l_i, a^i d_i l_i' : 0 \leq i \leq p-1 \},$$

or

$$A = \{ a^i l_i, a^i dl_i' : 0 \leq i \leq p-1 \}$$

(in case (i) $d = e$).

Consider a character χ of G for which $\chi(a) = \rho$. If $\chi(A) = 0$, then as before it follows (4). If

$$l_0^{-1} l_0' = l_1^{-1} l_1' = \cdots = l_{p-1}^{-1} l_{p-1}' = c,$$

then

$$l_0' = l_0 c, l_1' = l_1 c, \ldots, l_{p-1}' = l_{p-1} c$$

and so c and cd is a period of A respectively. Thus we may assume that $l_i^{-1}l_i' \neq l_j^{-1}l_j'$ for some i and j. Now $c = l_i^{-1}l_i'l_j(l_j')^{-1}$ is not the identity element. We show that $\chi(c) = 1$. Indeed in case (i) $\chi(l_i) = \chi(l_i')$; in case (ii) $\chi(l_i^{-1}l_i') = 1$; in case (iii) $\chi(l_i) = 1$ and $\chi(l_i') = \chi(l_0')$ for each i, $0 \leq i \leq p-1$. Therefore in each case $\chi(c) = 1$.

Summing up our consideration we may say that $\chi(A) \neq 0$ and so $\chi(B) = 0$ for each character χ of G with $\chi(a) \neq 1$ and $\chi(c) \neq 1$. By Lemma 3.2.7, there are $X, Y \subset G$ such that $B = X\langle a \rangle \cup Y\langle c \rangle$, where the products are direct and the union is disjoint. For the cardinalities this gives $2q = xp + y2$, where $x = |X|$ and $y = |Y|$ are nonnegative integers. As x is an even number we get $q = (x/2)p + y$. As $p > q$ this implies $x = 0$. Therefore c is a period of B.

This completes the proof.

Lemma 8.2.4. *Let p be a prime and let H be a subgroup of type (p, p) of a finite abelian group G. Let $L_0, \ldots, L_p$ be all the maximal subgroups of H. If $G = L_i A$ is a factorization of G for each i, $0 \leq i \leq p-1$, then the elements of $L_p \setminus \{e\}$ are periods of A.*

Proof. Let x, y be basis elements of H and let $L_p = \langle y \rangle$, $L_i = \langle xy^i \rangle$, $0 \leq i \leq p-1$. By Lemma 3.2.6, it is enough to show that $\chi(A) = 0$ for each character χ of G with $\chi(y) \neq 1$. Choose a character χ of G for which $\chi(y) = \rho$, where ρ is a primitive pth root of unity. Clearly, χ is not the principal character of G. Now $\chi(x) = \rho^j$ for some j, $0 \leq j \leq p-1$. There is a k such that $0 \leq k \leq p-1$ and $\chi(xy^k) = 1$. So $\chi(\langle xy^k \rangle) = p$. Therefore from $0 = \chi(G) = \chi(\langle xy^k \rangle)\chi(A)$ it follows that $\chi(A) = 0$.

This completes the proof.

Theorem 8.2.4. *Let p be an odd prime. Groups of type $(p^3, 2, 2)$ have the Hajós 2-property.*

Proof. Let G be a group of type $(p^3, 2, 2)$. Clearly G is a direct product of its subgroups $\langle a \rangle$ and H, where $|a| = p^3$ and H is of type $(2, 2)$. Consider the factorization $G = AB$. We would like to show that A or B is periodic. Only the $|A| = 2p$, $|B| = 2p^2$ case is not covered by earlier results. Let

$$A = \{a^{\alpha(i)}h_i : 1 \leq i \leq 2p\},$$

$$B = \{a^{\beta(j)}k_j : 1 \leq j \leq 2p^2\},$$

where $0 \leq \alpha(i), \beta(j) \leq p^3 - 1$, $h_i k_j \in H$. If the $\alpha(i)$ numbers are distinct, then we can use the argument in the proof of Theorem 5.4.1 to show that A or B is periodic. Thus we may assume that the $\alpha(i)$ numbers are not distinct. Similarly if the $\beta(j)$ numbers are distinct, then we can conclude that A or B is periodic. Thus we may assume that the $\beta(j)$ numbers are not distinct. Choose a primitive (p^3)th root of unity ρ and consider the characters χ_1, χ_2, χ_3 defined by

$$\chi_1(a) = \rho, \quad \chi_1(h) = 1 \quad \text{for each} \quad h \in H,$$

$$\chi_2(a) = \rho^p, \quad \chi_2(h) = 1 \ \text{ for each } \ h \in H,$$

$$\chi_3(a) = \rho^{p^2}, \quad \chi_3(h) = 1 \ \text{ for each } \ h \in H.$$

Each of χ_1, χ_2, χ_3 must annihilate A or B. We claim that exactly one of χ_1, χ_2, χ_3 annihilates A and the remaining two annihilate B. In order to prove the claim assume the contrary say, $\chi_1(B) = \chi_2(B) = \chi_3(B) = 0$. Using cyclotomic polynomials we get the contradiction $p^3 \mid |B| = 2p^2$. Hence one of χ_1, χ_2, χ_3 cannot annihilate B and so annihilates A. If two of χ_1, χ_2, χ_3 annihilate A, then we get the contradiction that $p^2 \mid |A| = 2p$.

Let A', B' be the multisets consisting of the p-parts of elements of A and B respectively. The product $A'B'$ is a 4-fold factorization of $\langle a \rangle$. It follows that each $\alpha(i)$ occurs with the same multiplicity. Then this multiplicity must be 2. Each $\beta(j)$ occurs with the same multiplicity and this common multiplicity must be 2. We may represent A and B in the forms

$$A = \{a^{\alpha(i)} h_i, a^{\alpha(i)} h_i' : 1 \le i \le p\},$$

$$B = \{a^{\beta(j)} k_j, a^{\beta(j)} k_j' : 1 \le j \le p^2\}.$$

(Here we reused $\alpha(i)$ and $\beta(j)$.) Set

$$A_i = A(a^{\alpha(i)} h_i)^{-1},$$

$$B_j = B(a^{\beta(j)} k_j)^{-1}.$$

From the factorization $G = A_i B_j$ and

$$e, h_i^{-1} h_i' \in A_i, \quad e, k_j^{-1} k_j' \in B_j$$

it follows that the product $\{e, h_i^{-1} h_i'\}\{e, k_j^{-1} k_j'\}$ is direct and is equal to H. If

$$h_1^{-1} h_1' = \cdots = h_p^{-1} h_p' = d,$$

then

$$h_1' = h_1 d, \ldots, h_p' = h_p d$$

and so A is periodic with period d. We may assume that $h_r^{-1} h_r' \ne h_s^{-1} h_s'$ for some r, s. Therefore both

$$\langle h_r^{-1} h_r' \rangle \langle k_j^{-1} k_j' \rangle \quad \text{and} \quad \langle h_s^{-1} h_s' \rangle \langle k_j^{-1} k_j' \rangle$$

are factorizations of H. Lemma 8.2.4 with the $p = 2$ choice implies $k_j^{-1} k_j' = (h_r^{-1} h_r')(h_s^{-1} h_s') = d$. This holds for each j, $1 \le j \le p^2$. Therefore B is periodic with period d.

This completes the proof.

Theorem 8.2.5. *Let p be an odd prime. Groups of type $(p^2, 2, 2, 2)$ have the Hajós 2-property.*

Proof. The proof will be similar to the previous proof. Let G be a group of type $(p^2, 2, 2, 2)$. Clearly G is a direct product of its subgroups $\langle a \rangle$ and H, where $|a| = p^2$ and H is of type $(2, 2, 2)$. Consider the factorization $G = AB$. We would like to show that A or B is periodic. Only the $|A| = 2p$, $|B| = 4p$ case is not covered by earlier results. Let

$$A = \{a^{\alpha(i)} h_i : 1 \le i \le 2p\},$$

$$B = \{a^{\beta(j)} k_j : 1 \le j \le 4p\},$$

where $0 \le \alpha(i), \beta(j) \le p^2 - 1$, $h_i k_j \in H$. We may assume that the $\alpha(i)$ numbers are not distinct and the $\beta(j)$ numbers are not distinct. Choose a primitive (p^2)th root of unity ρ and consider the characters χ_1, χ_2 defined by

$$\chi_1(a) = \rho, \quad \chi_1(h) = 1 \quad \text{for each} \quad h \in H,$$

$$\chi_2(a) = \rho^p, \quad \chi_2(h) = 1 \quad \text{for each} \quad h \in H.$$

We claim that exactly one of χ_1, χ_2 annihilates A and the other annihilates B.

Let A', B' be the multisets consisting of the p-parts of elements of A and B respectively. The product $A'B'$ is an 8-fold factorization of $\langle a \rangle$. It follows that each $\alpha(i)$ occurs with the same multiplicity. Then this multiplicity must be 2. Each $\beta(j)$ occurs with the same multiplicity and this common multiplicity must be 4. We may represent A and B in the forms

$$A = \{a^{\alpha(i)} h_i, a^{\alpha(i)} h_i' : 1 \le i \le p\},$$

$$B = \{a^{\beta(i)} k_{i,j}, a^{\beta(i)} k_{i,j}' : 1 \le i \le p, 1 \le j \le 4\}.$$

If

$$h_1^{-1} h_1' = \cdots = h_p^{-1} h_p' = d,$$

then

$$h_1' = h_1 d, \ldots, h_p' = h_p d$$

and so A is periodic with period d. We may assume that $h_r^{-1} h_r' \ne h_s^{-1} h_s'$ for some r, s. Set

$$A_i = A(a^{\alpha(i)} h_i)^{-1}, \quad B_j = B(a^{\beta(j)} k_j)^{-1},$$

$$C_i = \{e, h_i^{-1} h_i'\}, \quad D_j = \{e, k_{j,1}^{-1} k_{j,2}', k_{j,1}^{-1} k_{j,3}', k_{j,1}^{-1} k_{j,4}'\}.$$

From the factorization $G = A_i B_j$ we get that $C_r D_j = H$, $C_s D_j = H$ are factorizations of H. By the $p = 2$ case of Lemma 8.2.4, $d = h_r^{-1} h_r' (h_s^{-1} h_s')$ is a period of D_j. This holds for each j, $1 \le j \le p$. Therefore B is periodic with period d.

This completes the proof.

With Theorem 8.2.5 we completed our survey of groups with the Hajós 2-property. In the remaining part we will consider a group with the Hajós 2-property factored into many factors.

Lemma 8.2.5. *Let $q_1 = p_1 \cdots p_s$, where $p_1, \ldots, p_s$ are primes. If $(q_1, \ldots, q_n)$ is a periodicity forcing factorization type for an abelian group, then so is $(p_1, \ldots, p_s, q_2, \ldots, q_n)$.*

Proof. Suppose that $(q_1, \ldots, q_n)$ is a periodicity forcing factorization type for the finite abelian group G and consider a normalized factorization

$$G = B_1 \cdots B_s A_2 \cdots A_n, \tag{6}$$

where

$$|B_1| = p_1, \ldots, |B_s| = p_s, |A_2| = q_2, \ldots, |A_n| = q_n.$$

We would like to show that at least one of the factors $B_1, \ldots, B_s$, $A_2, \ldots, A_n$ is periodic. If one of the factors is periodic then there is nothing to prove so we assume that none of the factors is periodic.

If the order of each element in B_i is a power of p_i, then we define C_i to be B_i. Assume that there are elements in B_i whose order is not a power of p_i. In factorization (6), by Lemma 1.4.3, B_i can be replaced by a normalized subset C_i such that the order of each element in C_i is a product of at most two distinct primes and C_i is not a subgroup of G. Note that as $|C_i|$ is a prime and $e \in C_i$, C_i is periodic if and only if C_i is a subgroup of G. Setting $C = C_1 \cdots C_s$ we get a factorization $G = C A_2 \cdots A_n$. The type of this factorization is $(q_1, q_2, \ldots, q_n)$. So by our assumption one of the factors $C, A_2, \ldots, A_n$ is periodic. If one of $A_2 \ldots, A_n$ is periodic, then we are done, so we assume that C is periodic. Now a periodic subset C of G is factored into subsets $C_1, \ldots, C_s$. In addition, there is a subset A of G such that $G = CA$ is a factorization. Simply set $A = A_2 \cdots A_n$. The hypotheses of Theorem 5.4.2 are satisfied, therefore this theorem gives that at least one of the factors $C_1, \ldots, C_s$ must be periodic.

This contradiction completes the proof.

Lemma 8.2.6. *Let G be a finite abelian group with the Hajós 2-property and let $q_1 = p$ be a prime such that either $p = 2$ or $p \geq 3$ and the p-component of G is cyclic. Then (q_1, q_2, q_3) is a periodicity forcing factorization type for G.*

Proof. Let $G = A_1 A_2 A_3$ be a normalized factorization of G such that $|A_1| = q_1$, $|A_2| = q_2$, $|A_3| = q_3$. We would like to show that one of the factors A_1, A_2, A_3 is periodic. If one of A_1, A_2, A_3 is periodic, then there is nothing to prove. So we assume that none of the factors is periodic. Assume that $p \geq 3$. By Lemma 1.4.3, A_1 can be replaced by $A_1' = \{e, a, a^2, \ldots, a^{p-1}\}$ for each $a \in A_1 \setminus \{e\}$. If A_1' is a subgroup of G for each $a \in A_1 \setminus \{e\}$, then A_1 is equal to the unique subgroup of G that has p elements. But A_1 is not a subgroup. Thus for some $a \in A_1 \setminus \{e\}$ the subset A_1' is not a subgroup. This is equivalent to $a^p \neq e$. In the $p = 2$ case we set $A_1' = A_1 = \{e, a\}$.

From the factorization $G = (A_1' A_2) A_3$ it follows that $A_1' A_2$ is periodic, say with period g, that is $A_1' A_2 g = A_1' A_2$. We claim that $\chi(A_2) = 0$ holds for all characters χ of G with $\chi(g) \neq 1$ and $\chi(a^p) \neq 1$. Indeed, from $\chi(a^p) \neq 1$ it follows that $\chi(A_1') \neq 0$. As $\chi(g) \neq 1$ from $\chi(A_1'A_2)\chi(g) = \chi(A_1'A_2)$ we get $0 = \chi(A_1'A_2) = \chi(A_1')\chi(A_2)$ and so $\chi(A_2) = 0$. Note that $\chi(a^p) \neq 1$ and $\chi(g) \neq 1$ implies $\chi(A_2) = 0$ is equivalent to

$$\operatorname{Ann}(\langle a^p \rangle) \cap \operatorname{Ann}(\langle g \rangle) \subset \operatorname{Ann}(A_2).$$

By Lemma 3.2.7, there are subsets X, Y of G such that

$$A_2 = X\langle a^p \rangle \cup Y\langle g \rangle,$$

where the products are direct and the union is disjoint. Similarly, from the factorization $G = A_2(A_1'A_3)$ it follows that $A_1'A_3$ is periodic, say with period h. Then there are subsets U, V of G such that

$$A_3 = U\langle a^p \rangle \cup V\langle h \rangle,$$

where the products are direct and the union is disjoint.

If $X = \emptyset$, then A_2 is periodic with period g. If $U = \emptyset$, then A_3 is periodic with period h. So we may assume that $X \neq \emptyset$ and $V \neq \emptyset$. Choose $x \in X$, $u \in U$. Multiply the factorization $G = A_1A_2A_3$ by $g = x^{-1}u^{-1}$ to get the factorization

$$G = Gg = A_1(A_2x^{-1})(A_3u^{-1}).$$

Now $\langle a^p \rangle \subset A_2x^{-1}$, $\langle a^p \rangle \subset A_3u^{-1}$ contradict the definition of the factorization. This completes the proof.

Theorem 8.2.6. *Let $G = A_1 \cdots A_n$ be a factorization of the finite abelian group G. If G has the Hajós 2-property, then at least one of the factors $A_1, \ldots, A_n$ is always periodic.*

Proof. We inspect groups with their possible factorization types in a case by case manner.

(1) Let G be of type (p^α, q), where p, q are distinct primes. In the factorization $G = A_1 \cdots A_n$ at most one of the factors is not of a prime power order. By Theorem 6.2.1, one of the factors is periodic.

(2) Let G be of type $(2^\alpha, 2)$. Theorem 7.2.1 takes care of this case.

(3) Let G be of type

$$(p^3, 2, 2), \quad (p^2, 2, 2, 2), \quad (p, 2, 2, 2, 2).$$

Note that $|G|$ is a product of five primes. If the factorization type is $(q_1, \ldots, q_5)$, then each q_i is a prime and Rédei's theorem applies. If the factorization type is $(q_1, \ldots, q_4)$, then only one q_i can be composite, say q_1, and Lemma 8.2.5 applies. If the factorization type is (q_1, q_2, q_3), then at least one of the q_i's is a prime, say q_1, and Lemma 8.2.6 applies.

(4) Let G be of type

$$(p^2, q^2), \quad (p^2, q, r), \quad (p, q, r, s),$$
$$(p, 4, 2), \quad (p, q, 2, 2), \quad (2^2, 2^2).$$

Note that $|G|$ is a product of four primes. If the factorization type is (q_1, q_2, q_3, q_4), then Rédei's theorem works. If the factorization type is (q_1, q_2, q_3), then Lemma 8.2.5 is applicable.

(5) The case when G is of type $(p, 3, 3)$, $(3^2, 3)$, (p, p), that is, when $|G|$ is a product of at most three primes is trivial.

This inspection completes the proof.

Chapter 9

The Rédei property

9.1 Groups of type (p, p, p) and Latin squares

By Rédei's theorem, if a finite abelian group is factored into normalized subsets of prime cardinality, then at least one of the factors is a subgroup of the group. The special case when G is of type (p, p) plays an important part of the proof of the general case and has interesting geometric and combinatorial applications. In connection with this result, Rédei formulated the next conjecture which also appears as Problem 5 in the Open Problems section of his book:

Let p be a prime and G be a group of type (p, p, p). Consider a normalized factorization $G = AB$ of G, where $|A| = p$ and $|B| = p^2$. Then either $\langle A \rangle \neq G$ or $\langle B \rangle \neq G$.

If Rédei's conjecture holds, then we can easily describe all normalized factorizations of groups of type (p, p, p) where p is prime. To see why, suppose $G = AB$ is a normalized factorization such that $|A| = p$ and $|B| = p^2$. It follows that either $\langle A \rangle \neq G$ or $\langle B \rangle \neq G$. If $\langle A \rangle$ is of order p, say $\langle A \rangle = \langle x \rangle$, then B must be a complete set of representatives modulo $\langle x \rangle$ which contains the identity element. If $\langle A \rangle$ is of order p^2, say $\langle A \rangle = \langle x, y \rangle$, then G is a partition of the cosets modulo $H = \langle x, y \rangle$, say

$$G = eH \cup zH \cup z^2 H \cup \cdots \cup z^{p-1} H.$$

Set $B_i = z^i H \cap B$ for each i, $0 \leq i \leq p - 1$. As

$$
\begin{aligned}
A(z^{-i} B_i) &= A(H \cap z^{-i} B) \\
&= AH \cap z^{-i} AB \\
&= H \cap G \\
&= H
\end{aligned}
$$

and as the product AB_i is direct, it follows that $A(z^{-i} B_i)$ is a factorization of H. By Rédei's theorem either A is a subgroup of G or $z^{-i} B_i$ is a coset modulo a subgroup of H. (Keep in mind that the factor $z^{-i} B$ is not necessarily normalized.)

Since A is not a subgroup, $z^{-i}B_i$ is a coset modulo subgroup of order p. Thus there are subgroups $K_0, K_1, \ldots, K_{p-1}$ of H such that $AK_0, AK_1, \ldots, AK_{p-1}$ are factorizations of H and

$$z^0 h_0 K_0 \cup z^1 h_1 K_1 \cup z^2 h_2 K_2 \cup \cdots \cup z^{p-1} h_{p-1} K_{p-1}$$

is a partition of B, where $h_0, h_1, h_2, \ldots, h_{p-1}$ are elements of H. The final case to consider is when $\langle B \rangle$ is of order p^2, say $\langle B \rangle = \langle x, y \rangle$. It follows that A must be a complete set of representatives modulo $\langle x, y \rangle$ that contains the identity element.

In the following we describe how Latin squares can be used to investigate this conjecture of L. Rédei. Using this bookkeeping tool we verify the conjecture for $p \leq 11$.

Theorem 9.1.1. *Let p be a prime and let G be a group of type (p, p, p). If $G = AB$ is a factorization and $p \leq 11$, then either $\langle A \rangle \neq G$ or $\langle B \rangle \neq G$.*

Proof. The normalized subset B of the finite abelian group is said to be *reducible* if there are normalized subsets C, D of G such that $|C| < |B|$, $|D| < |B|$ and B is factored into CD. We claim that the factors must be irreducible in a counterexample to Rédei's conjecture. To prove this claim, consider the factorization $G = AB$, where $|A| = p$ and $|B| = p^2$. Clearly A is irreducible. Assume that B can be factored into CD, where $|C| = |D| = p$. From the factorization $G = ACD$ it follows that one of the factors is a subgroup of G. If this factor is A, then $\langle A \rangle \neq G$ and we are done. Thus we may assume that either C or D is a subgroup of G. For the sake of concreteness assume that D is a subgroup of G. From the factorization $G = ACD$ we get the factorization $G/D = \big((AD)/D\big) \cdot \big((CD)/D\big)$ of the factor group G/D. Again by Rédei's theorem, one of the factors is a subgroup of G/D. Hence either AD or CD is a subgroup of G and so either $\langle A \rangle \neq G$ or $\langle B \rangle \neq G$. This proves the claim.

We assume that $\langle A \rangle \neq G$. As a consequence, there are elements x, y, z of A such that $\langle x, y, z \rangle = G$. In the normalized factorization $G = AB$, the factor A can be replaced by

$$A' = \{e, z, z^2, \ldots, z^{p-1}\}$$

to get the normalized factorization $G = A'B$. Multiplying by g we get the factorization $G = (gA')B$ for each g of G. This latter factorization is not necessarily normalized. As the two factors of a factorization can have at most one element in common, we have that each of the p^2 cosets modulo subgroup $\langle z \rangle$ intersects the factor B in at most one element. On the other hand, these cosets form a partition of G and so each element b of B belongs to one of the cosets. Consequently, each coset contains precisely one element from B.

Elements of B are of the form $x^i y^j z^k$, where i, j, k are integers in the range $0, 1 \ldots, p-1$. Actually, i and j run through the range $0, 1, \ldots, p-1$ independently. We can code the element $x^i y^j z^k$ by the ordered triple (i, j, k). These p^2 triples can be recorded conveniently as a p by p table whose ith row, jth column entry is k. We can repeat the previous argument with $\langle x \rangle$ in place of $\langle z \rangle$. As before, we get that each coset modulo subgroup $\langle x \rangle$ contains exactly one element from B. Fixing i, we can see that each row of the table is a permutation of the elements $0, 1, \ldots, p-1$. Similarly, as each coset modulo subgroup $\langle y \rangle$ contains exactly one element from B, each column of the table is a permutation of the elements $0, 1, \ldots, p-1$. Thus if $\langle A \rangle = G$, then the factor B can be described using a p by p Latin square. In fact this can be done in three ways since the roles of the elements x, y, z are symmetric.

The subgroups $\langle x \rangle$, $\langle y \rangle$, and $\langle z \rangle$ are not the only ones restricting the structure of B. As $y, z \in A$, each coset modulo subgroup $\langle z^{-1}y \rangle$ contains exactly one element from B. Thus if a and b are entries in the uth and vth columns in the same row, then

$$b - a \not\equiv (-1)(v - u) \pmod{p}. \tag{1}$$

Similarly, since $x, z \in A$ it follows that if a and b are entries in the uth and vth rows in the same column, then (1) holds. This means that the Latin square describing B can contain only a certain restricted type of permutation in its rows and columns.

The Latin square corresponding to B has p positions which contain a given fixed element, k, one from each row and one from each column. Consider the permutation on $0, 1, \ldots, p-1$ which maps i to j precisely when k is the entry in row i, column j. We refer to this permutation as the *kth transversal*. The fact that $x, y \in A$ implies that the transversals satisfy those restrictions discussed above in connection with the rows and columns. This is because each coset modulo subgroup $\langle y^{-1}x \rangle$ contains exactly one element from B. Thus if the ath and the bth positions in the uth and vth columns contain the same element, then (1) holds.

A permutation f of the elements of a finite abelian group H is called a *complete mapping* of H if $a \to af(a)$, $a \in H$ is again a permutation of the elements of H. Using this terminology we may say that the rows, columns and transversals of Latin squares corresponding to the factor B are complete mappings of the additive group modulo p. For instance the expression that the ith row contains the complete map f simply means that the triples $(i, j, f(j))$, $0 \le j \le p-1$ belong to B. The triple $(i, j, f(j))$ can be viewed as a point in the 3-dimensional affine space over the modulo p finite field. In case the points $(i, j, f(j))$, $0 \le j \le p-1$ form a straight line we call the map f *linear* as well.

Consider the Latin square that corresponds to factor B. Suppose that each transversal is linear. The transversals are disjoint sets and consequently the straight lines forming the transversals are parallel. Because the roles of rows, columns, and transversals in a Latin square are symmetric it follows that if each row is linear, then these rows are parallel.

We now discuss some further ideas that will be used systematically later. Each entry in the Latin square corresponding to factor B is the intersection of a column, a row and a transversal. We claim that if there is a counterexample for Rédei's conjecture then there is a corresponding Latin square which contains an entry such that the column, row and transversal containing it are not linear. Since factor B can be replaced by $b^{-1}B$ for each $b \in B$, we may assume that this happens at the top left entry of the Latin square. The argument is the following. If each column of the Latin square is a straight line, then B factors into a product of two smaller subsets. Namely, B is the product of the elements of the first column and the first row. Hence B is reducible. Thus there is a nonlinear column, say the first one. Similarly, there must be a nonlinear row, say the first one. The intersection of the first column and row contains the zero entry. We may assume that the 0th transversal is linear since otherwise we are done. Let u, v, w be the permutations in the first column, first row and the 0th transversal respectively. If each transversal is linear, then factor B is reducible. Thus there is an i such that the $u(i)$th transversal is not linear. We may assume that the ith row is linear since otherwise we are done. Let P be the plane spanned by the triples in the ith row and the 0th transversal.

Consider $u(j)$ the entry in the jth position in the first column. Again either the jth row or the $u(j)$th transversal is linear since otherwise we are done. In case the jth row is linear, it must be parallel to the ith row. Furthermore this row intersects the 0th transversal. Hence the triples representing the jth row lie in plane P. Consequently, the triple $\big(j, 0, u(j)\big)$ lies in P. In case the $u(j)$th transversal is linear, it must be parallel to the 0th transversal. Furthermore this transversal intersects the ith row. Hence the triples representing the $u(j)$th transversal lie in plane P. Consequently, $\big(j, 0, u(j)\big)$ lies in P. Since j was arbitrary, we get the contradiction that the first column is linear.

When $p = 2$ or $p = 3$ the factor A contains only one or two nonidentity elements. On the other hand, G cannot be generated by less than three elements. Therefore $\langle A \rangle \neq G$.

Let us turn to the case $p = 5$. If $\langle A \rangle \neq G$, then we are done so we suppose that $x, y, z \in A$ such that $\langle x, y, z \rangle = G$. Consider the 5 by 5 Latin square associated with the factor B. We have seen that we might assume that the first row, first column and 0th transversal represent non-linear complete mappings. When $p = 5$, the complete mappings are all straight lines and so in this case there is no counterexample for Rédei's conjecture.

Let $p = 7$ and assume that $G = AB$ is a counterexample to Rédei's conjecture. Consider all the complete mappings on the additive group modulo 7. These constitute the candidates for the rows, columns and transversals of the Latin square corresponding to factor B. We need to focus only on those mappings which fix 0, since each of the other ones is the result of adding an arbitrary constant modulo p to one of these mappings. There are 19 such complete mappings. Five of them are linear; we list the remaining ones, $\gamma_1, \ldots, \gamma_{14}$, in Table 1.

The automorphism ϕ of G defined by $\phi(x) = x^3$, $\phi(y) = y^3$, $\phi(z) = z^3$ takes the factorization $G = AB$ to the factorization $G = \phi(G) = \phi(A)\phi(B)$. This ϕ induces a map $(i, j, k) \to (3i, 3j, 3k)$ on the triples, where the multiplication is taken modulo 7. This acts on the set of 7 by 7 Latin squares. If u is the first column of the Latin square corresponding to factor B, then $u'(i) = 3u(3^{-1}i)$ is the first column of the Latin square that codes factor $\phi(B)$. Here the operations are taken modulo 7. The automorphism ϕ generates a permutation of the listed

γ_1	0 1 4 2 6 3 5	γ_8	0 4 1 3 5 6 2
γ_2	0 2 6 3 1 4 5	γ_9	0 4 2 5 6 1 3
γ_3	0 5 2 6 1 3 4	γ_{10}	0 3 1 5 2 4 6
γ_4	0 2 4 1 5 3 6	γ_{11}	0 5 1 2 4 6 3
γ_5	0 2 3 6 4 1 5	γ_{12}	0 4 6 1 2 5 3
γ_6	0 3 4 6 1 5 2	γ_{13}	0 2 4 5 1 6 3
γ_7	0 1 3 5 2 6 4	γ_{14}	0 4 1 6 2 3 5

Table 1

complete mappings. The corresponding permutation decomposed into cycles is

$$(\gamma_1\gamma_2\gamma_3\gamma_4\gamma_5\gamma_6)(\gamma_7\gamma_8\gamma_9\gamma_{10}\gamma_{11}\gamma_{12})(\gamma_{13}\gamma_{14}). \tag{2}$$

The automorphism ψ of G defined by $\psi(x) = x$, $\psi(y) = z$, $\psi(z) = y$ takes the factorization $G = AB$ to the factorization $G = \psi(G) = \psi(A)\psi(B)$. If u is the first column of the Latin square corresponding to factor B, then u^{-1} is the first column of the Latin square corresponding to factor $\psi(B)$. Here u^{-1} denotes the inverse of u as a permutation. The automorphism ψ induces a permutation of the complete mappings. This permutation as a product of cycles is

$$(\gamma_1\gamma_7)(\gamma_2\gamma_8)(\gamma_3\gamma_9)(\gamma_4\gamma_{10})(\gamma_5\gamma_{11})(\gamma_6\gamma_{12})(\gamma_{13}\gamma_{14}). \tag{3}$$

Permutations (2) and (3) generate a permutation group. This permutation group splits the listed complete mappings into two transitivity classes.

We make some useful observations. Let u, v, and w be the complete mappings represented by the the 1st column, 1st row and the 0th transversal respectively. For a given i, suppose $w(i) = j$. Let β_i and γ_j denote the complete mappings represented by row i and column j respectively. We have $\beta_i(0) = u(i)$, $\gamma_j(0) = v(j)$, and $\beta_i(j) = \gamma_j(i) = 0$. Hence, $u(i) = \beta_i(0) - \beta_i(j) \neq j$ since β_i is a complete mapping. Thus for each i, $u(i) \neq w(i)$. Furthermore, $v(j) = \gamma_j(0) - \gamma_j(i) \neq i$ since γ_j is a complete mapping. Thus for each i, $v\big(w(i)\big) \neq i$.

We now consider 7 by 7 Latin squares, in search of one corresponding to a counterexample. We may assume that the complete mappings u, v, and w are not linear and so they belong to the mappings $\gamma_1, \ldots, \gamma_{14}$. Considering the permutation group discussed above, we may suppose that u is either γ_1 or γ_{13}. First consider the case where $u = \gamma_1$. Only the $v = \gamma_8$, $v = \gamma_{10}$, and $v = \gamma_{12}$ choices satisfy $u(i) \neq v(i)$, $1 \leq i \leq p - 1$. In the case $u = \gamma_1$, $v = \gamma_8$ as $u(i) \neq w(i)$, w can only be γ_8, γ_{10}, γ_{12}. Since $v\big(w(i)\big) \neq i$, these possibilities are excluded by $v\big(w(3)\big) = 3$, $v\big(w(5)\big) = 5$, $v\big(w(2)\big) = 2$ respectively. In the case $u = \gamma_1$, $v = \gamma_{10}$ as $u(i) \neq w(i)$, w can only be γ_8, γ_{10}, γ_{12}. Since $v\big(w(i)\big) \neq i$, these possibilities are excluded by $v\big(w(4)\big) = 4$, $v\big(w(6)\big) = 6$, $v\big(w(3)\big) = 3$ respectively. In the case $u = \gamma_1$, $v = \gamma_{12}$ the only possibilities for w are γ_8, γ_{10}, γ_{12} which are excluded by $v\big(w(6)\big) = 6$, $v\big(w(1)\big) = 1$, $v\big(w(5)\big) = 5$ respectively.

Now consider the case $u = \gamma_{13}$. Since $u(i) \neq v(i)$, $1 \leq i \leq p - 1$ the only choice for v is γ_{14}. Since $u(i) \neq w(i)$, we must have $v = \gamma_{14}$. At this point, there is no consistent choice available for the second column.

This contradiction shows that there is no counterexample to Rédei's conjecture when $p = 7$.

Finally, let $p = 11$. Suppose there is a counterexample to Rédei's conjecture for $p = 11$. Consequently, there is an 11 by 11 Latin square representing factor B which has complete mappings for its rows, columns and transversals. Furthermore there exists one in which the initial row, column and 0th transversal are nonlinear and fix 0. Any such Latin square will be called *qualified*. Finally, a counterexample to Rédei's conjecture implies the existence of a non-reducible qualified Latin square. We will show that no such Latin square exists but the numbers involved require the use of a computer to assist in the search.

We will regard a Latin square as a collection of triples (i, j, k) where k is the entry in row i, column j. Consider the permutation on triples which takes (i, j, k) to $(3i, 3j, 3k)$ where multiplication is performed modulo 11. This acts on the set of qualified Latin squares and preserves the property of non-reducibility. We note that if u is the complete mapping representing the initial column of a Latin square, then the initial column of the permuted one, u', will satisfy $u'(i) = 3u(3^{-1}i)$.

In a similar fashion the transformation taking (i, j, k) to (i, k, j) acts on the set of all qualified Latin squares, preserves non-reducibility and changes the initial column, u, of a Latin square to u^{-1}. The group generated by both of these permutations partitions the 3441 complete mappings that fix 0 into 23 transitivity classes of nonlinear mappings and six classes containing linear mappings. Thus, we need to consider only 23 cases; each case uses a representative nonlinear mapping from a transitivity class as the initial column of the Latin square representing factor B. There are exactly 389 complete mappings that fix 1 as well as zero and at least one of these belongs to each of the 23 transitivity classes. Thus, we chose as our representative a complete mapping that fixes both 0 and 1.

C. Ward and S. Szabó performed an exhaustive computer search for all qualified Latin squares having an initial column equal to one of the 23 representatives discussed above. The search was performed in a depth-first fashion. At stage 1 the initial row was chosen from the nonlinear complete mappings that fix 0. Subsequently, the columns were chosen, one per stage. At each stage, backtracking occurred when a candidate column or row was inconsistent with the previously established parts of the Latin square.

The search revealed exactly 50 qualified Latin squares meriting further consideration as possible counterexamples. We needed to check each of these to see if it corresponded to a non-reducible factorization of B. Suppose there exist integers, α, β, ω, and δ, and complete mappings f and g on the additive group modulo p such that the Latin square triples all satisfy the relationship

$$k = f(\alpha i + \beta j) + g(\omega i + \delta j),$$

$f(2)$ $\cdots$ $f(10)$	(β, δ)
2 4 5 7 10 3 6 8 9	$(8,6)$ $(0,6)$ $(9,0)$
2 5 6 9 3 10 4 7 8	$(5,4)$ $(4,0)$ $(0,8)$
2 5 10 4 6 3 8 9 7	$(4,4)$ $(8,7)$ $(3,0)$ $(1,4)$ $(0,6)$ $(0,7)$ $(9,0)$ $(7,0)$
3 4 8 5 9 2 6 10 7	$(7,0)$ $(0,3)$ $(3,2)$
3 5 6 4 8 10 7 9 2	$(8,7)$ $(0,6)$ $(9,0)$
3 5 6 4 9 7 10 8 2	$(4,0)$ $(0,8)$ $(5,4)$
3 5 6 10 8 2 4 9 7	$(7,0)$ $(0,3)$ $(3,2)$
3 5 10 7 4 2 9 6 8	$(9,8)$ $(8,0)$ $(0,4)$ $(8,7)$ $(0,6)$ $(9,0)$ $(3,3)$ $(3,10)$
4 7 5 9 2 8 10 3 6	$(5,4)$ $(7,6)$ $(1,3)$ $(4,0)$ $(3,0)$ $(0,7)$ $(0,8)$ $(8,10)$
7 5 3 10 4 9 6 8 2	$(7,0)$ $(0,3)$ $(0,4)$ $(9,8)$ $(8,0)$ $(3,2)$ $(1,7)$ $(4,10)$

Table 2

where the operations are taken modulo p. If the matrix

$$\begin{bmatrix} \alpha & \beta \\ \omega & \delta \end{bmatrix} \quad \text{has an inverse, say} \quad \begin{bmatrix} a & b \\ c & d \end{bmatrix}$$

modulo p, then the collection of triples in the Latin square is the direct sum of

$$\left\{ (am, cm, f(m)) : 0 \leq m \leq p - 1 \right\}$$

and

$$\left\{ (bn, dn, g(n)) : 0 \leq n \leq p - 1 \right\}.$$

All 50 qualified Latin squares under investigation satisfied a special case of this, namely

$$k = f(i + \beta j) + \delta j.$$

Table 2 gives the complete mappings f and the (β, δ) pairs that generate these 50 Latin squares.

Thus, there is no counterexample to Rédei's conjecture when $p = 11$ by virtue of the fact that all qualified Latin squares are reducible.

This completes the proof.

9.2 The Rédei property

Let G be a finite abelian group. If from each factorization $G = AB$ it follows that either $\langle A \rangle \neq G$ or $\langle B \rangle \neq G$, then we say that G has the *Rédei property*. Adopting this terminology we can rephrase the problem of Section 9.1 such that groups of type (p, p, p) do not have the Rédei property. In coding theory the problem of which finite elementary 2-groups possess the Rédei property is considered in the context of tiling finite dimensional affine spaces over the field GF(2). In 1995 in connection with a problem in Fourier analysis, R. Tijdeman conjectured that every finite cyclic group has the Rédei property. (This conjecture will be disproved later in this section.) Thus there are three areas of mathematics where the Rédei property naturally appeared. In this section we will show that only a tiny portion of the finite abelian groups has the Rédei property.

The next two lemmas help to extend factorizations from subgroups to a group. We state the first lemma without proof only for the sake of easy reference.

Lemma 9.2.1. *If two groups do not have the Rédei property, then nor does their direct product.*

Lemma 9.2.2. *Let the finite abelian group G be the internal direct product of its subgroups H and K, where K is cyclic. If $H = AB$ is a normalized factorization of H with $\langle A \rangle = \langle B \rangle = H$ and if there is an element $a \in A \setminus \{e\}$ such that $\langle A \setminus \{a\} \rangle = H$, then there is a normalized factorization $G = A'B'$ of G with $\langle A' \rangle = \langle B' \rangle = G$.*

Proof. Let k be a generator element of K, that is, let $K = \langle k \rangle$ and let $A^* = A \setminus \{a\}$. Define the subsets A', B' of G by

$$A' = \big(A \setminus \{a\}\big) \cup \{ak\} = A^* \cup \{ak\},$$

$$B' = BK = B\langle k \rangle.$$

We claim that $G = A'B'$ is a factorization of G. First we verify that $A'B' = G$. Indeed,

$$
\begin{aligned}
A'B' &= \big(A^* \cup \{ak\}\big)(BK) \\
&= A^*BK \cup akBK \\
&= A^*BK \cup aBkK \\
&= A^*BK \cup aBK \\
&= \big(A^* \cup \{a\}\big)BK \\
&= ABK \\
&= HK \\
&= G.
\end{aligned}
$$

Secondly, we show that $|A'||B'| = |G|$. Indeed,

$$
\begin{aligned}
|A'||B'| &= |A||B||K| \\
&= |H||K| \\
&= |G|.
\end{aligned}
$$

Thus $G = A'B'$ is a factorization of G and obviously the factorization is normalized.

L_1	a	a	a	b
L_2	a	a	b	b
L_3	a	b	b	b

Table 1

We claim that $\langle A' \rangle = \langle B' \rangle = G$. From $a \in H = \langle A \setminus \{a\} \rangle \subset \langle A' \rangle$ and $ak \in \langle A' \rangle$ it follows that $k \in \langle A' \rangle$. Now $H = \langle A \setminus \{a\} \rangle \subset \langle A' \rangle$ and $k \in \langle A' \rangle$ imply that $G = \langle H, k \rangle \subset \langle A' \rangle$ and so $\langle A' \rangle = G$. From $B' = B\{e, k, \ldots\}$ it is clear that $B \subset B'$ and $k \in B'$. Hence $H = \langle B \rangle \subset \langle B' \rangle$ and $k \in \langle B' \rangle$. From this it follows that $G = \langle H, k \rangle \subset \langle B' \rangle$ and so $\langle B' \rangle = G$.

This completes the proof.

Let $p \geq 3$ be a prime. We say that a p-group is bad if it is a direct product of L_1, L_2, L_3 and each L_i is one of the following types:

(a) (p^α), where $\alpha \geq 2$ is an integer,

(b) (p, p).

Let F_p be the family of groups whose type is one of

$$(p, p, p, p, p), \quad (p^\alpha, p, p, p), \quad (p^\alpha, p^\beta, p),$$

where $\alpha, \beta \geq 2$ are integers or a subgroup of such a group. In other words F_p contains the groups with the above types and F_p is closed under the operation of forming subgroups. It is plain that members of F_p do not have any bad subgroups and F_p is the complete list of groups that do not have any bad subgroups.

Theorem 9.2.1. *Let p be an odd prime. If G is a finite abelian p-group that is not a member of the F_p family, then G does not have the Rédei property.*

Proof. First we will show that bad groups do not have the Rédei property. There are four choices for the types of L_1, L_2, L_3 shown in Table 1.

(1) Suppose G is an elementary p-group of rank 6. Let x_1, x_2, y_1, y_2, z_1, z_2 be basis elements of G. Set

$$K_1 = \langle x_1 \rangle, \quad H_1 = \langle x_2 \rangle, \quad A_1 = \left(K_1 \setminus \{x_1^{p-1}\} \right) \cup \{x_1^{p-1} x_2\},$$

$$K_2 = \langle y_1 \rangle, \quad H_2 = \langle y_2 \rangle, \quad A_2 = \left(K_2 \setminus \{y_1^{p-1}\} \right) \cup \{y_1^{p-1} y_2\},$$

$$K_3 = \langle z_1 \rangle, \quad H_3 = \langle z_2 \rangle, \quad A_3 = \left(K_3 \setminus \{z_1^{p-1}\} \right) \cup \{z_1^{p-1} z_2\},$$

$$H = H_1 H_2 H_3, \quad A = A_1 A_2 A_3.$$

The product $HA_1A_2A_3$ is a factorization of G. Indeed,

$$HA_1A_2A_3 = (H_1 A_1)(H_2 A_2)(H_3 A_3) = \langle x_1, x_2 \rangle \langle y_1, y_2 \rangle \langle z_1, z_2 \rangle = G.$$

Next we construct a subset B from H by adding and removing certain subsets. Remove

$$x_2^{p-1} H_2, \quad y_2^{p-1} H_3, \quad z_2^{p-1} H_1$$

from H. Add

$$x_2^{p-1} H_2 y_1, \quad y_2^{p-1} H_3 z_1, \quad z_2^{p-1} H_1 x_1$$

to H and denote the resulting set by B. We claim that

$$x_2^{p-1} H_2 A = x_2^{p-1} y_1 H_2 A, \tag{1}$$

$$y_2^{p-1} H_3 A = y_2^{p-1} z_1 H_3 A, \tag{2}$$

$$z_2^{p-1} H_1 A = z_2^{p-1} x_1 H_1 A. \tag{3}$$

Let us verify (1):

$$
\begin{aligned}
x_2^{p-1} y_1 H_2 A &= x_2^{p-1} y_1 H_2 A_1 A_2 A_3 \\
&= x_2^{p-1} y_1 (H_2 A_2) A_1 A_3 \\
&= x_2^{p-1} y_1 \langle y_1, y_2 \rangle A_1 A_3 \\
&= x_2^{p-1} \langle y_1, y_2 \rangle A_1 A_3 \\
&= x_2^{p-1} (H_2 A_2) A_1 A_3 \\
&= x_2^{p-1} H_2 A_1 A_2 A_3 \\
&= x_2^{p-1} H_2 A.
\end{aligned}
$$

Equations (2) and (3) can be verified in a similar way. Next we claim that the sets

$$x_2^{p-1} H_2 A, \quad y_2^{p-1} H_3 A, \quad z_2^{p-1} H_1 A \tag{4}$$

are disjoint. In order to prove the claim assume the contrary, that

$$x_2^{p-1} H_2 A_1 A_2 A_3 \cap y_2^{p-1} H_3 A_1 A_2 A_3 \neq \emptyset.$$

This is equivalent to

$$x_2^{p-1} \langle y_1, y_2 \rangle A_1 A_3 \cap y_2^{p-1} \langle z_1, z_2 \rangle A_1 A_2 \neq \emptyset.$$

The x_1, x_2 component of an element of $x_2^{p-1} \langle y_1, y_2 \rangle A_1 A_3$ must be in one of the forms

$$x_1^\alpha x_2, \quad x_1, \quad 0 \leq \alpha \leq p - 2.$$

The x_1, x_2 component of an element of $y_2^{p-1} \langle z_1, z_2 \rangle A_1 A_2$ must be in one of the forms

$$x_1^\beta, \quad x_1^{p-1} x_2, \quad 0 \leq \beta \leq p - 2.$$

There are four ways to equate these elements but none of them is possible. The remaining case of the claim can be verified in a similar way. Finally we claim that in the $G = AB$ factorization $\langle A \rangle = G$ and $\langle B \rangle = G$. To verify the claim note that $A_1, A_2, A_3 \subset A$ and

$$x_1, x_1^{p-1} x_2 \in A_1, \quad y_1, y_1^{p-1} y_2 \in A_2, \quad z_1, z_1^{p-1} z_2 \in A_3$$

which gives $\langle A \rangle = G$. Then we can see that $x_2, y_2, z_2 \in B$ and

$$x_2^{p-1} y_1, \, y_2^{p-1} z_1, \, z_2^{p-1} x_1 \in B$$

and so $\langle B \rangle = G$.

(2) Let G be a group of type $(p^\alpha, p^\beta, p^\gamma)$, where $\alpha, \beta, \gamma \geq 2$ are integers. Let x, y, z be basis elements of G with $|x| = p^\alpha$, $|y| = p^\beta$, $|z| = p^\gamma$. Set

$$A_1 = \{e, x, x^2, \ldots, x^{p-1}\}, \quad H_1 = \langle x^p \rangle,$$

$$A_2 = \{e, y, y^2, \ldots, y^{p-1}\}, \quad H_2 = \langle y^p \rangle,$$

$$A_3 = \{e, z, z^2, \ldots, z^{p-1}\}, \quad H_3 = \langle z^p \rangle,$$

$$H = H_1 H_2 H_3, \quad A = A_1 A_2 A_3.$$

The product HA is a factorization of G as the computation

$$\begin{aligned} HA \; &= \; (H_1 A_1)(H_2 A_2)(H_3 A_3) \\ &= \; \langle x \rangle \langle y \rangle \langle z \rangle \\ &= \; G \end{aligned}$$

shows. Construct B from H by removing

$$x^{p^\alpha - p} H_2, \quad y^{p^\beta - p} H_3, \quad z^{p^\gamma - p} H_1$$

from H and adding

$$x^{p^\alpha - p} H_2 y, \quad y^{p^\beta - p} H_3 z, \quad z^{p^\gamma - p} H_1 x$$

to H. Now

$$\begin{aligned} x^{p^\alpha - p} y H_2 A \; &= \; x^{p^\alpha - p} y H_2 A_1 A_2 A_3 \\ &= \; x^{p^\alpha - p} y (H_2 A_2) A_1 A_3 \\ &= \; x^{p^\alpha - p} y \langle y \rangle A_1 A_3 \\ &= \; x^{p^\alpha - p} \langle y \rangle A_1 A_3 \\ &= \; x^{p^\alpha - p} (H_2 A_2) A_1 A_3 \\ &= \; x^{p^\alpha - p} H_2 A_1 A_2 A_3 \\ &= \; x^{p^\alpha - p} H_2 A. \end{aligned}$$

Similarly

$$y^{p^\beta - p} z H_3 A = y^{p^\beta - p} H_3 A, \quad z^{p^\gamma - p} x H_1 A = z^{p^\gamma - p} H_1 A.$$

We claim that the sets

$$x^{p^\alpha - p} H_2 A, \quad y^{p^\beta - p} H_3 A, \quad z^{p^\gamma - p} H_1 A$$

are disjoint. Let us check, say that the sets

$$x^{p^\alpha - p} H_2 = x^{p^\alpha - p} \langle y \rangle A_1 A_3 \tag{5}$$

and

$$y^{p^\beta - p} H_3 A = y^{p^\beta - p} \langle z \rangle A_1 A_2 \tag{6}$$

are disjoint. The exponent of the x component of an element of (5) is at least $p^\alpha - p$. The exponent of the x component of an element of (6) is at most $p - 1$. This proves the claim.

Using these one can verify that $G = AB$ is a factorization of G. Note that $x, y, z \in A$, $x^p, y^p, z^p \in B$,

$$x^{p^\alpha - p} y, \ y^{p^\beta - p} z, \ z^{p^\gamma - p} \in B$$

and so $\langle A \rangle = \langle B \rangle = G$. Thus G does not have the Rédei property.

(3) Let G be a group of type $(p^\alpha, p^\beta, p, p)$, where $\alpha, \beta \geq 2$ are integers. Let x, y, z_1, z_2 be basis elements of G with $|x| = p^\alpha$, $|y| = p^\beta$, $|z_1| = |z_2| = p$. Set

$$A_1 = \{e, x, x^2, \ldots, x^{p-1}\}, \quad H_1 = \langle x^p \rangle,$$

$$A_2 = \{e, y, y^2, \ldots, y^{p-1}\}, \quad H_2 = \langle y^p \rangle,$$

$$K_3 = \langle x_1 \rangle, \quad H_3 = \langle z_2 \rangle, \quad A_3 = \left(K_3 \setminus \{z_1^{p-1}\} \right) \cup \{z_1^{p-1} z_2\},$$

$$H = H_1 H_2 H_3, \quad A = A_1 A_2 A_3.$$

One can verify that $G = HA$ is a factorization of G. Construct B from H by removing

$$x^{p^\alpha - p} H_2, \quad y^{p^\beta - p} H_3, \quad z_2^{p-1} H_1$$

from H and adding

$$x^{p^\alpha - p} H_2 y, \quad y^{p^\beta - p} H_3 z_1, \quad z_2^{p-1} H_1 x$$

to H. Then $G = AB$ is a factorization of G and $\langle A \rangle = \langle B \rangle = G$.

(4) Let G be a group of type (p^α, p, p, p, p), where $\alpha \geq 2$ is an integer. Let x, y_1, y_2, z_1, z_2 be basis elements of G with $|x| = p^\alpha$, $|y_1| = |y_2| = |z_1| = |z_2| = p$. Set

$$A_1 = \{e, x, x^2, \ldots, x^{p-1}\}, \quad H_1 = \langle x^p \rangle,$$

$$K_2 = \langle y_1 \rangle, \quad H_2 = \langle y_2 \rangle, \quad A_2 = \left(K_2 \setminus \{y_1^{p-1}\} \right) \cup \{y_1^{p-1} y_2\},$$

$$K_3 = \langle z_1 \rangle, \quad H_3 = \langle z_2 \rangle, \quad A_3 = \left(K_3 \setminus \{z_1^{p-1}\} \right) \cup \{z_1^{p-1} z_2\},$$

$$H = H_1 H_2 H_3, \quad A = A_1 A_2 A_3.$$

One can verify that $G = HA$ is a factorization of G. Construct B from H by removing

$$x^{p^\alpha - p} H_2, \quad y_2^{p-1} H_3, \quad z_2^{p-1} H_1$$

from H and adding

$$x^{p^\alpha - p} H_2 y_1, \quad y_2^{p-1} H_3 z_1, \quad z_2^{p-1} H_1 x$$

to H. Then $G = AB$ is a factorization of G and $\langle A \rangle = \langle B \rangle = G$.

(5) Let G be a group of type

$$(p^{\alpha(1)}, \ldots, p^{\alpha(m)}, p^{\beta(1)}, \ldots, p^{\beta(n)}),$$

where $\alpha(1), \ldots, \alpha(m) \geq 2$, $\beta(1) = \cdots = \beta(n) = 1$.

Suppose first that $m = 0$, $n \geq 6$, that is, G is an elementary p-group of rank n. Divide n by 6 with remainder

$$n = 6u + v, \quad 0 \leq v \leq 5.$$

Here $u \geq 1$ as $k \geq 6$. Clearly G is a direct product of subgroups $H_1, \ldots, H_u$, $K_1, \ldots, K_v$ such that each H_i is of rank 6 and each K_j is cyclic of order p. The construction described in step (1) applied to H_u shows that H_u does not have the Rédei property. In the construction A has $p^3 \geq 27$ elements. There are at least 20 elements $a \in A \setminus \{e\}$ such that $\langle A \setminus \{a\} \rangle = H_u$ and so by Lemma 9.2.2, $H_u \times K_1$ does not have the Rédei property. Continuing in this way we get that $M = H_u \times K_1 \times \cdots \times K_v$ does not have the Rédei property. As $H_1, \ldots, H_{u-1}$ do not have the Rédei property by Lemma 9.2.1, $L = H_1 \times \cdots \times H_{u-1}$ does not have the Rédei property. Then finally $G = L \times M$ does not have the Rédei property.

Next suppose that $m = 1$, $n \geq 4$. Divide $n - 4$ by 6 with remainder

$$n - 4 = 6u + v, \quad 0 \leq v \leq 5.$$

Now G is a direct product of subgroups

$$H_1, H_2, \ldots, H_{u+1}, K_1, \ldots, K_v$$

such that H_1 is of type (p^α, p, p, p, p), H_i is an elementary p-group of rank 6 for $2 \leq i \leq u+1$ and K_j is a cyclic group of order p. Lemma 9.2.2 gives that $H_{u+1} \times K_1 \times \cdots \times K_v$ does not have the Rédei property. By Lemma 9.2.1, $H_1 \times \cdots \times H_u$ does not have the Rédei property. Finally G does not have the Rédei property.

In the $m = 2$, $n \geq 2$ case divide $n - 2$ by 6 and argue as before to show that G does not have the Rédei property.

In the $m \geq 3$ case divide m by 3 and divide n by 6 with remainder to get a suitable direct product decomposition of G into subgroups to verify that G does not have the Rédei property.

This completes the proof.

We say that a 2-group is bad if it is a direct product of the groups L_1, L_2, L_3 whose types are one of the following:

(a) (2^α), where $\alpha \geq 3$ is an integer,

(b) $(2, 2, 2, 2, 2)$.

(When no factor of type (b) is involved, then the condition $\alpha \geq 3$ can be relaxed to $\alpha \geq 2$.) Let F_2 be the family of 2-groups whose type is one of

$$(\ \underbrace{2, \ldots, 2}_{14}\ ), \quad (2^\alpha, \underbrace{2, \ldots, 2}_{9}\ ), \quad (2^\alpha, 2^\beta, \underbrace{2, \ldots, 2}_{4}\ ),$$

where $\alpha \geq 3$, $\beta \geq 2$ are integers or a subgroup of such a group. The F_2 family is constructed such that its members do not have any bad subgroups and F_2 contains each 2-group that has no bad subgroups.

Theorem 9.2.2. *Let G be a finite abelian 2-group that is not a member of the F_2 family. Then G does not have the Rédei property.*

Proof. First we show that bad groups do not have the Rédei property. There are four choices for the types of L_1, L_2, L_3 in a bad group. The cases are listed in Table 1.

(1) Suppose G is an elementary 2-group of rank 15. Let

$$x_1, \ldots, x_5, \quad y_1, \ldots, y_5, \quad z_1, \ldots, z_5$$

be basis elements of G. Set

$$K_1 = \langle x_1, x_2, x_3 \rangle, \quad H_1 = \langle x_4, x_5 \rangle,$$

$$K_2 = \langle y_1, y_2, y_3 \rangle, \quad H_2 = \langle y_4, y_5 \rangle,$$

$$K_3 = \langle z_1, z_2, z_3 \rangle, \quad H_3 = \langle z_4, z_5 \rangle,$$

$$A_1 = \big(K_1 \setminus \{x_2 x_3, x_1 x_2 x_3\}\big) \cup \big\{(x_2 x_3)x_4, (x_1 x_2 x_3)x_5\big\},$$

$$A_2 = \big(K_2 \setminus \{y_2 y_3, y_1 y_2 y_3\}\big) \cup \big\{(y_2 y_3)y_4, (y_1 y_2 y_3)y_5\big\},$$

$$A_3 = \big(K_3 \setminus \{z_2 z_3, z_1 z_2 z_3\}\big) \cup \big\{(z_2 z_3)z_4, (z_1 z_2 z_3)z_5\big\},$$

$$A = A_1 A_2 A_3, \quad H = H_1 H_2 H_3.$$

Note that $G = HA$ is a factorization of G. Indeed,

$$\begin{aligned} HA_1 A_2 A_3 &= (H_1 A_1)(H_2 A_2)(H_3 A_3) \\ &= \langle x_1, \ldots, x_5 \rangle \langle y_1, \ldots, y_5 \rangle \langle z_1, \ldots, z_5 \rangle \\ &= G. \end{aligned}$$

We construct a subset B from H by adding and removing subsets. Remove

$$\begin{array}{lll} x_4 x_5 H_2, & x_4 x_5 H_2 z_4, & x_4 x_5 H_2 z_5, \\ y_4 y_5 H_3, & y_4 y_5 H_3 x_4, & y_4 y_5 H_3 x_5, \\ z_4 z_5 H_1, & z_4 z_5 H_1 y_4, & z_4 z_5 H_1 y_5. \end{array}$$

Add

$$\begin{array}{lll} x_4 x_5 H_2 y_1, & x_4 x_5 z_4 H_2 y_2, & x_4 x_5 z_5 H_2 y_3, \\ y_4 y_5 H_3 z_1, & y_4 y_5 x_4 H_3 z_2, & y_4 y_5 x_5 H_3 z_3, \\ z_4 z_5 H_1 x_1, & z_4 z_5 y_4 H_1 x_2, & z_4 z_5 y_5 H_1 x_3. \end{array}$$

We claim that $x_4 x_5 H_2 A = x_4 x_5 y_1 H_2 A$. Indeed,

$$\begin{aligned}
x_4 x_5 y_1 H_2 A &= x_4 x_5 y_1 H_2 A_1 A_2 A_3 \\
&= x_4 x_5 y_1 (H_2 A_2) A_1 A_3 \\
&= x_4 x_5 y_1 \langle y_1, \ldots, y_5 \rangle A_1 A_3 \\
&= x_4 x_5 \langle y_1, \ldots, y_5 \rangle A_1 A_3 \\
&= x_4 x_5 (H_2 A_2) A_1 A_3 \\
&= x_4 x_5 H_2 A.
\end{aligned}$$

Each of the removed subsets has a counterpart that is added to M. Similar equations hold for the remaining eight removed and added subsets. One can check that the sets

$$\begin{array}{lll}
x_4 x_5 H_2 A, & x_4 x_5 z_4 H_2 A, & x_4 x_5 z_5 H_2 A, \\
y_4 y_5 H_3 A, & y_4 y_5 x_4 H_3 A, & y_4 y_5 x_5 H_3 A, \\
z_4 z_5 H_1 A, & z_4 z_5 y_4 H_1 A, & z_4 z_5 y_5 H_1 A
\end{array}$$

are disjoint. This can be used to show that $G = AB$ is a factorization of G. Also one can check that $\langle A \rangle = \langle B \rangle = G$.

(2) Let G be a group of type $(2^\beta, 2^\gamma, 2^\delta)$, where $\beta, \gamma, \delta \geq 2$ are integers. Let x, y, z be basis elements of G with $|x| = 2^\beta$, $|y| = 2^\gamma$, $|z| = 2^\delta$. Set

$$A_1 = \{e, x\}, \quad A_2 = \{e, y\}, \quad A_3 = \{e, z\},$$

$$H_1 = \langle x^2 \rangle, \quad H_2 = \langle y^2 \rangle, \quad H_3 = \langle z^2 \rangle,$$

$$H = H_1 H_2 H_3, \quad A = A_1 A_2 A_3.$$

Now $G = AH$ is a factorization of G. Construct B from H by removing

$$x^{2^\beta - 2} H_2, \quad y^{2^\gamma - 2} H_3, \quad z^{2^\delta - 2} H_1$$

and adding

$$x^{2^\beta - 2} H_2 y, \quad y^{2^\gamma - 2} H_3 z, \quad z^{2^\delta - 2} H_1 x.$$

Then $G = AB$ is a factorization of G and $\langle A \rangle = \langle B \rangle = G$. (When the reader verifies $\langle B \rangle = G$ we advise doing first the $\beta, \gamma, \delta \geq 3$ case then separately the $\beta = \gamma = \delta = 2$ case, $\beta = \gamma = 2$, $\delta \geq 3$ case, $\beta = 2$, $\gamma, \delta \geq 3$ case.)

(3) Let G be a group of type $(2^\alpha, 2^\beta, 2, 2, 2, 2, 2)$, where $\alpha \geq 3$, $\beta \geq 2$. Let x, y, $z_1, \ldots, z_5$ be basis elements of G with $|x| = 2^\alpha$, $|y| = 2^\beta$, $|z_1| = \cdots = |z_5| = 2$. Set

$$A_1 = \{e, x\}, \quad A_2 = \{e, y\},$$

$$H_1 = \langle x^2 \rangle, \quad H_2 = \langle y^2 \rangle,$$

$$K_3 = \langle z_1, z_2, z_3 \rangle, \quad H_3 = \langle z_4, z_5 \rangle,$$

$$A_3 = \left(K_3 \setminus \{z_2 z_3, z_1 z_2 z_3\} \right) \cup \left\{ (z_2 z_3) z_4, (z_1 z_2 z_3) z_5 \right\},$$

$$H = H_1 H_2 H_3, \quad A = A_1 A_2 A_3.$$

We can see that $G = AH$ is a factorization of G. Let us construct B from H by removing

$$x^{2^{\alpha}-2}H_2,$$
$$y^{2^{\beta}-2}H_3, \qquad y^{2^{\beta}-2}H_3x^{2^{\alpha}-4}, \qquad y^{2^{\beta}-2}H_3x^{2^{\alpha}-6},$$
$$z_4z_5H_1$$

and adding

$$x^{2^{\alpha}-2}H_2y,$$
$$y^{2^{\beta}-2}H_3z_1, \qquad y^{2^{\beta}-2}x^{2^{\alpha}-4}H_3z_2, \qquad y^{2^{\beta}-2}x^{2^{\alpha}-6}H_3z_3,$$
$$z_4z_5H_1x$$

One can verify that $G = AB$ is a factorization of G and $\langle A \rangle = \langle B \rangle = G$.

(4) Let G be a group of type

$$(2^{\alpha}, \; \underbrace{2, \ldots, 2}_{10} \;),$$

where $\alpha \geq 3$. Let $x, y_1, \ldots, y_5, z_1, \ldots, z_5$ be basis elements of G such that $|x| = 2^{\alpha}$, $|y_1| = \cdots = |y_5| = 2$, $|z_1| = \cdots = |z_5| = 2$. Set

$$A_1 = \{e, x\}, \quad H_1 = \langle x^2 \rangle,$$

$$K_2 = \langle y_1, y_2, y_3 \rangle, \quad H_2 = \langle y_4, y_5 \rangle,$$

$$K_3 = \langle z_1, z_2, z_3 \rangle, \quad H_3 = \langle z_4, z_5 \rangle,$$

$$A_2 = \big(K_2 \setminus \{y_2y_3, y_1y_2y_3\}\big) \cup \big\{(y_2y_3)y_4, (y_1y_2y_3)y_5\big\},$$

$$A_3 = \big(K_3 \setminus \{z_2z_3, z_1z_2z_3\}\big) \cup \big\{(z_2z_3)z_4, (z_1z_2z_3)z_5\big\},$$

$$A = A_1A_2A_3, \quad H = H_1H_2H_3.$$

One can check that $G = AH$ is a factorization of G. Construct B from H by removing

$$x^{2^{\alpha}-2}H_2, \qquad x^{2^{\alpha}-2}H_2z_4, \qquad x^{2^{\alpha}-2}H_2z_5,$$
$$y_4y_5H_3, \qquad y_4y_5H_3x^{2^{\alpha}-4}, \qquad y_4y_5H_3x^{2^{\alpha}-6},$$
$$z_4z_5H_1$$

and adding

$$x^{2^{\alpha}-2}H_2y_1, \qquad x^{2^{\alpha}-2}z_4H_2y_2, \qquad x^{2^{\alpha}-2}z_5H_2y_3,$$
$$y_4y_5H_3z_1, \qquad y_4y_5x^{2^{\alpha}-4}H_3z_2, \qquad y_4y_5x^{2^{\alpha}-6}H_3z_3,$$
$$z_4z_5H_1x.$$

Then $G = AB$ is a factorization of G and $\langle A \rangle = \langle B \rangle = G$.

L_1	a	a	a	a	a	a	b	b	b	c
L_2	a	a	a	b	b	c	b	b	c	c
L_3	a	b	c	b	c	c	b	c	c	c

Table 2

(5) Let G be a group of type

$$(2^{\alpha(1)}, \ldots, 2^{\alpha(m)}, 2^{\beta(1)}, \ldots, 2^{\beta(n)}),$$

where $\alpha(1), \ldots, \alpha(m) \geq 2$, $\beta(1) = \cdots = \beta(n) = 1$.

Suppose first that $m = 0$, $n \geq 15$, that is, G is an elementary 2-group of rank n. Divide n by 15 with remainder

$$n = 15u + v, \quad 0 \leq v \leq 14.$$

Then G is a direct product of subgroups $H_1, \ldots, H_u, K_1, \ldots, K_v$, where H_i is an elementary 2-group of rank 15 and K_j is a cyclic subgroup of order 2. We have seen in step (1) that H_i does not have the Rédei property. By Lemma 9.2.2, $L = H_1 \times \cdots \times H_{u-1}$ does not have the Rédei property. Applying the construction presented in step (1) to H_u gives that H_u does not have the Rédei property. The subset A has $8^3 = 512$ elements. There are 466 elements $a \in A \setminus \{e\}$ such that $\langle A \setminus \{a\} \rangle = H_u$. By Lemma 9.2.2, $H_u \times K_1 \times \cdots \times K_v$ does not have the Rédei property. Then by Lemma 9.2.1, $G = L \times M$ does not have the Rédei property.

Then we can settle the $m = 1$, $n \geq 10$ case, the $m = 2$, $n \geq 5$ case, and the $n \geq 3$ case using similar arguments.

This completes the proof.

We say that a finite abelian group G is bad if $G = L_1 \times L_2 \times L_3$ and each L_i is one of the following types:

(a) (uv), where $u \geq 2$, $v \geq 4$,
(b) (p, p), where $p \geq 3$ is a prime,
(c) $(2, 2, 2, 2, 2)$.

(When no factor of type (c) is involved, the condition $v \geq 4$ can be relaxed to $v \geq 2$.)

Theorem 9.2.3. *Let G be a bad group. Then G does not have the Rédei property.*

Proof. Let $G = L_1 \times L_2 \times L_3$. There are ten choices for the types of L_1, L_2, L_3 listed in Table 2.

In each case we show that G does not have the Rédei property. The constructions are similar to those we have seen in connection with p-groups. The 7th and 10th cases have already been settled. Also note that the 5th, 8th, 9th cases can be reduced to the 1st, 2nd, 3rd cases respectively. Working out two more cases will illustrate the procedure sufficiently.

(1) Let L_i be of type $\big(u(i)v(i)\big)$, $u(i) \geq 2$, $v(i) \geq 4$, with basis element x_i for $1 \leq i \leq 3$. Set

$$A_i = \{e, x_i, x_i^2, \ldots, x_i^{u(i)-1}\}, \quad H_i = \langle x_i^{u(i)} \rangle, \quad 1 \leq i \leq 3,$$

$$H = H_1 H_2 H_3, \quad A = A_1 A_2 A_3.$$

Note that $G = AH$ is a factorization of G. Let us construct a set B from H by removing

$$x_1^{[v(1)-1]u(1)} H_2, \quad x_2^{[v(2)-1]u(2)} H_3, \quad x_3^{[v(3)-1]u(3)} H_1$$

from H and adding

$$x_1^{[v(1)-1]u(1)} H_2 x_2, \quad x_2^{[v(2)-1]u(2)} H_3 x_3, \quad x_3^{[v(3)-1]u(3)} H_1 x_1$$

to H. We can check that $G = AB$ is a factorization of G and $\langle A \rangle = \langle B \rangle = G$.

(2) Let L_1 be of type (u, v), $u \geq 2$, $v \geq 4$. Let L_2 be of type (p, p), where $p \geq 3$ is a prime. Let L_3 be of type $(2, 2, 2, 2, 2)$. Let the basis elements of L_1, L_2, L_3 be x, y_1, y_2, $z_1, \ldots, z_5$ respectively. Set

$$A_1 = \{e, x, x^2, \ldots, x^{u-1}\}, \quad H_1 = \langle x^u \rangle,$$

$$K_2 = \langle y_1 \rangle, \quad H_2 = \langle y_2 \rangle, \quad A_2 = \big(K_2 \setminus \{y_1^{p-1}\}\big) \cup \{y_1 y_2^{p-1}\},$$

$$K_3 = \langle z_1, z_2, z_3 \rangle, \quad H_3 = \langle z_4, z_5 \rangle,$$

$$A_3 = \big(K_3 \setminus \{z_2 z_3, z_1 z_2 z_3\}\big) \cup \big\{(z_2 z_3)z_4, (z_1 z_2 z_3)z_5\big\},$$

$$A = A_1 A_2 A_3, \quad H = H_1 H_2 H_3.$$

One can check that $G = AH$ is a factorization of G. Construct B from H by removing

$$x^{(v-1)u} H_2,$$
$$y_2^{p-1} H_3, \qquad y_2^{p-1} H_3 x^{(v-2)u}, \qquad y_2^{p-1} H_3 x^{(v-3)u},$$
$$z_4 z_5 H_1$$

and adding

$$x^{(v-1)u} H_2 y_1,$$
$$y_2^{p-1} H_3 z_1, \qquad y_2^{p-1} x^{(v-2)u} H_3 z_2, \qquad y_2^{p-1} x^{(v-3)u} H_3 z_3,$$
$$z_4 z_5 H_1 x$$

to H. It can be verified that $G = AB$ is a factorization of G and $\langle A \rangle = \langle B \rangle = G$. This completes the proof.

Let F_{odd} be the family of groups whose type is in Table 3 or a subgroup of such a group. Here p, q, r, s, t are distinct odd primes and $\alpha, \beta \geq 2$ are integers.

(p,p,p,p,p)	(p^{α},p,p,p)	(p^{α},p^{β},p)
(p,p,p,p,q)	(p,p,p,q,q)	(p^{α},p,p,q)
(p^{α},p^{β},q)	(p^{α},p,q,q)	(p^{α},p,q^{β})
(p,p,q,q,r)	(p,p,p,q,r)	(p^{α},q,q,r)
(p^{α},q^{β},r)	(p^{α},p,q,r)	
(p,p,q,r,s)	(p^{α},q,r,s)	
(p,q,r,s,t)		

Table 3

Theorem 9.2.4. *Let G be a finite abelian group of odd order. If G has the Rédei property, then G must be a member of the F_{odd} family.*

Proof. Let G be a finite abelian group of odd order and suppose that G has the Rédei property. Let $p(1),\ldots,p(n)$ be all the distinct prime divisors of $|G|$ and let G_i be the $p(i)$-component of G, $1 \le i \le n$. We have seen in the proof of Theorem 9.2.2 that G_i can be written in the form $H_i \times K_i$ such that H_i is a direct product of bad groups and K_i belongs to the $F_{p(i)}$ family. Consider $K = K_1 \times \cdots \times K_n$ and remember that K is a direct product of cyclic subgroups of prime power order. Group these cyclic factors together to form as many bad groups as possible. Then $K = L \times M$, where L is a direct product of bad groups and from the cyclic direct factors of M it is not possible to form any bad group. It follows that $|M|$ has at most five distinct prime divisors, since otherwise using six cyclic factors of distinct prime power order one could form a bad group.

There are distinct primes p, q, r, s, t such that M is a direct product of five groups each belonging to the $F_p,\ldots,F_t$ families respectively. As a consequence M is a direct product of at most 25 cyclic groups of prime power order. Clearly G is a direct product of a number of bad groups and M. Choose a bad component G' of G. We have seen in the proof of Theorem 9.2.3 that there is a factorization $G' = AB$ such that $\langle A \rangle = \langle B \rangle = G'$. We can also read off from the form of A that there are at least 20 elements $a \in A \setminus \{e\}$ for which $\langle A \setminus \{a\} \rangle = G'$. Also there are at least 20 elements $b \in B \setminus \{e\}$ such that $\langle B \setminus \{b\} \rangle = G'$. By Lemma 9.2.2, $G' \times M$ does not have the Rédei property. (Reversing the roles of A and B if it is necessary.) Then the direct product of $G' \times M$ and all the remaining bad direct factors of G does not have the Rédei property. In other words G does not have the Rédei property. This is contrary to our assumption. It follows that G has no bad direct factor, that is, $G = M$.

It remains to show that M belongs to the F_{odd} family. Now M is a direct product of groups belonging to the $F_p,\ldots,F_t$ families. We can list the types of groups of these families and sort out the combinations in which a bad subgroup occurs.

This completes the proof.

Naturally there is a list of group types similar to the one in Theorem 9.2.4 for finite abelian groups of even order. We do not attempt to compile this list. It looks reasonable to get more detailed results on 2-groups before turning to groups of even order.

9.3 Some groups with the Rédei property

This section presents a few instances when finite abelian groups have the Rédei property. By Theorem 9.1.1, if $p \leq 11$ is a prime and G is of type (p, p, p) then G possesses the Rédei property. Next we show that groups of type (p^α, q^β) have the Rédei property.

Theorem 9.3.1. *If p and q are distinct primes, then the cyclic group G of order $p^\alpha q^\beta$ has the Rédei property.*

Proof. Let $G = AB$ be a factorization of G where $\langle A \rangle = G$ and $\langle B \rangle = G$. As $\langle A \rangle = G$, either A contains an element of order $p^\alpha q^\beta$ or elements a and a' of orders $p^\alpha q^\gamma$, $\gamma < \beta$, and $p^\delta q^\beta$, $\delta < \alpha$. In the second case $a^{-1}a'$ has order $p^\alpha q^\beta$. Thus, in each case AA^{-1} contains an element x of order $p^\alpha q^\beta$. Similarly $\langle B \rangle = G$ implies that BB^{-1} contains an element y of order $p^\alpha q^\beta$.

Let ϕ be an automorphism of G defined by $\phi(x) = y$. There is an integer k such that $y = x^k$. Clearly k is relatively prime to $p^\alpha q^\beta$. Note that $\phi(A) = A^k$. By Lemma 3.2.5, it follows that in the factorization $G = AB$ the factor A can be replaced by $\phi(A)$ to get the factorization $G = \phi(A)B$ of G and so the product $\phi(A)B$ is direct. Hence

$$\phi(A)\left[\phi(A)\right]^{-1} \cap BB^{-1} = \{e\}.$$

But we know that

$$y \in \phi(A)\left[\phi(A)\right]^{-1} \cap BB^{-1}.$$

This contradiction completes the proof.

Theorem 9.3.2. *Let p, q be distinct primes. Groups of type (p, p, q) have the Rédei property.*

Proof. Let G be a group of type (p, p, q) and let x, y, z be basis elements of G with $|x| = |y| = p$ and $|z| = q$. Let $G = AB$ be a normed factorization of G. We will show that $\langle A \rangle \neq G$ or $\langle B \rangle \neq B$

Assume first that $|A| = p$ and $|B| = pq$. If A contains only p-elements, then $A \subset \langle x, y \rangle$ and so $\langle A \rangle \neq G$. We may assume that A contains not only p-elements. By Lemma 6.1.1, A can be replaced by the A' in the factorization $G = AB$, where A' consists of the p-parts of the elements of A. From the factorization $G = A'B$

it follows that the p-parts of the elements of A are distinct. This gives that the p-parts of the elements of $A \setminus \{e\}$ have order p. (The point is that none of them can be e.) There is an element $a \in A$ such that $|a| = pq$. It follows from Lemma 1.4.3 that in the factorization $G = AB$ the factor A can be replaced by the cyclic subset $A'' = \{e, a, a^2, \ldots, a^{p-1}\}$ to get the factorization $G = A''B$. By Lemma 2.1.1, B is periodic. There is a subset C of B such that $B = HC$, where the product is direct and $H = \langle a^p \rangle$. From the factorization $G = AHC$, by considering the factor group G/H, we get the factorization $G/H = (AH)/H \cdot (CH)/H$. As $|H| = |a^p| = q$ it follows that G/H is of type (p, p). By Rédei's theorem $(AH)/H$ or $(CH)/H$ is a subgroup of G/H. If $(AH)/H$ is a subgroup of G/H, then AH is a subgroup of G of order pq. Hence $\langle A \rangle \neq G$. If $(CH)/H$ is a subgroup of G/H, then CH is a subgroup of G of order pq. But $B = CH$ and so $\langle B \rangle \neq G$. The case when $|A| = q$ and $|B| = p^2$ can be settled in a similar way.

This completes the proof.

Lemma 9.3.1. *The Hajós 2-property implies the Rédei property.*

Proof. Let G be a finite abelian group with the Hajós 2-property. We show that if $G = AB$ is a normalized factorization of G, then either $\langle A \rangle \neq G$ or $\langle B \rangle \neq G$. We proceed by induction on n, the number of not necessarily distinct prime factors of $|G|$, that is $n = \nu(|G|)$. If $n = 1$, then since $|G| = |A||B|$ it follows that either $A = G$, $B = \{e\}$ or $A = \{e\}$, $B = G$. Hence either $\langle A \rangle \neq G$ or $\langle B \rangle \neq G$. Suppose that $n \geq 2$. As G has the Hajós 2-property, either A or B is periodic. Assume that B is periodic, that is, $B = HC$, where the product is direct, H is a proper subgroup and C is a subset of G. From the factorization $G = AB = A(HC)$ we get the factorization $G/H = (AH)/H \cdot (BH)/H$ of the factor group G/H. Note that $|G/H| < |G|$ and that G/H also has the Hajós 2-property because of its type so by the inductive assumption either $\langle (AH)/H \rangle \neq G/H$ or $\langle (BH)/H \rangle \neq G/H$. From this it follows that either $\langle A \rangle \neq G$ or $\langle B \rangle \neq G$.

This completes the proof.

The proof of the next theorem is based on counting characters that annihilate a factor in a factorization.

Theorem 9.3.3. *A group of type $(3, 3, 3, 3)$ has the Rédei property.*

Proof. Let G be a group of type $(3, 3, 3, 3)$. Assume the contrary, that $G = AB$ is a normalized factorization of G such that $\langle A \rangle = \langle B \rangle = G$. We may assume that $|A| \leq |B|$ since this is only a matter of relabelling the factors. So either $|A| = 3$ or $|A| = 9$. In the $|A| = 3$ case, $A \setminus \{e\}$ has only two elements that cannot span the whole of G. Thus it is enough to consider the $|A| = |B| = 9$ case only.

If χ is a nonprincipal character of G, then $\chi(G) = 0$. From the factorization $G = AB$ it follows that $0 = \chi(G) = \chi(AB) = \chi(A)\chi(B)$ and so either $\chi(A) = 0$ or $\chi(B) = 0$. There are $3^4 - 1 = 80$ nonprincipal characters of G. Hence either $\chi(A) = 0$ holds for at least 40 nonprincipal characters χ of G or $\chi(B) = 0$ holds for at least 40 nonprincipal characters χ of G. By relabelling, we may assume that $\chi(A) = 0$ holds for at least 40 nonprincipal characters χ of G.

As $\langle A \rangle = G$, there are elements x, y, u, v of A such that x, y, u, v form a basis for G. Indeed, x can be any element of $A \setminus \{e\}$. If $A \setminus \langle x \rangle = \emptyset$, then $\langle A \rangle \subset \langle x \rangle \neq G$. So $A \setminus \langle x \rangle \neq \emptyset$ and y can be any element of this set. If $A \setminus \langle x, y \rangle = \emptyset$, then $\langle A \rangle \subset \langle x, y \rangle \neq G$. Hence $A \setminus \langle x, y \rangle \neq \emptyset$ and u can be any element of this set. Finally, we can choose v to be any element of $A \setminus \langle x, y, u \rangle$.

Let $A = \{e, x, y, u, v, a_1, \ldots, a_4\}$. Here the elements $a_1, \ldots, a_4$ are distinct and they belong to $G \setminus \{e, x, y, u, v\}$. There are

$$\binom{76}{4} = 1\ 282\ 975$$

choices for $a_1, \ldots, a_4$. For each choice we count the number of characters χ of G for which $\chi(A) = 0$. A computer assisted inspection reveals that $\chi(A) = 0$ can hold for at most 38 characters χ of G.

This contradiction completes the proof.

Theorem 9.3.4. *A group of type $(3^2, 3^2)$ has the Rédei property.*

Proof. Let G be a group of type $(3^2, 3^2)$. Assume the contrary, that there is a normalized factorization $G = AB$ of G with $\langle A \rangle = \langle B \rangle = G$. We claim that neither A nor B is periodic. In order to prove the claim assume that A is periodic. Now $A = CH$, where C is a subset and $H \neq \{e\}$ is a subgroup of G and the product CH is direct. From the factorization $G = AB$, by considering the factor group G/H we get the factorization $G/H = (AH)/H \cdot (BH)/H$ of G/H. Because of the type of G/H it has the Hajós 2-property and so the Rédei property. One of the factors does not span G/H and so one of A and B does not span G.

We may assume that $|A| \leq |B|$. Hence $|A| = 3$ or $|A| = 9$. In the $|A| = 3$ case by Lemma 8.2.1 it follows that A or B is periodic. Thus it is enough to consider the $|A| = |B| = 9$ case.

If χ is a nonprincipal character of G, then $\chi(A) = 0$ or $\chi(B) = 0$. There are $3^4 - 1 = 80$ nonprincipal characters of G. By symmetry we may assume that $\chi(A) = 0$ holds for at least 40 characters χ of G.

As $\langle A \rangle = G$, there are elements x, y of A such that x, y form a basis for G. Indeed, there is a basis u, v of G with $|u| = |v| = 9$. The elements of order 3 of G span the subgroup $\langle u^3, v^3 \rangle$ of order 9. Thus if A does not contain an element of order 9, then $\langle A \rangle \neq G$. Let x be any element of A with $|x| = 9$. In the $x = u^\alpha v^\beta$, $0 \leq \alpha, \beta \leq 8$ basis representation of x at least one of α and β must be relatively prime to 9. We assume that α is prime to 9. Note that $\langle x \rangle \cap \langle v \rangle = \{e\}$ and so x, v also form a basis for G. If $A \setminus \langle x \rangle = \emptyset$, then $\langle A \rangle \subset \langle x \rangle \neq G$. Thus $A \setminus \langle x \rangle \neq \emptyset$. Let $y = x^\gamma v^\delta$, $0 \leq \gamma, \delta \leq 8$ be an element of $A \setminus \langle x \rangle$. If for each choice of y $3 \mid \delta$, then $\langle A \rangle \subset \langle x, v^3 \rangle \neq G$. So there is a y for which δ is prime to 3. Now $\langle x \rangle \cap \langle y \rangle = \{e\}$ and x, y form a basis for G.

Let $A = \{e, x, y, a_1, \ldots, a_6\}$, where $a_1, \ldots, a_6$ are distinct elements of $G \setminus \{e, x, y\}$. There are

$$\binom{78}{6} = 256\ 851\ 595$$

choices for $a_1, \ldots, a_6$. For each choice we count the number of characters χ of G with $\chi(A) = 0$. We do this by computing $\chi(A)$. With a little extra work in certain cases we can recognize if A or B is periodic. Namely, if A is periodic with period g, then we may assume that $|g|$ is a prime, that is, $|g| = 3$. In fact we may assume that g is one of x^3, y^3, x^3y^3, x^6y^3. If $\chi(A) \neq 0$ implies $\chi(g) = 1$, then A is periodic with period g. If $\chi(A) = 0$ implies $\chi(g) = 1$, then B is periodic with period g.

A computer assisted inspection shows that $\chi(A) = 0$ holds for at most 38 characters χ of G if A and B are not periodic.

This contradiction completes the proof.

Theorem 9.3.5. *A group of type* $(8, 4)$ *has the Rédei property.*

Proof. Let G be a group of type $(8, 4)$. Assume the contrary, that there is a normalized factorization $G = AB$ of G such that $\langle A \rangle = \langle B \rangle = G$. We claim that it may be assumed that there are elements x, y in A such that $|x| = 8$, $|y| = 4$ and x, y form a basis for G. In order to verify this claim let u, v be a basis for G with $|u| = 8$, $|v| = 4$. The elements of order less than or equal to 4 span the subgroup $\langle u^2, v \rangle$ of order 16. Note that A must contain an element x with $|x| = 8$. Let $x = u^\alpha v^\beta$, where $0 \leq \alpha \leq 7$, $0 \leq \beta \leq 3$. Since $|x| = 8$, it follows that α is odd. We claim that $\langle x \rangle \cap \langle v \rangle = \{e\}$. In order to prove the claim pick an element g from $\langle x \rangle \cap \langle v \rangle$. Now $g = x^\gamma$ and $g = v^\delta$, where $0 \leq \gamma \leq 7$, $0 \leq \delta \leq 3$. From $x^\gamma = v^\delta$ we get $(u^\alpha v^\beta)^\gamma = v^\delta$. It follows that $8 \mid \alpha\gamma$. Since α is odd we get $8 \mid \gamma$. Therefore $g = x^\gamma = e$. So $\langle x \rangle \cap \langle v \rangle = \{e\}$ and x, v form a basis for G. Let $y = x^\gamma v^\delta$, $0 \leq \gamma \leq 7$, $0 \leq \delta \leq 3$ be an element of $A \setminus \langle x \rangle$. If for each choice of y $2 \mid \delta$, then $\langle A \rangle \subset \langle x, v^2 \rangle \neq G$. So there is a y with odd δ. Now $\langle x, y \rangle = G$. If $|y| = 4$, then x, y form a basis for G and we are done. So we may assume that $|y| = 8$, that is, γ is odd. Multiplying the factorization $G = AB$ by x^{-1} we get the factorization $G = Gx^{-1} = (Ax^{-1})B$. Here $e, x^{-1}, yx^{-1} \in Ax^{-1}$. Note that $\langle x^{-1}, yx^{-1} \rangle = \langle x, y \rangle$. Further $|x^{-1}| = 8$ and $|yx^{-1}| < |x| = 8$ since $yx^{-1} = x^{\gamma-1}v^\delta$ and $\gamma - 1$ is even.

Similarly, it may be assumed that there are elements u, v in B such that $|u| = 8$, $|v| = 4$ and u, v is a basis for G.

We may assume that $|A| \leq |B|$. So either $|A| = 2$ or $|A| = 4$. Therefore we may restrict our attention to the $|A| = 4$, $|B| = 8$ case. The proper subgroups of G have the Hajós 2-property and so the Rédei property. Thus we may assume that neither A nor B is periodic.

If χ is a nonprincipal character of G, then either $\chi(A) = 0$ or $\chi(B) = 0$. There are $2^5 - 1 = 31$ nonprincipal characters of G. Let $A = \{e, x, y, a\}$, where x, y is a basis for G with $|x| = 8$, $|y| = 4$ and $a \in G \setminus \{e, x, y\}$. There are $2^5 - 3 = 29$ choices for a. An inspection shows that $\chi(A) = 0$ holds for at most 11 characters χ of G. So $\chi(B) = 0$ must hold for at least 20 characters χ of G.

Let $B = \{e, u, v, b_1, \ldots, b_5\}$, where u, v is a basis for G, with $|u| = 8$, $|v| = 4$ and $b_1, \ldots, b_5$ are distinct elements of $G \setminus \{e, u, v\}$. There are

$$\binom{29}{5} = 118\ 755$$

choices for $b_1, \ldots, b_5$. A computer assisted inspection shows that $\chi(B) = 0$ holds for at most 19 characters χ of G when A and B are not periodic.

This contradiction completes the proof.

Theorem 9.3.6. *A group of type $(2^3, 2^3)$ has the Rédei property.*

Proof. Let G be a group of type $(2^3, 2^3)$. Assume the contrary, that there is a normalized factorization $G = AB$ of G such that $\langle A \rangle = \langle B \rangle = G$. The proper subgroups of G have the Rédei property so we may assume that neither A nor B is periodic. Further we may assume that $|A| \leq |B|$ and so $|A| = 2$ or $|A| = 4$ or $|A| = 8$. In the $|A| = 2$ case clearly $\langle A \rangle \neq G$. In the $|A| = 4$ case Theorem 7.2.7 gives the contradiction that B is periodic.

In the $|A| = 8$ case $|B| = 8$ and so we may assume that $\chi(A) = 0$ holds for at least 32 nonprincipal characters χ of G. As $\langle A \rangle = G$, there are elements x, y in A such that x, y is a basis for G. Let $A = \{e, x, y, a_1, \ldots, a_5\}$, where $a_1, \ldots, a_5$ are distinct elements of $G \setminus \{e, x, y\}$. There are

$$\binom{61}{5} = 5\ 949\ 147$$

choices for $a_1, \ldots, a_5$. An inspection shows that if A and B are not periodic, then $\chi(A)$ holds for at most 31 characters χ of G. This contradiction sorts out the $|A| = 8$ case.

This completes the proof.

Lemma 9.3.2. *Let G be a finite abelian p-group, where p is a prime. Let $G = AB$ be a normalized factorization of G. If $|A| = p$, then either $\langle A \rangle$ is an elementary p-group or B is periodic.*

Proof. If each $a \in A \setminus \{e\}$ is of order p, then $\langle A \rangle$ is of type $(p, \ldots, p)$, that is, $\langle A \rangle$ is an elementary group and so there is nothing to prove. We may assume that there is an $a \in A \setminus \{e\}$ with $|a| \geq p^2$. By Lemma 1.4.3, A can be replaced by the cyclic subset $A' = \{e, a, a^2, \ldots, a^{p-1}\}$ for each $a \in A \setminus \{e\}$. From the factorization $G = A'B$ by Lemma 2.1.1 we get that B is periodic.

This completes the proof.

Theorem 9.3.7. *Let G be a group of type (p^α, p^β), where p is a prime. If $G = AB$ is a normalized factorization of G such that $|A| = p$, then either $\langle A \rangle \neq G$ or $\langle B \rangle \neq G$.*

Proof. We may assume that $\alpha \geq \beta$. We proceed by induction on $n = \alpha + \beta$. If $n = 1$, then $|A| = p$ and $|B| = 1$. Hence $B = \{e\}$ and so $\langle B \rangle \neq G$. If $n = 2$, then

$|A| = |B| = p$ so by Rédei's theorem A or B is a subgroup of G and so $\langle A \rangle \neq G$ or $\langle B \rangle \neq G$. For the remaining part of the proof suppose that $n \geq 3$. Now $\alpha \geq 2$ and consequently G is not an elementary group. By Lemma 9.3.2, either $\langle A \rangle$ is an elementary group or B is periodic. If $\langle A \rangle$ is an elementary group, then $\langle A \rangle$ cannot be equal to G and so we are done. Assume that B is periodic, that is, $B = HC$, where the product is direct H is a proper subgroup and C is a subset of G. From the factorization $G = AB = A(HC)$ we get the factorization $G/H = (AH)/H \cdot (BH)/H$ of the factor group G/H. Since $|G/H| < |G|$ and since the type of G/H has not changed by the inductive hypothesis, either $\langle (AH)/H \rangle \neq G/H$ or $\langle (BH)/H \rangle \neq G/H$. From this it follows that either $\langle A \rangle \neq G$ or $\langle B \rangle \neq G$.

This completes the proof.

Corollary 9.3.1. *Groups of type (p^2, p) have the Rédei property for each prime p.*

Proof. Let p be a prime and let $G = AB$ be a normalized factorization, where G is a group of type (p^2, p). As $|G| = |A||B|$, by relabelling we may assume that $|A| = p$ and $|B| = p^2$. By Theorem 9.3.7, either $\langle A \rangle \neq G$ or $\langle B \rangle \neq G$.

This completes the proof.

Theorem 9.3.8. *Let G be a group of type $(p^\alpha, p^\beta, p^\gamma)$, where p is a prime $p \leq 11$. If $G = AB$ is a normalized factorization of G such that $|A| = p$, then either $\langle A \rangle \neq G$ or $\langle B \rangle \neq G$.*

Proof. We may assume that $\alpha \geq \beta \geq \gamma$. The $\gamma = 0$ case is covered by Theorem 9.3.1 so we assume that $\gamma \geq 1$. We proceed by induction on $n = \alpha + \beta + \gamma$. If $n = 3$, then $|A| = p$, $|B| = p^2$ and as $\gamma \geq 1$, G is of type (p, p, p). Hence by Theorem 9.1.1, $\langle A \rangle \neq G$ or $\langle B \rangle \neq G$. In the remaining part of the proof we suppose that $n \geq 4$. Now G cannot be an elementary group. By Lemma 9.3.2, either $\langle A \rangle$ is an elementary group or B is periodic. If $\langle A \rangle$ is an elementary group, then $\langle A \rangle \neq G$ and we are done. So we may assume that B is periodic, that is, $B = HC$, where the product is direct, H is a proper subgroup and C is a subset of G. From the factorization $G/H = (AH)/H \cdot (BH)/H$ of the factor group G/H in the way we have seen in the proof of Theorem 9.3.7, it follows that $\langle A \rangle \neq G$ or $\langle B \rangle \neq G$.

This completes the proof.

We prove two lemmas essentially about groups of types $(4, 4, 4)$ and $(4, 4, 2)$.

Lemma 9.3.3. *If $G = AB$ is a normalized factorization of the finite abelian group G, where $|A| = 4$ and $\langle A \rangle$ is of type $(4, 4, 4)$, then B is periodic.*

Proof. Let $A = \{e, x, y, z\}$. As $\langle A \rangle$ is of type $(4, 4, 4)$ it follows that $|x| = |y| = |z| = 4$ and $\langle x, y, z \rangle$ is of type $(4, 4, 4)$. Set $g = (xyz)^2$. We claim that $B = Bg$, that is B is periodic with period g. To prove the claim it is enough to verify that

$$\chi(B) = \chi(B)\chi(g) \tag{1}$$

for each character χ of G. If $\chi(B) = 0$ or $\chi(g) = 1$, then the equation (1) clearly holds. Suppose there is a character χ of G with $\chi(g) \neq 1$ and $\chi(B) \neq 0$. Now χ is not the principal character of G. Applying χ to the factorization $G = AB$ gives that $0 = \chi(G) = \chi(A)\chi(B)$. As $\chi(B) \neq 0$ we must have $\chi(A) = 0$, that is

$$0 = \chi(e) + \chi(x) + \chi(y) + \chi(z).$$

The four roots of unity lemma tells us that we may assume that $1 + \chi(x) = 0$ and $\chi(y) + \chi(z) = 0$. Using these we get

$$\begin{aligned}
\chi(g) &= \chi(x^2)\chi(y^2)\chi(z^2) \\
&= (-1)^2[\chi(y)]^2[-\chi(y)]^2 \\
&= [\chi(y)]^4 \\
&= 1.
\end{aligned}$$

This contradiction completes the proof.

Lemma 9.3.4. *Let $G = AB$ be a normalized factorization, where G is of type $(4, 4, 2)$, $|A| = 4$, $\langle A \rangle = G$. Then B is periodic.*

Proof. As $\langle A \rangle = G$, there is a basis x, y, z of G such that $|x| = |y| = 4$, $|z| = 2$ and $A = \{e, x, y, az\}$, where $a \in \langle x, y \rangle$. Set $g = (xyaz)^2$. In the way we have seen in the proof of Lemma 9.3.3 it follows that $B = Bg$. Thus if $g \neq e$, then B is periodic with period g. Let $a = x^\alpha y^\beta$, $0 \leq \alpha, \beta \leq 3$. Clearly $g = e$ only when $2 \mid (\alpha + 1)$, $2 \mid (\beta + 1)$. This leaves the following four possibilities for A.

$$A = \{e, x, y, xyz\}, \quad A = \{e, x, y, xy^3 z\},$$

$$A = \{e, x, y, x^3 yz\}, \quad A = \{e, x, y, x^3 y^3 z\}.$$

We will show that in these cases z is a period of B. It is enough to verify that

$$\chi(B) = \chi(B)\chi(z) \tag{2}$$

for each character χ of G. If $\chi(B) =$ or $\chi(z) = 1$, then (2) obviously holds. Assume that $\chi(B) \neq 0$ and $\chi(z) \neq 1$. Now χ is not the principal character of G. Applying χ to the factorization $G = AB$ we get $0 = \chi(G) = \chi(A)\chi(B)$. It follows that $\chi(A) = 0$. We claim that for all the four choices of A, $\chi(A) = 0$ implies $\chi(z) = 1$. To prove the claim assume that $A = \{e, x, y, xyz\}$. From

$$0 = 1 + \chi(x) + \chi(y) + \chi(xyz)$$

by the four roots of unity lemma we face the following possibilities

$$\chi(x) = -1, \quad \chi(y) = -\chi(xyz),$$

$$\chi(y) = -1, \quad \chi(x) = -\chi(xyz),$$

$$\chi(xyz) = -1, \quad \chi(x) = -\chi(y).$$

A routine computation shows that $\chi(z) = 1$ holds in each case. A similar argument works for the remaining three choices for A.

This completes the proof.

Theorem 9.3.9. *Let G be a group of type $(2^{\alpha(1)}, \ldots, 2^{\alpha(s)})$, where $1 \leq \alpha(i) \leq 2$ for each i, $1 \leq i \leq s$. If $G = AB$ is a normalized factorization of G, where $|A| = 2$ or $|A| = 4$, then either $\langle A \rangle \neq G$ or $\langle B \rangle \neq G$.*

Proof. We may assume that $\langle A \rangle = G$ since otherwise there is nothing to prove. If $|A| = 2$, then A contains only one nonidentity element. Hence G must be of type (2) or (4) by the assumption on the type of G. In these cases G has the Hajós 2-property and consequently the Rédei property.

Turn to the case when $|A| = 4$. A contains three nonidentity elements and so by the type of G it follows that the type of $\langle A \rangle = G$ is one of

$$(2, 2, 2), \quad (4, 2, 2), \quad (4, 4, 2), \quad (4, 4, 4).$$

In the first two cases G has the Hajós 2-property and so $\langle B \rangle \neq G$. In the last two cases by Lemma 9.3.4 and by Lemma 9.3.3, B is periodic, that is, $B = HC$, where the product is direct and H is a proper subgroup of G. Considering the factorization $G/H = (AH)/H \cdot (BH)/H$ of the factor group G/H, completes the proof in the known way.

Corollary 9.3.2. *A group of type $(4, 4, 2)$ has the Rédei property.*

Proof. Let G be a group of type $(4, 4, 2)$ and let $G = AB$ be a normalized factorization of G. We may assume that $|A| \leq |B|$ and so either $|A| = 2$ or $|A| = 4$. Further we may assume that $\langle A \rangle = G$ since otherwise there is nothing to prove. When $|A| = 2$, A has only one nonidentity element and in this case A cannot span the whole of G. So $|A| = 4$. Now by Lemma 9.3.4, B is periodic and considering a factor group completes the proof.

If G is an elementary 2-group with $|G| \leq 2^5$, then G has the Hajós 2-property and so by Lemma 9.3.1 the Rédei property. Next we show that elementary 2-groups of rank 6 and 7 have the Rédei property.

Theorem 9.3.10. *If G is of type $(2, 2, 2, 2, 2, 2)$, then G has the Rédei property.*

Proof. Let $G = AB$ be a normalized factorization of G. We may assume that $|A| \leq |B|$. When $|A| \leq 4$, then A contains only three nonidentity elements and so $\langle A \rangle$ cannot be equal to G. Thus $|A| = |B| = 8$. G has 64 characters and so 63 nonprincipal characters. From $0 = \chi(G) = \chi(A)\chi(B)$ it follows that $\chi(A) = 0$ or $\chi(B) = 0$ holds for at least 32 characters χ of G. For the sake of definiteness we assume that $\chi(A) = 0$ holds for at least 32 characters of G. In other words $|\mathrm{Ann}(A)| \geq 32$. As $\langle A \rangle = G$, A is in the form $A = \{e, x_1, \ldots, x_6, a\}$, where $x_1, \ldots, x_6$ is a basis of G. The element a has the representation

$$a = x_1^{\alpha(1)} \cdots x_6^{\alpha(6)}, \quad 0 \leq \alpha(i) \leq 1.$$

Because of the symmetry in the roles of $x_1, \ldots, x_6$ we may assume that $a = x_1 \cdots x_s$ with some $2 \le s \le 6$.

Note that $\chi(A) = 0$ is equivalent to exactly four of

$$\chi(e), \chi(x_1), \ldots, \chi(x_6), \chi(a)$$

being equal to 1 and exactly four of them equal to -1. Let u be the number of -1's among $\chi(x_1), \ldots, \chi(x_s)$ and let v be the number of -1's among $\chi(x_{s+1}), \ldots, \chi(x_6)$. Clearly $0 \le u \le s$, $0 \le v \le 6 - s$ and $u + v = 4$, $2 \mid u$ if $\chi(a) = 1$ and $u + v = 3$, $2 \nmid u$ if $\chi(a) = -1$. The number of characters χ of G with $\chi(A) = 0$ is

$$w = \sum \binom{s}{u}\binom{6-s}{v},$$

where the summation extends over all possible (u, v) pairs.

Assume that $a = x_1 \cdots x_5$, that is, $s = 5$. If $\chi(a) = 1$, then $u = 4$, $v = 0$. If $\chi(a) = -1$, then $u = 3$, $v = 0$. Hence

$$w = \binom{5}{4}\binom{1}{0} + \binom{5}{3}\binom{1}{0} = 15.$$

Assume that $a = x_1 \cdots x_4$, that is, $s = 4$. If $\chi(a) = 1$, then $u = 2$, $v = 2$ or $u = 4$, $v = 0$. If $\chi(a) = -1$, then $u = 1$, $v = 2$ or $u = 3$, $v = 0$. Now

$$w = \binom{4}{2}\binom{2}{2} + \binom{4}{4}\binom{2}{0} + \binom{4}{1}\binom{2}{2} + \binom{4}{3}\binom{2}{0} = 15.$$

Assume that $a = x_1 x_2 x_3$, that is, $s = 3$. If $\chi(a) = 1$, then $u = 2$, $v = 2$. If $\chi(a) = -1$, then $u = 1$, $v = 2$ or $u = 3$, $v = 0$. Therefore

$$w = \binom{3}{2}\binom{3}{2} + \binom{3}{1}\binom{3}{2} + \binom{3}{3}\binom{3}{0} = 19.$$

Assume that $a = x_1 x_2$, that is, $s = 2$. If $\chi(a) = 1$, then $u = 0$, $v = 4$ or $u = 2$, $v = 2$. If $\chi(a) = -1$, then $u = 1$, $v = 3$ and so

$$w = \binom{2}{0}\binom{4}{4} + \binom{2}{2}\binom{4}{2} + \binom{2}{1}\binom{4}{2} = 19.$$

The above argument can be summarized such that we get the $32 \le |\mathrm{Ann}(A)| \le 19$ contradiction unless $a = x_1 \cdots x_6$. Thus

$$A = \{e, x_1, \ldots, x_6, x_1 \cdots x_6\}.$$

The fact that $G = AB$ is a factorization is equivalent to the condition $AA^{-1} \cap BB^{-1} = \{e\}$ and $|A| = |B| = 8$. We say that the quantity $\alpha(1) + \cdots + \alpha(6)$ is the weight of the element $x_1^{\alpha(1)} \cdots x_6^{\alpha(6)}$ of G. The set $AA^{-1} = AA$ consists of all

1	0	0	1	0	1
1	0	0	0	1	1
0	1	0	1	0	1
0	0	1	1	1	0
0	1	0	0	1	1
1	0	1	1	0	1
1	0	1	0	1	1
0	1	1	1	0	1
0	1	1	0	1	1

Table 1 (left):

1	1	0	0	1	1
1	0	1	0	1	1
1	0	0	1	1	1
0	1	1	0	1	1
0	1	0	1	1	1
0	0	1	1	1	1

Table 1   Table 2

the elements of G whose weight is $0, 1, 2, 5, 6$. Consequently the set $BB^{-1} = BB$ consists of all the elements of G whose weight is $0, 3, 4$. Since $e \in B$ it follows that $B \subset BB$ and the elements of $B \setminus \{e\}$ must have weight $3, 4$.

Let us define a graph Γ. The vertices of Γ are the elements of G with weight $3, 4$. The vertices $h, g \in G$ are connected if gh has weight 3 or 4. If with the $B = \{e, b_1, \ldots, b_7\}$ choice $G = AB$ is a factorization of G, then the nodes $b_1, \ldots, b_7$ of Γ are mutually connected. We will show that no such B exists. If each element $B \setminus \{e\}$ has weight 4, then we may assume that $b_1 = x_1 x_2 x_3 \in B$. Then note that there are exactly six nodes of weight 4 that are connected to b_1. These are listed in Table 1 using only the exponents to describe the elements.

Let χ_i be the character of G defined by $\chi_i(x_i) = -1$, $\chi_i(x_j) = 1$ if $i \neq j$. Clearly $\chi_i(A) \neq 0$. Now the elements of B are fixed and we get the $\chi_5(B) \neq 0$ contradiction. Thus B contains an element of weight 3, say $b_1 = x_1 x_2 x_3 \in B$. If each element of $B \setminus \{e\}$ has weight 3, then we may assume that $b_2 = x_1 x_4 x_5 \in B$. Note that there are four elements of weight 3 connected to both b_1 and b_2. Thus B contains an element of weight 4. We will assume that $b_1 = x_1 x_2 x_3$, $b_2 = x_1 x_2 x_4 x_5$ are elements of B. There are nine nodes of Γ that are connected to both b_1 and b_2. These elements are listed in Table 2. Some six of these elements must belong to B. Note that for each choice of these six elements $\chi_6(B) \neq 0$.

This completes the proof.

Theorem 9.3.11. *If G is an elementary 2-group of rank 7, then G has the Rédei property.*

Proof. Let $G = AB$ be a normalized factorization of G and assume the contrary, that $\langle A \rangle = G$, $\langle B \rangle = G$. We may assume that $|A| \leq |B|$. If $|A| = 2$ or $|A| = 4$, then $\langle A \rangle \neq G$. So we may assume that $|A| \geq 8$. Now $|A| = 8$, $|B| = 16$. Clearly A is in the form $A = \{e, x_1, \ldots, x_7\}$, where $x_1, \ldots, x_7$ is a basis for G. Set $g = x_1 \cdots x_7$. We claim that B is periodic with period g. We will prove that

$$\chi(B) = \chi(B)\chi(g) \qquad (3)$$

for each character χ of G. If $\chi(B) = 0$ or $\chi(g) = 1$, then (3) obviously holds.

Suppose there is a character χ of G such that $\chi(B) \neq 0$, $\chi(g) \neq 1$. Now χ is not the principal character of G. $0 = \chi(G) = \chi(A)\chi(B)$ implies that $\chi(A) = 0$.

$$0 = \chi(e) + \chi(x_1) + \cdots + \chi(x_7)$$

shows that four of $\chi(x_1), \ldots, \chi(x_7)$ are -1 and three of them are 1. Thus $\chi(g) = 1$.

Using the fact that B is periodic in the known way we can draw the conclusion that $\langle A \rangle \neq G$ or $\langle B \rangle \neq G$.

This contradiction completes the proof.

Chapter 10

Infinite groups

10.1 Infinite constructions

We extend the concept of factorization to the case of infinite abelian groups. We will restrict our attention to factorizations where each factor contains the identity element of the group. Let G be an abelian group. Let I be an index set and let A_i, $i \in I$ be a family of subsets of G such that $e \in A_i$ for each $i \in I$. If each element g of G can be expressed uniquely in the form

$$g = \prod_{i \in I} a_i, \quad a_i \in A_i,$$

where there are only finitely many among the elements a_i for which $a_i \neq e$, then we say that G is factored into its subsets A_i or

$$G = \prod_{i \in I} A_i$$

is a factorization of G. We will show that many of the results that hold for finite abelian groups do not hold for the infinite case or can hold only for a restricted range of infinite abelian groups.

We say that Hajós's theorem does not hold for a group G if G can be factored into (finite) cyclic subsets such that none of the factors is a subgroup of G. If G does not admit any factorization into finite cyclic subsets, or in every factorization of G into cyclic subsets at least one of the factors is always a subgroup of G, then Hajós's theorem holds for G.

Lemma 10.1.1. *If G is one of the types*
> *(1) (∞),*
> *(2) $\left[r(1)s(1), r(2)s(2), \ldots \right]$,*
where $r(1), s(1), r(2), s(2), \ldots$ are integers such that $2 \leq s(1) \leq r(2)$, $2 \leq s(2) \leq r(3), \ldots$ then Hajós's theorem does not hold for G.

Proof. Let G be an infinite cyclic group with generator element x. In other words G is of type (∞) and $G = \langle x \rangle$. We will show that Hajós's theorem does not hold for G. Set $A_i = \{e, x^{(-2)^i}\}$, $0 \leq i < \infty$. Plainly A_i is not a subgroup of G for any i, $0 \leq i < \infty$. We claim that $G = A_0 A_1 A_2 \cdots$ is a factorization of G. In order to prove the claim choose a $g \in G$. Now $g = x^\alpha$, where α is an integer. Every integer α can be represented uniquely in the form

$$\alpha = \alpha(0)(-2)^0 + \alpha(1)(-2)^1 + \cdots + \alpha(s)(-2)^s,$$

where $0 \leq \alpha(0), \ldots, \alpha(s) \leq 1$. (We are representing α in a number system where the base is (-2) instead of the commonly used 10.) Therefore

$$
\begin{aligned}
g &= x^\alpha \\
&= \left[x^{\alpha(0)}\right]\left[(x^{-2})^{\alpha(1)}\right] \cdots \left[(x^{(-2)^s})^{\alpha(s)}\right] \\
&= a_0 a_1 \cdots a_s,
\end{aligned}
$$

where $a_0 \in A_0, \ldots, a_s \in A_s$.

To prove the second half of the lemma consider a group G of type

$$\big(r(1)s(1), r(2)s(2), \ldots\big), \quad 2 \leq s(1) \leq r(2), 2 \leq s(2) \leq r(3), \ldots$$

with basis elements $x_1, x_2, x_3, \ldots$ such that

$$|x_1| = r(1)s(1), \ |x_2| = r(2)s(2), \ldots$$

Set

$$A_i = \left\{e, x_i, x_i^2, \ldots, x_i^{r(i)-1}\right\},$$

$$B_i = \left\{e, x_i^{r(i)} x_{i+1}^{-1}, (x_i^{r(i)} x_{i+1}^{-1})^2 \ldots, (x_i^{r(i)} x_{i+1}^{-1})^{s(i)-1}\right\}$$

for each i, $1 \leq i < \infty$. Obviously A_i, B_i are not subgroups of G for any i, $1 \leq i < \infty$. We claim that $G = A_1 B_1 A_2 B_2 \cdots$ is a factorization of G. To verify the claim choose a $g \in G \setminus \{e\}$ and consider the representation of g in the basis $x_1, x_2, \ldots$. There are indices i, j such that

$$g = x_i^{\alpha(i)} \cdots x_j^{\alpha(j)},$$

where

$$0 \leq \alpha(i) \leq r(i)s(i) - 1, \ldots, 0 \leq \alpha(j) \leq r(j)s(j) - 1$$

and $\alpha(i) \neq 0$, $\alpha(j) \neq 0$. We call $n = j - i + 1$ the length of g. We proceed by induction on n to prove that g can be represented in the form

$$g = \prod_{i=1}^{\infty} a_i b_i$$

such that only finitely many of the elements a_i, b_i are not e.

Assume first that $n = 1$. Divide $\alpha(i)$ by $r(i)$ with remainder to get

$$\alpha(i) = \beta + \gamma r(i), \quad 0 \le \beta \le r(i) - 1.$$

From $\alpha(i) < r(i)s(i)$ it follows that $\gamma < s(i)$. So

$$
\begin{aligned}
g &= x_i^{\alpha(i)} \\
&= x_i^{\beta} x_i^{\gamma r(i)} \\
&= \left[x_i^{\beta}\right]\left[\left(x_i^{r(i)} x_{i+1}^{-1}\right)^{\gamma}\right]\left[x_{i+1}^{\gamma}\right] \\
&= a_i b_i a_{i+1},
\end{aligned}
$$

where $a_i \in A_i$, $b_i \in B_i$, $a_{i+1} \in A_{i+1}$.

Next assume that $n \ge 2$. Divide $\alpha(i)$ by $r(i)$ with remainder to get

$$\alpha(i) = \beta + \gamma r(i), \quad 0 \le \beta \le r(i) - 1.$$

It follows that $\gamma \le s(i) - 1$. Hence

$$
\begin{aligned}
g &= x_i^{\alpha(i)} \cdots x_j^{\alpha(j)} \\
&= x_i^{\beta} x_i^{\gamma r(i)} x_{i+1}^{\alpha(i+1)} \cdots x_j^{\alpha(j)} \\
&= \left[x_i^{\beta}\right]\left[\left(x_i^{r(i)} x_{i+1}^{-1}\right)^{\gamma}\right]\left[x_{i+1}^{\alpha(i+1)+\gamma} \cdots x_j^{\alpha(j)}\right] \\
&= a_i b_i g',
\end{aligned}
$$

where $a_i \in A_i$, $b_i \in B_i$, and the length of g' is smaller than the length of g.

To prove the uniqueness of the representation of g assume that both

$$g = a_1 b_1 \cdots a_m b_m,$$

$$g = a_1' b_1' \cdots a_m' b_m'$$

are representations of g, where

$$a_1, a_1' \in A_1, \quad b_1, b_1' \in B_1, \ldots$$

Let i be the first index for which $a_i b_i \ne a_i' b_i'$. If $a_i \ne a_i'$, then we get the contradiction that in two representations of g in the basis $x_1, x_2, \ldots$ the exponents of x_i are not equal. If $b_i \ne b_i'$, then we get a similar contradiction with the exponents of x_{i+1}.

This completes the proof.

Exercise 10.1.1. Let A_i be the subset of an infinite cyclic group defined in the proof of Lemma 10.1.1. Compute the products $A_0 A_1$, $A_0 A_1 A_2$.

Exercise 10.1.2. Read Lemma 1.3.1 critically to check that it holds also for infinite abelian groups.

We say that Rédei's theorem does not hold for an abelian group G if G has a factoring into nonsubgroup normalized subsets of prime cardinalities. If G does not have any factorization into normalized subsets of prime cardinalities, or whenever G admits such a factorization it follows that at least one of the factors is always a subgroup, we say that Rédei's theorem holds for G.

Lemma 10.1.2. *Let G be a group of type $[r(1), r(2), \ldots]$, where $r(1), r(2), \ldots$ are integers such that $3 \le r(1), r(2), \ldots$. Then Rédei's theorem does not hold for G.*

Proof. Let $x_1, x_2, x_3, \ldots$ be a basis for G such that $|x_1| = r(1)$, $|x_2| = r(2), \ldots$. Set

$$A_i = \left\{ e, x_i, x_i^2, \ldots, x_i^{r(i)-2}, x_i^{r(i)-1} x_{i+1}^2 \right\}$$

for each i, $1 \le i < \infty$. Obviously A_i is not a subgroup of G. We claim that $G = A_1 A_2 A_3 \cdots$ is a factorization of G. To prove the claim choose a $g \in G \setminus \{e\}$ and consider the representation of g in the basis $x_1, x_2, \ldots$. There are indices i, j such that

$$g = x_i^{\alpha(i)} \cdots x_j^{\alpha(j)},$$

where

$$0 \le \alpha(i) \le r(i) - 1, \ldots, 0 \le \alpha(j) \le r(j) - 1$$

and $\alpha(i) \ne 0$, $\alpha(j) \ne 0$. We call $n = j - i + 1$ the length of g. Using an induction on n we show that g has a representation in the form

$$g = \prod_{i=1}^{\infty} a_i, \quad a_i \in A_i$$

such that only finitely many of the a_i elements are not e.

Assume first that $n = 1$. Now $g = x_i^{\alpha(i)}$. If $1 \le \alpha(i) \le r(i) - 2$, then with the $a_i = x^{\alpha(i)}$ choice $g = a_i$, $a_i \in A_i$. If $\alpha(i) = r(i) - 1$, then

$$
\begin{aligned}
g &= x_i^{p-1} \\
&= \left[x_i^{r(i)-1} x_{i+1}^2 \right] \left[x_{i+1}^{r(i+1)-2} \right] \\
&= a_i a_{i+1},
\end{aligned}
$$

where $a_i \in A_i$, $a_{i+1} \in A_{i+1}$.

Assume next that $n \ge 2$. If $1 \le \alpha(i) \le r(i) - 2$, then

$$g = \left[x_i^{\alpha(i)}\right]\left[x_{i+1}^{\alpha(i+1)} \cdots x_j^{\alpha(j)}\right]$$
$$= a_i g',$$

where $a_i \in A_i$ and the length of g' is smaller than the length of g. If $\alpha(i) = r(i)-1$, then

$$g = x_i^{r(i)-1} \cdots x_j^{\alpha(j)}$$
$$= \left[x_i^{r(i)-1} x_{i+1}^2\right]\left[x_{i+1}^{\alpha(i+1)-2} x_{i+2}^{\alpha(i+2)} \cdots x_j^{\alpha(j)}\right]$$
$$= a_i g',$$

where $a_i \in A_i$ and the length of g' is smaller than the length of g.

In order to verify the uniqueness of the representation of g assume that

$$g = a_1 \cdots a_m, \quad g = a_1' \cdots a_n'$$

are representations of g with $a_1, a_1' \in A_1$, $a_2, a_2' \in A_2, \ldots$ Let i be the first index for which $a_i \neq a_i'$. The possibilities for the representation of a_i in the basis $x_1, x_2, \ldots$ are

$$a_i = x_i^{r(i)-1} x_{i+1}^2, \quad a_i = x_i^{\alpha(i)}, \ 0 \leq \alpha(i) \leq r(i) - 2.$$

The possibilities for the basis representation of a_i' are

$$a_i' = x_i^{r(i)-1} x_{i+1}^2, \quad a_i' = x_i^{\beta(i)}, \ 0 \leq \beta(i) \leq r(i) - 2.$$

This gives that in two representations of g in the basis $x_1, x_2, \ldots$ the exponents of x_i are not equal.

This completes the proof.

A variant of Lemmas 1.4.1–1.4.3 hold for infinite abelian groups if we assume that A is a finite subset. In our applications $\langle A \rangle$ is a finite subgroup of G so the following result will be sufficient for our purposes.

Lemma 10.1.3. *Let G be an abelian group and let A, B be subsets of G such that $\langle A \rangle$ is finite. Assume that $e \in A$, $|A| = p$ is a prime and $G = AB$ is a factorization of G. Then A can be replaced by A' where either $A' = A^t$, p does not divide t or $A' = \{e, a, a^2, \ldots, a^{p-1}\}$, $a \in A \setminus \{e\}$.*

Proof. Let $H = \langle A \rangle$. The factorization $G = AB$ is equivalent to the sets $Ab, b \in B$ forming a partition of G. We can choose a complete set of representatives of G modulo H from B. Let b_i, $i \in I$ be a complete set of representatives of G modulo H chosen from B. Now

$$G = \bigcup_{i \in I} Hb_i = \bigcup_{i \in I} A(B \cap Hb_i).$$

Clearly $H = A(Bb_i^{-1} \cap H)$ is a factorization of H. As H is a finite group, Lemmas 1.4.2, 1.4.3 are applicable and we get that $H = A'(Bb_i^{-1} \cap H)$ is a factorization of

H. This gives that

$$G = \bigcup_{i \in I} A'(B \cap Hb_i) = A'B.$$

This completes the proof.

We say that the simulation theorem does not hold for an abelian group G if G can be factored into nonsubgroup (finite) simulated subsets. If G does not admit any factorization into finite simulated subsets or in each factorization of G into simulated subsets one of the factors is always a subgroup, then we say that the simulation theorem holds for G.

Lemma 10.1.2 shows that if G is of type $[r(1), r(2), \ldots]$ with $3 \leq r(1), r(2), \ldots$, then the simulation theorem does not hold for G.

Lemma 10.1.4. *If G is a group of type $(2, 2, 2, \ldots)$, then G can be factored into nonsubgroup simulated subsets.*

Proof. Let G be a group of type $(2, 2, 2, \ldots)$ with basis elements $x_1, x_2, x_3, \ldots$. Set

$$A_i = \{e, x_i, x_{i+1}, x_i x_{i+1} x_{i+2}\}, \quad 1 \leq i < \infty.$$

Clearly A_i is a nonsubgroup simulated subset of G. We claim that $G = A_1 A_3 A_5 \cdots$ is a factorization of G. To prove the claim choose a $g \in G \setminus \{e\}$ and consider the representation of g in the basis $x_1, x_2, x_3, \ldots$. There are indices i, j such that

$$g = x_i^{\alpha(i)} \cdots x_j^{\alpha(j)}, \quad 0 \leq \alpha(i), \ldots, \alpha(j) \leq 1$$

and $\alpha(i) = \alpha(j) = 1$. We call $n = j - i + 1$ the length of g. We proceed by induction on n to show that g can be represented in the form

$$g = \prod_{i=0}^{\infty} a_{2i+1}, \quad a_{2i+1} \in A_{2i+1},$$

where only finitely many a_{2i+1} is not e.

Suppose first that $n = 1$. This means that $g = x_i$. If i is odd, then $x_i \in A_i$. If i is even, then $x_i \in A_{i-1}$. Thus g has the desired representation.

Suppose next that $n = 2$. This means that $g = x_i x_{i+1}$. If i is odd, then

$$
\begin{aligned}
g &= x_i x_{i+1} \\
&= [x_i x_{i+1} x_{i+2}][x_{i+2}] \\
&= a_i a_{i+2},
\end{aligned}
$$

where $a_i \in A_i$, $a_{i+2} \in A_{i+2}$. If i is even, then $i + 1$ is odd and $g = x_i x_{i+1} = a_{i-1} a_{i+1}$, where $a_{i-1} \in A_{i-1}$, $a_{i+1} \in A_{i+1}$.

Assume next that $n = 3$. Now $g = x_i x_{i+1} x_{i+2}$ or $g = x_i x_{i+2}$. Suppose that $g = x_i x_{i+1} x_{i+2}$. If i is odd, then $g = a_i$, where $a_i \in A_i$. If i is even, then

$$g = [x_i][x_{i+1} x_{i+2}] = a_{i-1} g',$$

where $a_{i-1} \in A_{i-1}$ and the length of g' is smaller than the length of g. Suppose that $g = x_i x_{i+2}$. If i is odd, then $g = a_i a_{i+2}$ with $a_i \in A_i$, $a_{i+2} \in A_{i+2}$. If i is even, then $g = a_{i-1} a_{i+1}$, where $a_{i-1} \in A_{i-1}$, $a_{i+1} \in A_{i+1}$.

Assume finally that $n \geq 4$. If i is even, then $g = a_{i-1} g'$, where $a_{i-1} \in A_{i-1}$ and the length of g' is smaller than the length of g. For the remaining part we may assume that i is odd. If $\alpha(i+1) = 0$, then $g = a_i g'$, where $a_i \in A_i$ and the length of g' is smaller than the length of g. We may assume that $\alpha(i+1) = 1$. If $\alpha(i+2) = 0$, then

$$
\begin{aligned}
g &= x_i x_{i+1} x_{i+3}^{\alpha(i+3)} \cdots x_j \\
&= \left[x_i x_{i+1} x_{i+2} \right] \left[x_{i+2} x_{i+3}^{\alpha(i+3)} \cdots x_j \right] \\
&= a_i g',
\end{aligned}
$$

where $a_i \in A_i$ and the length of g' is smaller than the length of g. If $\alpha(i+3) = 1$, then

$$
\begin{aligned}
g &= x_i x_{i+1} x_{i+2} x_{i+3}^{\alpha(i+3)} \cdots x_j \\
&= \left[x_i x_{i+1} x_{i+2} \right] \left[x_{i+3}^{\alpha(i+3)} \cdots x_j \right] \\
&= a_i g',
\end{aligned}
$$

where $a_i \in A_i$ and the length of g' is smaller than the length of g.

To prove the uniqueness of the representation of g assume that

$$
g = a_1 \cdots a_{2m+1}, \quad g = a_1' \cdots a_{2n+1}'
$$

are representations of g with $a_1, a_1' \in A_1$, $a_3, a_3' \in A_3, \ldots$ Let i be the first index for which $a_i \neq a_i'$. The possibilities for the basis representation of both a_i and a_i' are

$$
x_i, \quad x_{i+1}, \quad x_i x_{i+1} x_{i+1}.
$$

This gives that in two representations of g in the basis $x_1, x_2, \ldots$ the exponents of x_i are not equal or the exponents of x_{i+1} are not equal.

This completes the proof.

10.2   Infinite abelian groups

We will show that some results on factoring finite abelian groups sometimes can be extended for the infinite case. By Lemma 10.1.1, Hajós's theorem can hold only for some special types of infinite abelian groups. By the next theorem, Hajós's theorem indeed holds for some infinite abelian groups. A cyclic group of order p will be denoted by Z_p. The roots of unity whose orders are p powers form an abelian group under multiplication. This is the Prüferian group and will be denoted by $Z(p^\infty)$.

Theorem 10.2.1. *Let G be a group of form*

$$G = F \times \left(\prod_m Z_p \right),$$

where F is a finite abelian group, m is an arbitrary cardinal number, p is a prime which does not divide $|F|$. Then Hajós's theorem holds for G.

Proof. Assume the contrary, that there is a factorization

$$G = \prod_{i \in I} A_i \tag{1}$$

of G such that each A_i is a nonsubgroup cyclic subset of G. Let A be a factor in (1). By Exercises 1.3.2, 1.3.3, we may assume that A has a prime cardinality, say $|A| = q$. We claim that q must be a divisor of $(|F|p)$. To prove the claim assume that q is not a divisor of $(|F|p)$. Set $t = |F|p$ and replace A by A^t in (1). The (t)th powers of the elements of A are all equal to e. Therefore A^t cannot be a member of a factorization. This contradiction shows that q divides $(|F|p)$.

Let

$$A = \{e, a, a^2, \ldots, a^{q-1}\}$$

and write a in the form $a = bc$, where b is the q-part of a and c is the q'-part of a. Set $t = |c|$ and replace A by A^t in (1). As A^t is a member of a factorization we must get that the (t)th powers of the elements of A are distinct.

We can transform a counterexample into a standard form. If $|b| = q^\alpha$, $\alpha \geq 2$, then we replace A by

$$A^* = \{e, b, b^2, \ldots, b^{q-1}\}.$$

Obviously A^* is not a subgroup of G. If $|b| = q$, then we must have $c \neq e$. In this case we can replace A by

$$A^* = \left\{e, bd, (bd)^2, \ldots, (bd)^{q-1}\right\},$$

where $|d| = r$ is a prime, $r \neq q$. Clearly A^* is not a subgroup of G. Then for notational convenience we rename A^* by A. Thus we may assume that if A is a factor in (1) with $|A| = q$, then A is in one of the following forms.

    (i) Each element of $A \setminus \{e\}$ has order q^α, $\alpha \geq 2$.

    (ii) Each element of $A \setminus \{e\}$ has order qr, where q, r are distinct primes.

In brief we will talk about type 1 and type 2 factors. For a factor A in (1) with $|A| = q$ we define A' to be the set of the q-parts of the elements of A. Each A can be replaced by A'. So

$$G = \prod_{i \in A_i} A_i = \left(\prod_{|A_i| \neq p} A_i' \right) \left(\prod_{|A_i| = p} A_i' \right)$$

is a factorization of G. Further

$$F = \Big(\prod_{|A_i| \neq p} A_i' \Big)$$

is a factorization of F and

$$\prod_m Z_p = \Big(\prod_{|A_i| = p} A_i' \Big)$$

is a factorization of the p-component of G. It follows that there are only finitely many factors in (1) whose cardinality is not p, say $B_1, \ldots, B_n$. We denote the remaining factors in (1) by C_j, $j \in J$.

The C_j factors are in the form

$$C_j = \big\{ e, x_j y_j, (x_j y_j)^2, \ldots, (x_j y_j)^{p-1} \big\},$$

where $|x_j| = p$, $|y_j|$ is a prime $y_j \in F$. Clearly x_j, $j \in J$ form a basis for the p-component of G. Each type 2 factor among $B_1, \ldots, B_n$ can be written in the form

$$B_i = \big\{ e, u_i v_i, (u_i v_i)^2, \ldots, (u_i v_i)^{q(i)-1} \big\},$$

where $|u_i| = q(i)$ and $|v_i|$ is a prime. Further either $v_i \in F$ or $|v_i| = p$. Let $B_1, \ldots, B_s$ be the type 2 factors among $B_1, \ldots, B_n$ for which the element v_i has order p. Each v_i, $1 \leq i \leq s$ can be represented as a product of finitely many x_j. There is an integer k such that

$$v_1, \ldots, v_s \in \langle x_1, \ldots, x_k \rangle.$$

Set $H = \langle F, x_1, \ldots, x_k \rangle$. We claim that

$$H = B_1 \cdots B_n C_1 \cdots C_k$$

is a factorization of H. The product $B_1 \cdots B_n C_1 \cdots C_k$ is direct because it is part of a factorization. The order of H is

$$|F||x_1| \cdots |x_k| = |B_1| \cdots |B_n||C_1| \cdots |C_k|$$

and obviously

$$B_1, \ldots, B_n, C_1, \ldots, C_k \subset H.$$

Now by the finite version of Hajós's theorem one of the factors $B_1, \ldots, B_n$, $C_1, \ldots, C_k$ is a subgroup of H and so of G.

This contradiction completes the proof.

Theorem 10.2.2. *Let $p, p_1, \ldots, p_s$ be distinct primes and let m be an arbitrary cardinal number. Let G be a group in the form*

$$G = F \times \Big(\prod_{i=1}^s Z(p_i^\infty) \Big) \times \Big(\prod_m Z_p \Big),$$

where F is a finite abelian group, none of the primes $p, p_1, \ldots, p_s$ divides $|F|$. Then Hajós's theorem holds for G.

Proof. Assume the contrary, that there is a factorization

$$G = \prod_{i \in I} A_i \tag{2}$$

such that each A_i is a nonsubgroup cyclic subset of G. Let A be a factor in (2). Using the ideas from the proof of Theorem 10.2.1 we may assume that $|A| = q$ is a prime and q is a divisor of $\big(|F|p_1 \cdots p_s p\big)$. Further we may assume that A is a type 1 or type 2 factor. For A we define A' to be the set of the q-parts of the elements of A. Each A can be replaced by A' and so we get that

$$F = \prod_{|A_i| \mid |F|} A_i'$$

is a factorization of F,

$$Z(p_i^\infty) = \prod_{|A_i| = p_i} A_i' \tag{3}$$

is a factorization of $Z(p_i^\infty)$, and

$$\prod_m Z_p = \prod_{|A_i| = p} A_i'$$

is a factorization of the p-component of G.

Let q be one of the primes $p_1, \ldots, p_s$. Let

$$A = \{e, a, a^2, \ldots, a^{q-1}\},$$

$$B = \{e, b, b^2, \ldots, b^{q-1}\}$$

be factors in (3) with $A \neq B$, $q = p_i$. We claim that $|a| \neq |b|$. To prove the claim assume the contrary, that $|a| = |b|$. Now $a, b \in Z(q^\infty)$ implies that $\langle a \rangle = \langle b \rangle$. There is an integer t such that $a = b^t$. By Lemma 1.3.1, B can be replaced by B^t. Note that $A = B^t$ which contradicts factorization (3).

Next we claim that there is a factor

$$A' = \{e, a, a^2, \ldots, a^{q-1}\}$$

in (3) with $q = p_i$ such that $|a| = q$. To prove the claim assume the contrary, that each nonidentity element of A_i' in (3) has order at least q^2. Choose an element $g \in Z(q^\infty)$ for which $|g| = q$. The element g can be expressed as a product of finitely many nonidentity elements coming from factors A_i' of (3). Say $g = h_1 \cdots h_u$. As the orders of $h_1, \ldots, h_u$ are distinct q powers we get the contradiction that

$$q = |g| = \max\big(|h_1|, \ldots, |h_u|\big) \geq q^2.$$

Each $Z(p_i^\infty)$ has a unique subgroup K_i of order p_i. There are finitely many factors in (2) whose cardinalities divide $|F|$. Let these be $B_1, \ldots, B_n$. Let $C_1, \ldots, C_s$ be the factors in (2) for which $|C_i| = p_i$ and each element of $C_i \setminus \{e\}$ has order p_i. Let D_j, $j \in J$ be all the remaining factors in (2). Since factorization (2) is in standard form the D_j factors are in form

$$D_j = \left\{ e, x_j y_j, (x_j y_j)^2, \ldots, (x_j y_j)^{p-1} \right\},$$

where $|x_j| = p$, $|y_j|$ is a prime, $y_j \in F K_1 \cdots K_s$. Clearly x_j, $j \in J$ form a basis of the p-component of G. A type 2 factor among $B_1, \ldots, B_n$ has the form

$$B_i = \left\{ e, a_i b_i, (a_i b_i)^2, \ldots, (a_i b_i)^{q(i)-1} \right\},$$

where $|a_i| = q(i)$, $|b_i|$ is a prime. Let $B_1, \ldots, B_u$ be all the type 2 factors among $B_1, \ldots, B_n$ for which the element b_i has order p. The factors $C_1, \ldots, C_s$ are all the type 2 factors in the form

$$C_i = \left\{ e, c_i d_i, (c_i d_i)^2, \ldots, (c_i d_i)^{q(i)-1} \right\},$$

where $|c_i| = q(i) = p_i$, $|d_i|$ is a prime. Let $C_1, \ldots, C_v$ be all the factors among $C_1, \ldots, C_s$ for which the element d_i has order p. There is an integer k such that

$$b_1, \ldots, b_u, d_1, \ldots, d_v \in \langle x_1, \ldots, x_k \rangle.$$

Set
$$H = \langle F, K_1, \ldots, K_s, x_1, \ldots, x_k \rangle.$$

We claim that
$$H = B_1 \cdots B_n C_1 \cdots C_s D_1 \cdots D_k$$

is a factorization of H.

To verify the claim note that the product

$$B_1 \cdots B_n C_1 \cdots C_s D_1 \cdots D_k$$

is direct. The order of H is equal to

$$|B_1| \cdots |B_n||C_1| \cdots |C_s||D_1| \cdots |D_k|$$

and finally
$$B_1, \ldots, B_n, C_1, \ldots, C_s, D_1, \ldots, D_k \subset H.$$

Now by the finite version of Hajós's theorem one of the factors is a subgroup of H and so of G.

This contradiction completes the proof.

By Lemma 10.1.2, Rédei's theorem can hold only for some special classes of infinite abelian groups. The next theorem shows that Rédei's theorem holds for certain infinite groups.

Theorem 10.2.3. *Let $p_1, \ldots, p_s$ be distinct primes and let m be an arbitrary cardinal number. Let G be a group in the form*

$$G = F \times \left(\prod_{i=1}^{s} Z(p_i^\infty) \right) \times \left(\prod_m Z_2 \right),$$

where F is a finite abelian group of odd order such that none of $p_1, \ldots, p_s$ divides $|F|$. Then Rédei's theorem holds for G.

Proof. Assume the contrary, that

$$G = \prod_{i=1}^{\infty} A_i \tag{4}$$

is a factorization of G such that each A_i is a nonsubgroup normalized subset of G with prime cardinality.

Let A be a factor in (4) with $|A| = q$, where q is a prime. Set $f = 2|F|p_1 \cdots p_s$. We claim that q divides f.

In order to prove the claim assume the contrary, that q does not divide f. Set $t = f$. By Lemma 10.1.4, A can be replaced by A^t. Note that the (t)th powers of elements of A are e. So A^t cannot be a member of a factorization. This contradiction proves that q divides f.

Let t be the least common multiple of the orders of the q'-parts of the elements of A. Replace A by A^t. As A^t is a member of a factorization it follows that the (t)th powers of the elements of A are distinct. Let A' be the set of the q-parts of the elements of A. Note that $A' = A^t$. Replacing all A by A' in (4) gives that

$$F = \prod_{|A_i| \mid |F|} A_i'$$

is a factorization of F,

$$Z(p_i^\infty) = \prod_{|A_i| = p_i} A_i' \tag{5}$$

is a factorization of $Z(p_i')$, and

$$\prod_m Z_2 = \prod_{|A_i| = 2} A_i'$$

is a factorization of the 2-component of G.

We transform (4) into a standard form. Let A be a factor in (4) with $|A| = q$. If $|A| = q = 2$, then write A in the form $A = \{e, ab\}$, where $|a| = 2$ and $|b|$ is odd. We can replace A by $A^* = \{e, ad\}$, where $|d|$ is an odd prime. Suppose $|A| = q$ is an odd prime. If there is an element $a \in A \setminus \{e\}$ such that the q-part of A has order q^α, $\alpha \geq 2$, then replace A by

$$A^* = \{e, b, b^2, \ldots, b^{q-1}\},$$

where $|b| = q^\alpha$, $\alpha \geq 2$. If for each $a \in A \setminus \{e\}$ the q-part of a has order q but there is an a for which the q'-part of a is not e, then we can replace A by

$$A^* = \{e, ab, (ab)^2, \ldots, (ab)^{q-1}\},$$

where $|a| = q$, $|b|$ is a prime. Finally if each $a \in A \setminus \{e\}$ has order q, then set $A^* = A$. Clearly A^* is not a subgroup of G. In other words we may assume that if A is a factor in (4), then A is in one of the following forms.

(i) Each element of $A \setminus \{e\}$ has order q^α, $\alpha \geq 2$.

(ii) Each element of $A \setminus \{e\}$ has order qr, where q, r are distinct primes.

(iii) Each element of $A \setminus \{e\}$ has order q.

Shortly we will talk about type 1, type 2, type 3 factors. Type 1 and type 2 factors are cyclic subsets and a type 3 factor is not a cyclic subset of G.

Let q be one of the primes $p_1, \ldots, p_s$. Let A, B be factors in (5) with $q = p_i$ such that $A \neq B$. We claim that there are no elements $a \in A \setminus \{e\}$, $b \in B \setminus \{e\}$ for which $|a| = |b|$. In order to prove the claim assume the contrary that $|a| = |b|$. Now $a, b \in Z(q^\infty)$ implies that $\langle a \rangle = \langle b \rangle$. There is an integer t such that $a = b^t$. By Lemma 10.1.3, A can be replaced by

$$A_1 = \{e, a, a^2, \ldots, a^{q-1}\}$$

and B can be replaced by

$$B_1 = \{e, b, b^2, \ldots, b^{q-1}\}.$$

Then B_1 can be replaced by B_1^t. Note that $A_1 = B_1^t$ which contradicts factorization (5).

Next we claim that there is a factor A' in (5) such that each element in $A' \setminus \{e\}$ has order q. To prove the claim replace each A'_i in (5) by

$$A'' = \{e, a_i, a_i^2, \ldots, a_i^{q-1}\},$$

where a_i is an element of A'_i with maximum order and so $|a_i| \geq q^2$. Assume the contrary claim that the nonidentity elements of A''_i has order at least q^2. Choose an element $g \in Z(q^\infty)$ for which $|g| = q$ and consider the representation $g = h_1 \cdots h_u$ of g as a finite product of nonidentity elements coming from A''_i. This leads to the

$$q = |g| = \max(|h_1|, \ldots, |h_u|) \geq q^2$$

contradiction. Therefore there is a factor A' in (5) such that each element of $A' \setminus \{e\}$ has order q.

The group $Z(q^\infty)$ with $q = p_i$ has a unique subgroup K_i of order $p_i = q$. As each nonidentity element of A' has order q it follows that $A' = K_i$.

There are finitely many factors in (4) whose cardinality divides $|F|$, say $B_1, \ldots, B_n$. The type 2 factors among $B_1, \ldots, B_n$ are in the following form

$$B_i = \{e, a_i b_i, (a_i b_i)^2, \ldots, (a_i b_i)^{q(i)-1}\},$$

where $|a_i| = q(i)$, $|b_i|$ is a prime. Let $B_1, \ldots, B_u$ be all the type 2 factors among $B_1, \ldots, B_n$ for which the element b_i has order p.

Let $C_1, \ldots, C_s$ be the factors in (4) for which $|C_i| = p_i$ and each nonidentity element of C_i has order p_i. Each of these factors are in the next form

$$C_i = \left\{ e, c_i d_i, (c_i d_i)^2, \ldots, (c_i d_i)^{q-1} \right\},$$

where $|c_i| = p_i = q$, $|d_i|$ is a prime. Let $C_1, \ldots, C_v$ be among $C_1, \ldots, C_s$ for which the element d_i has order p.

Let D_j, $j \in J$ be all the factors in (4) distinct from

$$B_1, \ldots, B_n, C_1, \ldots, C_s.$$

Each D_j, $j \in J$ is a type 2 factor in the form $D_j = \{e, x_j y_j\}$, where $|x_j| = 2$, $|y_j|$ is a prime, $y_j \in FK_1 \cdots K_s$. Now x_j, $j \in J$ is a basis for the 2-component of G. There is an integer k such that

$$b_1, \ldots, b_u, d_1, \ldots, d_v \in \langle x_1, \ldots, x_k \rangle.$$

Setting
$$H = \langle F, K_1, \ldots, K_s, x_1, \ldots, x_k \rangle$$

we get that
$$H = B_1 \cdots B_u C_1 \cdots C_s D_1 \cdots D_k$$

is a factorization of H. An appeal to the finite version of Rédei's theorem completes the proof.

We extend Theorem 7.4.1 for certain infinite abelian groups.

Theorem 10.2.4. *Let*

$$G = H \times \prod_{i=1}^{r} Z(p_i^\infty),$$

where H is a finite abelian group, $p_1, \ldots, p_r$ are distinct primes, and $p_i \nmid |H|$ for each i, $1 \le i \le r$. If

$$G = B \prod_{i=1}^{\infty} A_i \tag{6}$$

is a normalized factorization of G such that $|B| = 4$ and each $|A_i|$ is a prime, then one of the factors $B, A_1, A_2, \ldots$ is periodic.

Proof. Let $|H|$ be the product of the (not necessarily distinct) primes $p_{r+1}, \ldots, p_s$. By the assumptions of the theorem

$$\{p_1, \ldots, p_r\} \cap \{p_{r+1}, \ldots, p_s\} = \emptyset.$$

For a factor A_j of the factorization (6) with $|A_j| = p$, where p is a prime, let A_j' be the set of the p-parts of the elements of A_j. We claim that in the factorization (6),

A_j can be replaced by A'_j. In order to prove the claim let m be a common multiple of the orders of the p'-parts of the elements of A_j and let n be a common multiple of the orders of the p-parts of the elements A_j. Such m, n do exist since each element of G has a finite order. As m and n are relatively prime by the Chinese remainder theorem, the system of congruences

$$\begin{aligned} t &\equiv 0 \quad (\text{mod } m), \\ t &\equiv 1 \quad (\text{mod } n) \end{aligned}$$

is solvable. By Lemma 10.2.3, A_j can be replaced by A_j^t. As $A_j^t = A'_j$, the claim is proved.

The fact that A_j can be replaced by A'_j implies that the p-parts of the elements of A_j are distinct. Further we can conclude that if $|A_j| = p$, where p is a prime, then p must be one of the primes $p_1, \ldots, p_s$. It follows in the same manner that the 2-parts of the elements of B are distinct. These elements form a set B' and $|B'| = 4$.

From the factorization

$$G = B' \prod_{i=1}^{\infty} A'_i$$

we draw further conclusions. Let p be one of the primes $p_1, \ldots, p_r$. If p is odd, then the product of all the A'_j's with $|A'_j| = p$ forms a factorization of the subgroup $Z(p^\infty)$ of G. If $p = 2$, then the product of B' and all the A'_j's with $|A'_j| = p$ form a factorization of the subgroup $Z(p^\infty)$ of G. If $|H|$ is odd, then the product of all the A'_j's with $|A'_j| \mid |H|$ forms a factorization of H. If $|H|$ is even, then the product of B' and all the A'_j with $|A'_j| \mid |H|$ forms a factorization of H.

The subgroups of $Z(p_i^\infty)$ form a chain. For each integer $j \geq 0$ there is a unique subgroup of order p_i^j. Let $H_{i,0}, H_{i,1}, \ldots$ be all the subgroups of $Z(p_i^\infty)$. We assume that $|H_{i,j}| = p_i^j$. There are factors $A''_{i,1}, A''_{i,2}, \ldots$ among $A'_1, A'_2, \ldots$ such that

$$A''_{i,1} = H_{i,1}, \quad A''_{i,1} A''_{i,2} = H_{i,2}, \ldots$$

Note that each nonidentity element of $A''_{i,j}$ must have order p_i^j.

Now let us go back to factorization (6). To prove the theorem we assume the contrary, that none of the factors $B, A_1, A_2, \ldots$ is periodic. Choose a factor A_j and assume that $|A_j| = p$, where p is a prime. (We know that p is one of the primes $p_1, \ldots, p_s$.) The p-parts of the elements of A_j are distinct and form a set A'_j with $|A'_j| = p$. If A'_j is not a subgroup of G, then replace A_j by $C_j = A'_j$. If A'_j is a subgroup of G, then there is an element in A_j whose q-part is not the identity element since A_j is not a subgroup of G. Here q is a prime $p \neq q$. Now A_j can be replaced by C_j such that C_j is not a subgroup of G and the orders of the elements of C_j divide pq. In other words C_j is constructed from the subgroup A'_j by multiplying some elements of A'_j by some elements of order q. Let us consider the factorization

$$G = B \prod_{i=1}^{\infty} C_i.$$

Here none of the factors $B, C_1, C_2, \ldots$ is periodic. For each i, $1 \le i \le r$ there is an integer $\alpha(i)$ such that the orders of the p_i-parts of the elements of B are less than or equal to $p_i^{\alpha(i)}$ and $\alpha(i) \ge 1$. The elements of $Z(p_i^{\infty})$ whose order is less than or equal to $p_i^{\alpha(i)}$ form the unique subgroup $H_{i,\alpha(i)}$ of $Z(p_i^{\infty})$. Set

$$K = H H_{1,\alpha(1)} \cdots H_{r,\alpha(r)}.$$

Clearly, $B \subset K$. Let $D_1, \ldots, D_n$ be all the C_i factors for which $C_i \subset K$. We claim that $K = B D_1 \cdots D_n$ is a factorization of K. As $B, D_1, \ldots, D_n \subset K$, it is enough to verify that $|B||D_1| \cdots |D_n| = |K|$. In order to verify this equation let $D_1, \ldots, D_m$ be the factors among $D_1, \ldots, D_n$ whose cardinality is one of $p_{r+1}, \ldots, p_s$.

Assume first that $4 \mid |H|$. Note that $|B||D_{m+1}| \cdots |D_n| = |H|$. Further

$$
\begin{aligned}
|D_1| \cdots |D_m| &= \left(|A''_{1,1}| \cdots |A''_{1,\alpha(1)}| \right) \cdots \left(|A''_{r,1}| \cdots |A''_{r,\alpha(r)}| \right) \\
&= p_1^{\alpha(1)} \cdots p_r^{\alpha(r)} \\
&= |H_{1,\alpha(1)}| \cdots |H_{r,\alpha(r)}|.
\end{aligned}
$$

Assume next that $4 \mid |H_{1,\alpha(1)}| \cdots |H_{r,\alpha(r)}|$. Note that

$$|D_{m+1}| \cdots |D_n| = |H|$$

and

$$|B||D_1| \cdots |D_m| = |H_{1,\alpha(1)}| \cdots |H_{r,\alpha(r)}|.$$

Thus $K = B D_1 \cdots D_n$ is a factorization of the finite abelian group K. By Theorem 7.4.1, one of the factors is periodic.

This contradiction completes the proof.

Chapter 11

Further topics

11.1 Some applications

In this section we present some applications of the factorizations of abelian groups to the geometry of tilings to estimating Ramsey numbers in combinatorics. We sketch the connection with coding theory with variable length codes and error correcting codes. We will describe a construction related to Hilbert's 18th problem.

B. Gordon showed that for each $n \geq 3$, $k \geq 2$ there is an n-dimensional k-fold (not necessarily lattice-like) cube tiling without twin cubes.

It is enough to construct two 3-dimensional cube tilings without twins. One with multiplicity 2 and another one with multiplicity 3. Superpositions of these tilings provide 3-dimensional tilings without twins for each multiplicity k, $k \geq 2$. Then one can construct suitable 4-dimensional tilings, 5-dimensional tilings etc.

If there is an abelian group G, a subset B of G, cyclic subsets $A_1, \ldots, A_n$ of G such that

$$A_i = \{e, a_i, a_i^2, \ldots, a_i^{r(i)-1}\},$$

the product $BA_1 \cdots A_n$ is a k-fold factorization of G and

$$BB^{-1} \cap \{a_1^{r(1)}, \ldots, a_n^{r(n)}\} = \emptyset,$$

then there is an n-dimensional k-fold cube tiling without twin cubes. The connection between the algebra and geometry is based on similar ideas as in the case of Minkowski's, Furtwängler's, and Keller's conjectures.

The constructions we need are the following. Let the group G be generated by the elements x, y, z with $|x| = |y| = 4$, $|z| = 2$. Set

$$B = \{e, xy^2, x^2yz, x^2y, y^2z, x^3y^3, xy^3z, x^3z\},$$

$$\begin{aligned}
a_1 &= x, & r(1) &= 2, \\
a_2 &= y, & r(2) &= 2, \\
a_3 &= xyz, & r(3) &= 2.
\end{aligned}$$

One can verify that the product $BA_1A_2A_3$ is a 2-fold factorization of G and $a_i^{r(i)} = b^{-1}b'$ does not hold for any choices of $b, b' \in B$ and $1 \leq i \leq 3$.

Next let the group G be of type $(4, 2, 3)$ with basis elements x, y, z, where $|x| = 4$, $|y| = 2$, $|z| = 3$. Set

$$B = \{e, xy, x^2z^2, x^3yz^2\},$$

$$
\begin{aligned}
a_1 &= x, & r(1) &= 2, \\
a_2 &= yz, & r(2) &= 3, \\
a_3 &= x^2yz, & r(3) &= 3.
\end{aligned}
$$

A routine computation shows that the product $BA_1A_2A_3$ is a 3-fold factorization of G and $a_i^{r(i)} = b^{-1}b'$ does not hold for any choices of $b, b' \in B$ and $1 \leq i \leq 3$.

Next we have a look at Ramsey numbers. A complete graph K_n on n vertices has $\binom{n}{2}$ edges. We colour the edges of K_n using m colours. If K_n does not contain a complete subgraph K_r whose edges are coloured with the same colour, then we say that K_n is monochrome K_r free. The minimal integer n for which K_n has a colouring with m colours and K_n is monochrome K_r free is denoted by $R_m(r)$. The numbers $R_m(r)$ are called Ramsey numbers. Let H be a finite abelian group written additively. Set $H^* = H \setminus \{0\}$. A subset S of H^* is called symmetric if $x \in S$ implies $-x \in S$. A subset S is called r-free if there are no elements $x_1, y_1, \ldots, x_r, y_r \in H$ such that $x_1 \neq y_1, \ldots, x_r \neq y_r$ and $x_1 - y_1, \ldots, x_r - y_r$ all belong to S.

Observation 11.1.1. *If H is a finite abelian group and the sets $A_1, \ldots, A_m$ form a partition of H^* such that each A_i is symmetric and r-free, then $R_m(r) \geq |H| + 1$.*

Proof. Let $n = |H|$ and consider the complete graph K_n whose vertices are the elements of H. If $x - y \in A_i$, then we colour the vertex (x, y) of K_n with the colour A_i. This provides a monochrome K_r free colouring of K_n with m colours.

The proof is complete.

Let F denote the Galois field of p^h elements, that is, $F = \mathrm{GF}(p^h)$. Since F is a vector field over $\mathrm{GF}(p)$ the elements of F can be represented uniquely in the form

$$x_0 + x_1\alpha + \cdots + x_{h-1}\alpha^{h-1},$$

where $x_0, x_1, \ldots, x_{h-1}$ are integers between 0 and $p - 1$. We simply identify this element with its coordinate vector $(x_0, x_1, \ldots, x_{h-1})$ in the basis $1, \alpha, \ldots, \alpha^{h-1}$. The additive part of F is an abelian group of type $(p, \ldots, p)$.

Observation 11.1.2. *Let $n = (p^h - 1)/2m$ and let $Z_{mn} = A + B$ be a factorization with $|A| = m$, $|B| = n$. If*

$$X = \{\alpha^b, -\alpha^b : b \in B\}$$

bound	$R_m(r)$	$\mathrm{GF}(p^h)$
6	$R_2(3)$	$\mathrm{GF}(5)$
18	$R_2(4)$	$\mathrm{GF}(17)$
17	$R_3(3)$	$\mathrm{GF}(2^4)$
42	$R_2(5)$	$\mathrm{GF}(41)$
102	$R_2(6)$	$\mathrm{GF}(101)$
126	$R_2(7)$	$\mathrm{GF}(5^3)$
50	$R_4(3)$	$\mathrm{GF}(7^2)$
128	$R_3(4)$	$\mathrm{GF}(127)$

Table 1

is an r-free subset of F^*, then $R_m(r) \geq |F| + 1$.

Proof. Let
$$A = \{a(1), \ldots, a(m)\}, \qquad B = \{b(1), \ldots, b(n)\}.$$
Since $\alpha^{(p^h-1)/2} = -1$ we have

$$
\begin{aligned}
F^* &= \bigcup_{i=0}^{(p^h-3)/2} \{\alpha^i, -\alpha^i\} \\
&= \bigcup_{i=1}^{m} \bigcup_{j=1}^{n} \{\alpha^{a(i)+b(j)}, -\alpha^{a(i)+b(j)}\} \\
&= \left[\bigcup_{j=1}^{n} \{\alpha^{b(i)}, -\alpha^{b(i)}\} \right] \alpha^{a(i)} \\
&= X\alpha^{a(i)}.
\end{aligned}
$$

As X is plainly a symmetric set and since X is r-free, it follows that $X\alpha^{a(i)}$ is symmetric and r-free. Observation 11.1.1 gives that $R_m(r) \geq |F| + 1$.

This completes the proof.

Using Observation 11.1.2, R. Hill and R. W. Irving embarked on a computer search to provide lower bounds for Ramsey numbers. Their result is summarized in Table 1.

We will detail only the $50 \leq R_4(3)$ case as an illustration.

$$\mathrm{GF}(7^2) = \{0, 1, \alpha, \ldots, \alpha^{47}\},$$

where $\alpha^2 = 2\alpha + 2$. Identifying the element $x_0 + x_1\alpha$ of $\mathrm{GF}(7^2)$ with its coordinate vector (x_0, x_1) in the basis $1, \alpha$ we get

$$\begin{aligned}
&\alpha^0 = (1,0), &&\alpha^6 = (4,1), &&\alpha^{12} = (4,3), &&\alpha^{18} = (1,1),\\
&\alpha^1 = (0,1), &&\alpha^7 = (2,6), &&\alpha^{13} = (6,3), &&\alpha^{19} = (2,3),\\
&\alpha^2 = (2,2), &&\alpha^8 = (5,0), &&\alpha^{14} = (6,5), &&\alpha^{20} = (6,1),\\
&\alpha^3 = (4,6), &&\alpha^9 = (0,5), &&\alpha^{15} = (3,2), &&\alpha^{21} = (2,1),\\
&\alpha^4 = (5,2), &&\alpha^{10} = (3,3), &&\alpha^{16} = (4,0), &&\alpha^{22} = (2,4),\\
&\alpha^5 = (4,2), &&\alpha^{11} = (6,2), &&\alpha^{17} = (0,4), &&\alpha^{23} = (1,3).
\end{aligned}$$

Set $n = (7^2 - 1)/2 \cdot 6 = 4$. We can see that

$$\begin{aligned}
Z_{6\cdot 4} &= \{0,1,2,3,4,5\} + \{0,6,12,18\}\\
&= \{0,1,14,3,10,5\} + \{0,6,12,18\}\\
&= A + B
\end{aligned}$$

is a factorization of $Z_{6\cdot 4}$. Then a computer assisted inspection reveals that

$$X = \{\alpha^b, -\alpha^b : b \in B\} =$$

$$\{(1,0),(6,0),(0,1),(0,6),(6,5),(1,2),$$

$$(4,6),(3,1),(3,3),(4,4),(4,2),(3,5)\}$$

is a 3-free set. Observation 11.1.2 gives that $R_4(3) \geq 50$.

We turn our attention to coding theory. Let $X = \{x, y\}$ be a binary alphabet. Let X^* be the set of all possible words on X. The concatenation of words is an associative operation on X^*. The empty word ε is the identity element. A subset C of X^* is called a code if from each equation

$$c_1 \cdots c_u = d_1 \cdots d_v, \quad c_1, \ldots, c_u, d_1, \ldots, d_v \in C$$

it follows that $u = v$, that is, $c_1 \cdots c_u = d_1 \cdots d_u$ and then it follows that $c_1 = d_1, \ldots, c_u = d_u$. This simply means that any message can be decoded if we use the words in C as code words.

Let

$$C_1 = \{xy, xyy, xyyy, \ldots\}, \quad C_2 = \{x, y, xy\}.$$

The reader can verify that C_1 is a code and C_2 is not a code.

A code is called maximal if it is not contained by a strictly larger code. A standard application of the Zorn lemma gives that for each code C there is a maximal code D such that $C \subset D$. We also express this fact saying that D is a completion of C. Finite codes with finite completion are related to (additive) factorizations of the integers modulo n. C. de Felice showed that if the group Z_n has the Hajós 2-property and $Z_n = A + B$ is a factorization, then

$$\{x^n\} \cup \{yx^a : a \in A\} \cup \{x^b y : b \in B\}$$

is a code. In addition this code has a finite completion.

For instance the words

$$xxxx, \; y, \; yx, \; y, \; xxy$$

are constructed from the factorization $Z_4 = \{0, 1\} + \{0, 2\}$.

N. H. Lam proved the following similar result. If Z_n has the Hajós 2-property and $Z_n = A + B$ is a factorization of Z_n, then

$$\{x^n\} \cup \{x^a y x^b : a \in A, b \in B\}$$

is a code and this code has a finite completion.

The reader can verify that the words

$$xxxx, \; y, \; yxx, \; xxxy, \; xxxyxx$$

are constructed from the factorization $Z_4 = \{0, 3\} + \{0, 2\}$ following Lam's recipe.

The open problem below is motivated by completing codes.

Problem 11.1.1. Let $A_1 = \{0, p, k\}$, $B_1 = \{0, 1\}$ be subsets of the non-negative integers, where p is a prime that does not divide k and the sum $A_1 + B_1$ is direct in the set of integers. Is there an integer n such that (i) $p, k < n$, (ii) $Z_n = A + B$ is a factorization of Z_n, (iii) $A_1 \subset A$, $B_1 \subset B$. (A. Restivo, S. Salemi, T. Sportelli conjecture that the answer is "no".)

If A is a finite set, then A^n denotes the set of n-tuples with entries from A. We can make A an abelian group and then consider A^n as a direct product of groups. If $u = (u_1, \ldots, u_n)$, $v = (v_1, \ldots, v_n)$ are elements of A^n, then the Hamming distance $d(u, v)$ is equal to the number of i's for which $u_i \neq v_i$. One can verify that d is a distance and so A^n is a metric space. The sphere of center u and radius e is the set $S_e(u) = \{v \in F^n : d(u, v) \leq e\}$. A perfect e-error correcting code is a subset of C of A^n such that A^n is a disjoint union of spheres $S_e(u)$, $u \in C$. This means that the product $[C][S_e(0)]$ is direct and gives A^n, that is, $A^n = [C][S_e(0)]$ is a factorization of A^n. A code is called a subgroup code if C is a subgroup of A^n. B. Lindstrom asked if every perfect code is periodic. The next result of A. D. Sands gives an answer in the case of binary codes.

Theorem 11.1.1. *Let C be a 1-error correcting perfect binary code of length n. If $n \geq 3$, then C is periodic.*

Proof. Let G be an elementary 2-group of rank n. Let $x_1, \ldots, x_n$ be a basis for G. Each element g of G can be written uniquely in the form

$$g = x_1^{\alpha(1)} \cdots x_n^{\alpha(n)},$$

where $\alpha(1), \ldots, \alpha(n)$ are integers equal to 0 or 1. We can identify g with the n tuple $(\alpha(1), \ldots, \alpha(n))$ and so we can identify G with $\{0, 1\}^n$. If C is a perfect 1-error correcting binary code of length n, then $G = CB$ is a factorization of

G, where $B = \{e, x_1, \ldots, x_n\}$. Clearly B corresponds to the Hamming sphere of radius 1 in $\{0, 1\}^n$ centered at $(0, \ldots, 0)$. In the group ring $Z(G)$ the element

$$\overline{B} = \left[\sum_{i=1}^{n} (e + x_i) \right] - (n - 1)e \tag{1}$$

corresponds to the subset B.

The factorization $G = CB$ implies that

$$2^n = |G| = |C||B| = |C|(n + 1),$$

that is, $n + 1$ divides 2^n. There is an integer m such that $n + 1 = 2^m$, and so $n = 2^m - 1$. Set $h = x_1 \cdots x_n$. We want to show that $C = Ch$, that is,

$$\chi(C) = \chi(C)\chi(h) \tag{2}$$

holds for each character χ of G. If $\chi(C) = 0$ or $\chi(h) = 1$, then (2) clearly holds. We assume that $\chi(C) \neq 0$ and $\chi(h) \neq 1$. Now χ is not the principal character of G. Applying χ to the factorization $G = CB$ gives that

$$0 = \chi(G) = \chi(C)\chi(B).$$

As $\chi(C) \neq 0$ we get $\chi(B) = 0$. There is an integer k such that $\chi(x_i) = 1$ holds for k values of i and $\chi(x_i) = -1$ holds for $n - k$ values of i. From (1) we get

$$0 = \chi(B) = k \cdot 2 - (n - 1)$$

and so $k = (n - 1)/2$. Hence

$$k = \frac{n - 1}{2} = \frac{2^m - 2}{2} = 2^{m-1} - 1.$$

Note that $n = 2^m - 1$ is odd since $n = 0$ is not possible. Further if $n \geq 3$, then $m \geq 2$ and so $k = 2^{m-1} - 1$ is odd. Consequently $n - k$ is even. The

$$\chi(h) = (1)^k(-1)^{n-k} = 1$$

contradiction shows that h is a period of C.

This completes the proof.

We describe an instance of how the techniques developed studying error correcting codes help in a factorization construction. In this sense this is an application of coding theory to the factorization theory.

Let F be the field of integers modulo p, where p is a prime and let F^n be the vector space of all ordered n-tuples of elements of F. The additive structure of F^n is an abelian group G of type $(p, \ldots, p)$. The subspaces of F^n and the subgroups of G correspond to each other. A perfect e-error correcting code is a subset of C

of F^n such that F^n is a disjoint union of spheres $S_e(u)$, $u \in C$. This means that $G = C + S_e(0)$ is an additive factorization of G. The code C is called linear when it is a subspace of F^n.

The next exercise shows that under certain conditions factoring of the homomorphic image of a group can be lifted to get a factoring of the group.

Exercise 11.1.1. Let $\phi : G \rightarrow G'$ be a homomorphism from G onto G' and let $G' = A'B'$ be a factorization of G'. Prove that if A is a subset of G such that the restriction of ϕ to A is one-to-one between A and A', then G can be factored in the form $G = A(\phi^{-1}(B'))$.

We work out a numerical case in detail. Let $p = 5$ and let us consider the linear map $\phi : F^6 \rightarrow F^2$ defined by

$$
\begin{aligned}
\phi(1,0,0,0,0,0) &= (0,1), \\
\phi(0,1,0,0,0,0) &= (1,0), \\
\phi(0,0,1,0,0,0) &= (1,1), \\
\phi(0,0,0,1,0,0) &= (1,2), \\
\phi(0,0,0,0,1,0) &= (1,3), \\
\phi(0,0,0,0,0,1) &= (1,4).
\end{aligned}
$$

Let C be the kernel of ϕ. This C is a 4-dimensional subspace of F^6 which is spanned by the vectors c_1, c_2, c_3, c_4 given as the columns of Table 2. Now $F^6 = C + S_1(0)$ is an additive factorization of the additive part of F^6. As C is a subspace of F^6 this factorization is quasi-periodic. (For the definition of quasi-periodic factorization see Section 11.2 Subsection 4.) The construction can be modified to get less regular factorizations. Let $\pi_1, \ldots, \pi_n$ be permutations of the elements of F. The map $\pi : F^n \rightarrow F^n$ defined by

$$
\pi(u_1, \ldots, u_n) = \big(\pi_1(u_1), \ldots, \pi_n(u_n)\big)
$$

preserves the Hamming distance. Therefore the image of a perfect e-error correcting code is again a perfect e-error correcting code.

Returning to our example choose $\pi_1 = \pi_2 = (2,3)$ and let $\pi_3, \pi_4, \pi_5, \pi_6$ be the identity map. We claim that $\pi(C)$ spans F^6. Indeed, $\pi(c_1), \pi(c_2), \pi(c_3)$, $\pi(c_4), \pi(c_1 + c_2), \pi(c_3 + c_4)$ belong to $\pi(C)$ and they are linearly independent. This can be verified by computing the determinant of the matrix given in Table 3 whose columns are these vectors.

Now $F^6 = \pi(C) + \pi\big(S_1(0)\big)$ is a factorization, where $\pi(C)$ is not contained in any proper subspace of F^6. Further $\pi\big(S_1(0)\big) = S_1(0)$ contains

$$
(0, \ldots, 0), \quad (i, 0, \ldots, 0), \ldots, (0, \ldots, 0, i)
$$

for each i, $1 \leq i \leq p - 1$ and so it is not contained in any proper subspace of F^6.

1	1	1	1
1	2	3	4
-1	0	0	0
0	-1	0	0
0	0	-1	0
0	0	0	-1

Table 2

1	1	1	1	3	3
1	3	2	4	2	3
-1	0	0	0	-1	0
0	-1	0	0	-1	0
0	0	-1	0	0	-1
0	0	0	-1	0	-1

Table 3

The above argument shows that if G is a group of type

$$(5, 5, 5, 5, 5, 5),$$

then there is a factorization $G = AB$ of G such that $|A| = 5^2$, $|B| = 5^4$ and $\langle A \rangle = \langle B \rangle = G$. In general, if p is a prime $p \geq 5$ and G is a group of type $(p, \ldots, p)$ with $p + 1$ direct components, then there is a factorization $G = AB$ such that $|A| = p^2$, $|B| = p^{p-1}$ and $\langle A \rangle = \langle B \rangle = G$.

Problem 11.1.2. Let G be a group of type $(p, \ldots, p)$, where p is a prime. Let $G = AB$ be a factorization of G, where $\langle A \rangle = \langle B \rangle = G$. Is there such a factorization when $|A| = p$?

S. K. Stein constructed a 10-dimensional centrally symmetric star body whose translated copies tile the 10-dimensional space but translates by a lattice do not tile the 10-space. The construction was motivated by Hilbert's 18th problem. D. Hilbert asked if congruent copies of a polyhedron P tile the space, then does it follow that copies of P under a group of motions also tile the space.

Let $e_1, \ldots, e_n$ be the coordinate unit vectors of an orthogonal coordinate system in the n-dimensional space. The union of translated copies of an n-dimensional unit cube whose centers are

$$je_i, \quad 1 \leq i \leq n, \quad -k \leq j \leq k$$

is called an (n, k)-cross. Here n is the dimension and k is the length of the arms of the cross. Obviously an (n, k)-cross is a centrally symmetric star body.

Using the field $F = \mathrm{GF}(3^2)$ one can construct a 1-error correcting perfect code of length 10 over the alphabet F. The code C is a subset of F^{10} such that $F^{10} = C + B$ is an additive factorization of F^{10}, where B is the Hamming sphere of radius 1 centered at $(0, \ldots, 0)$. We know that the additive part of F^{10} is an elementary 3-group of rank 20. However the 1-error correcting perfect code of length 10 over the alphabet $A = \{0, 1, \ldots, 8\}$ gives rise to a factorization $G = C_1 + B_1$, where G is a group of type $(9, \ldots, 9)$ with basis elements $x_1, \ldots, x_{10}$ and

$$B_1 = \{jx_i : 1 \leq i \leq 10, \ 0 \leq j \leq 8\}.$$

The set B_1 corresponds to a Hamming sphere of radius 1. Note that $G = C_1 + B_2$ is also a factorization, where

$$B_2 = \{jx_i : 1 \le i \le 10, \ -4 \le j \le 4\}.$$

The set B_2 is associated with a Hamming sphere of radius 1 and also with a $(10, 4)$-cross. Filling the 10-dimensional space with 10-dimensional cubes of side length 9 meeting face to face and using the set C_1, we can see that translated copies of a $(10, 4)$-cross tile the 10-dimensional space.

Our next aim is to show that translated copies of a $(10, 4)$-cross by a lattice do not tile the 10-dimensional space. To do this we modify the cross by gluing small pyramids to the facets of a cross and cutting out small pyramids from the facets of a cross. One can modify the cross such that the cross still remains a centrally symmetric star body and also make sure that cubes forming the crosses meet face to face, in addition the end of an arm of a cross meets with an end of an arm of another cross. In other words if the vector v points to the center of a cross then so does the $v + 9e_i$ for each i, $1 \le i \le 10$.

Let L_1 be the lattice spanned by the vectors $e_1, \ldots, e_{10}$. Suppose there is a lattice L_2 such that translates of a $(10, 4)$-cross by the elements of L_2 form a tiling. Because of the pyramid shape dents and bumps $L_2 \subset L_1$. Let

$$B_3 = \{je_i : 1 \le i \le 10, \ -4 \le j \le -4\}.$$

Clearly B_3 consists of the centers of the unit cubes that form a $(10, 4)$-cross. Each element l_1 of L_1 can be written uniquely in the form

$$l_1 = b_3 + l_2, \quad b_3 \in B_3, \quad l_2 \in L_2.$$

In other words $L_1 = B_3 + L_2$ is a factorization. Considering the factor group $H = L_1/L_2$ we get that $|H| = |B_3| = 81$. Set $h_i = e_i + L_2$. Each element of $H \setminus \{0\}$ can be written uniquely in the form

$$ih_j, \quad 1 \le i \le 10, \quad 1 \le |j| \le 4. \tag{3}$$

It follows that the order of h_j is at least 9. On the other hand, since the end of an arm meets with the end of another arm, it follows that the order of h_j is 9. Consequently $H \setminus \{0\}$ has $6 \cdot 10 = 60$ elements of order 9, $2 \cdot 10 = 20$ elements of order 3. Note that the number of elements of order 3 in a finite abelian 3-groups is $3^k - 1$ for some positive integer k. As 20 is not of this form it follows that elements of $H \setminus \{0\}$ cannot be represented in form (3).

The argument above shows that translates of a suitably modified $(10, 4)$-cross tile the 10-space and there is no tiling if the translations form a lattice.

11.2 Miscellaneous results

This section briefly mentions some results about factorizations which are not covered by the earlier chapters.

1. Z-good groups. The subset A of a finite abelian group G is defined to be a *Z-set* if for all $a \in A$ and $n \in Z$ it follows that $a^n \in A$. A subgroup is closed under multiplication. A Z-set is closed only under powering elements. We call the group G, *Z-good* if in every factorization $G = AB$ with A and B are Z-sets, at least one of the factors is periodic. Otherwise G is called *Z-bad*.

In 1975, C. Okuda in his Ph.D thesis has classified all finite abelian groups into Z-good and Z-bad groups. The summary of his work is the following.

(1) A finite abelian group is Z-good if and only if at least one of its Sylow p-subgroups is Z-good.

This reduces the problem to p-groups.

(2) A p-group of rank ≤ 3 is Z-good.

(3) A group of type $(2^\alpha, 2^\beta, 2^\gamma, 2)$, $\alpha \geq \beta \geq \gamma \geq 2$ is Z-good.

(4) A group of type $(2^\alpha, 2, 2, 2, 2)$ is Z-good.

(5) A p-group of rank ≥ 4 is Z-bad if p is an odd prime.

(6) A 2-group of rank ≥ 6 is Z-bad.

(7) A group of type $(2^\alpha, 2^\beta, 2^\gamma, 2^\delta)$, $\alpha \geq \beta \geq \gamma \geq \delta \geq 2$, or of type $(2^\alpha, 2^\beta, 2^\gamma, 2^\delta, 2^\mu)$, $\alpha \geq \beta \geq 2$ is Z-bad.

The rank of a p-group is n if it is of type $(p^{\alpha(1)}, \ldots, p^{\alpha(n)})$. A Z-subset of a finite abelian group must have a very specific structure.

Exercise 11.2.1. Let A be a Z-subset of a finite abelian group G. Show that A must be a union of subgroups of G.

The next exercise deals with a Z-bad group.

Exercise 11.2.2. Let G be a group of type (p, p, p, p) with basis x, y, u, v, where p is a prime $p \geq 3$. Set

$$A = \big(\langle x, y \rangle \setminus \langle y \rangle\big) \cup \langle yu \rangle,$$

$$B = \Big(\langle u, v \rangle \setminus \big(\bigcup_{i=1}^{p-1} \langle uv^i \rangle\big)\Big) \cup \big(\bigcup_{i=1}^{p-1} \langle yuv^i \rangle\big).$$

Verify that $G = AB$ is a factorization of G and furthermore, A and B are nonperiodic Z-subsets.

2. Infinite abelian groups. In connection with infinite abelian groups the additive notation is most customary. Z denotes the infinite cyclic group; Q the additive group of the rationals; Z_n the cyclic group of order n; $Z(p^\infty)$ the Prüferian, where p is a prime. As in the finite case we say that the abelian group G has the Hajós 2-property if from each factorization $G = A + B$, $A \neq \{0\}$, $B \neq \{0\}$ it follows that either A or B is periodic. But the restriction is imposed that one of the factors has finitely many elements. It is not known if there is any infinite group with Hajós 2-property without this restriction. It was shown by Hajós in 1949 that Z has the Hajós 2-property. In 1961, A. D. Sands gave a complete classification of infinite abelian groups possessing the Hajós 2-property. Namely,

$$Z(p^\infty) + Z_q, \quad Z(2^\infty) + Z_2, \quad Q + Z_p$$

and their subgroups have the Hajós 2-property. All other groups are shown not to possess this property. The constructions of factorization with nonperiodic factors is mainly based on the ideas used in the finite case.

By Hajós's theorem, if a finite abelian group is factored into cyclic subsets, then at least one of the factors is a subgroup of the group. In 1958, L. Fuchs turned to the possible extensions of Hajós's theorem for infinite abelian groups. He suggested studying factorizations of infinite abelian groups by cyclic subsets of a finite number of elements. To extend the concept of factorization to the case when the number of components is not finite, we need a little care. As an infinite sum of elements does not necessarily exist, we assume that all infinite sums occurring in the definition of a direct sum of subsets contain only finitely many nonzero terms.

When G is a finite abelian group (now written additively) in the factorization $G = A + B$ the factor A can be replaced by $-A$. In connection with a geometrical problem, S. K. Stein asked if A can be replaced by $(-A)$ in any factorization $G = A + B$ of the abelian group G. When A is finite the answer is "yes" and in general the answer is "no". Let $G = Z(2^\infty)$. The elements of G are the roots of unity $e^{2\pi i(k/n)}$, where k is an integer and n is a 2-power. (The operation is the multiplication of complex numbers.) For the sake of brevity we denote $e^{2\pi i(k/n)}$ by (k/n). Let A, B be

$$\{(0/1), (1/2), (1/4), (1/8), (1/16), \ldots\},$$

$$\{(0/1), \underbrace{(3/4), (5/8), (7/8)}, \underbrace{(9/16), (11/16), (13/16), (15/16)}, \ldots\}$$

respectively. Now $G = A + B$ is a factorization of G. The inverses of

$$(1/4), (1/8), (1/16), \ldots$$

are in order $(3/4), (7/8), (15/16), \ldots$. Therefore neither $(-A) + B$ nor $A + (-B)$ is direct.

3. Noncommutative groups. In 1968, H. J. Bernstein generalized Hajós's theorem in a different direction by dropping the condition that the group involved is commutative. He proved the following three theorems.

(1) Let G be a finite group in which every cyclic subgroup of composite order is normal. Let G be a direct product of cyclic subsets. Then at least one of the factors must be a subgroup of G.

For example groups of prime exponent, dihedral and Hamiltonian groups satisfy the hypothesis of this theorem.

(2) Let G be a finite group in which the subgroup generated by the elements of composite order is a proper subgroup of G. Let G be a k-fold product of cyclic subsets. Then at least one of the factors is a subgroup of G.

Groups of prime exponent and dihedral groups are examples of groups that satisfy the conditions of this theorem.

(3) If a finite group is a k-fold product of not more than three cyclic subsets, then at least one of the factors is a subgroup.

The proofs are based on the ideas developed for the commutative case.

Besides Hajós's theorem, a result on factoring factoring finite abelian groups by simulated subsets was considered in the case of certain finite noncommutative groups, by K. Corrádi, P. Z. Hermann and S. Szabó.

4. Quasi-periodic factorizations. Analyzing the available factorizations $G = AB$ in 1950, G. Hajós observed that one of the factors, say B, always can be partitioned in the form $B = B_1 \cup \cdots \cup B_n$ such that

$$AB_1 h_1 = AB_2 h_2 = \cdots = AB_n h_n,$$

where $H = \{h_1, h_2, \ldots, h_n\}$ is a subgroup of G. Let us call a factorization $G = AB$ with this property a *quasi-periodic* factorization.

As an example consider the factorization $G = AB$, where G is of type $(4, 4, 3)$ with basis elements x, y, z, $|x| = |y| = 4$, $|z| = 3$, $A = \{e, x, y, xy\}$,

$$B = B_1 \cup B_2 \cup B_3.$$

Here

$$
\begin{aligned}
B_1 &= \{e, x^2 y, x^2 y^3, y^2\}, \\
B_2 &= \{xz, x^3 z, x^2 y^2 z, y^2 z\}, \\
B_3 &= \{x^2 z^2, x^2 y^2 z^2, y^2 z^2, z^2\}
\end{aligned}
$$

and $H = \{e, z, z^2\}$.

Let $G = AB$ be a factorization and assume that B is periodic. Now B can be factored in the form $B = HC$, where H is a nonidentity subgroup of G. If $H = \{h_1, \ldots, h_n\}$, then B can be partitioned in the form $B = B_1 \cup \cdots \cup B_n$ and

$$AB_1 h_1 = AB_2 b_2 = \cdots = AB_n b_n,$$

that is, the factorization $G = AB$ is quasi-periodic.

When Hajós discovered that neither factor in a factorization of certain finite abelian groups need be periodic, he asked the following question. Is every factorization $G = AB$ of a finite abelian group G quasi-periodic? A. D. Sands noticed that the factorization $G = A_1 A_2$ described by construction (2) in Lemma 2.3.2 answers Hajós's question in the negative. With the choice $A = A_2$, $B = A_1$ the factorization $G = AB$ is not quasi-periodic. If the factorization $G = AB$ is quasi-periodic, then one of the factors splits as a disjoint union of n subsets of equal order, where $n \geq 2$. These subsets can have order 1 only if one of the factors is periodic. Since neither A nor B is periodic and A has prime order, we see that the only possibility is that B splits as a union of p subsets each of order p. Let such a splitting occur and let $AB_1 = AB_i h_i$. Then the subgroup H has order p. Hence H must be contained in the subgroup $K = \langle x^p, y \rangle$. The product AH is direct and so $|AH| = p^2$. Now $A, H \subset K$. It follows that $K = AH$ is a factorization of K. From $G = AB$ we have

$$G = A(\cup B_i) = \cup(AB_i) = AB_1 H = AB_i h_i H = AB_i H = KB_i.$$

Hence each set B_i must be a set of coset representatives for G modulo K. There-fore each set B_i contains one element from $\langle x^p \rangle$ and one element from $\{e, y, y^2, \ldots, y^{p-1}\}x^r$, for each r such that $1 \leq r \leq p-1$. Let $x^{pa(1)}, y^{b(1)}x \in B_1$ and $x^{pa(2)}, y^{b(2)}x \in B_2$. Then $AB_1 = AB_2 h_2$ implies that

$$(AB_1) \cap K = (AB_2 h_2) \cap K.$$

Therefore $Ax^{pa(1)} = Ax^{pa(2)} h_2$. Since A is not periodic, we have $h_2 = x^{p(a(1)-a(2))}$. Similarly

$$(AB_1) \cap (Kx) = (AB_2 h_2) \cap (Kx)$$

implies that $Ay^{b(1)}x = Ay^{b(2)}xh_2$. Thus $h_2 = y^{b(1)-b(2)}$. This gives

$$x^{p(a(1)-a(2))} = y^{b(1)-b(2)}.$$

As G is a direct product of $\langle x \rangle$ and $\langle y \rangle$ it follows that $x^{pa(1)} = x^{pa(2)}$. This is impossible as $B_1 \cap B_2 = \emptyset$. Therefore the factorization $G = AB$ is not quasi-periodic.

Problem 11.2.1. Are there finite cyclic groups with not quasi-periodic factoriza-tions?

Under certain conditions a factorization must be quasi-periodic. Such a sit-uation is the content of the next exercise.

Exercise 11.2.3. Let $G = AB$ be a factorization of the finite abelian group G. Let us assume that A is contained in a proper subgroup K of G such that G is a direct product of K and a subgroup H. Show that $G = AB$ is a quasi-periodic factoring. (Hint: Set $B_i = B \cap (Kh_i)$ for each $h_i \in H$.)

Exercise 11.2.4. Let G be a group of type $(p, \ldots, p)$, where p is a prime and let $G = AB$ be a factorization of G such that $\langle A \rangle \neq G$. Show that the factorization is quasi-periodic. (Hint: $\langle A \rangle = K$ is a direct factor component of G.)

If the factorization $G = AB$ is quasi-periodic, then the equation $G = A(B_1 H)$ is also a factorization of G. In other words in the factorization $G = AB$ the factor B has been replaced by the periodic factor $B_1 H$. So we have the following problem.

Problem 11.2.2. Is it always possible to replace one factor in a factorization of a finite abelian group by a periodic factor?

5. Simulated factorizations. Consider a normalized factorization $G = A_1 \cdots A_n$ of a finite abelian group G. Suppose that each A_i is obtained from a subgroup H_i of G changing at most k elements in H_i. We call such a factorization a *simulated factorization*. We ask for which integers k it is true that, if each A_i differs from a subgroup H_i with $|A_i| = |H_i|$ in at most k elements, then one of the subsets must be actually a subgroup. By Theorem 1.2.1, if $k = 1$, then one factor must be a subgroup. Let p be the least prime factor of $|G|$. It follows from Theorem 5.3.1

type of G	$k(G)$	remark
(p)	∞	
(p^2)	∞	
(p, q)	∞	
(p, p)	∞	
$(4, 2)$	∞	
$(2, 2, 2)$	∞	
$(2, 2, 2, 2)$	∞	
(p^α)	$p^2 - p - 1$	$\alpha \geq 3$
(p, q^α)	$q - 1$	$\alpha \geq 2,\ p < q$
$(3^\alpha, 3)$	2	$\alpha \geq 2$
$(3, 3, 3)$	2	
$(3, 3, p^\alpha)$	2	$\alpha \geq 1,\ p \geq 5$

Table 1

that for $k = p - 2$ one factor must be a subgroup. For each finite abelian group G there will be a best possible value, say $k(G)$, for which this result holds. These values were determined by A. D. Sands.

There are certain finite abelian groups G such that in any normalized factorization $G = A_1 \cdots A_n$ of G one factor must be a subgroup. The list of these groups is known for us from Exercises 4.1.2–4.1.4. For these groups $k(G)$ can be set to be infinity. (See Table 1.)

Suppose G is not one of the types listed in Table 1 and the least prime divisor of $|G|$ is p.

If the p-component of G is cyclic, then $k(G) = p - 1$.

If the p-component of G is not cyclic, then $k(G) = p - 2$.

6. Replacing a factor by a subgroup. L. Fuchs asked if in a factorization $G = AB$ of a finite abelian group G, A or B can always be replaced by a subgroup of G. The next exercise shows that the answer is negative in general.

Exercise 11.2.5. Let p and q be distinct primes and let $G = \langle x \rangle \times \langle y \rangle$, where $|x| = p^2$, $|y| = q^2$. Define the subsets A and B of G in the following way.

$$A = \{e, x^p, x^{2p}, \ldots, x^{(p-1)p}\}\{e, y, y^2, \ldots, y^{q-1}\},$$
$$B = \{e, x, x^2, \ldots, x^p\}\{e, y^q, y^{2q}, \ldots, y^{(q-1)q}\}.$$

Show that $G = AB$ is a factorization of G. Then verify that neither A nor B can be replaced by a subgroup in this factorization. (Hint: Note that as G is cyclic, if it has subgroups of a given order it must be unique.)

A. D. Sands gave a positive answer for Fuchs's question in the case of finite cyclic groups. We present here only some direct consequences of our replacement results.

Theorem 11.1.1. *If $G = AB$ is a normalized factorization of the finite abelian group G, where $|A|$ is a prime power and is relatively prime to $|B|$, then B can be replaced by a subgroup.*

Proof. As $|A|$ and $|B|$ are relatively prime, there are subgroups K and L of G of orders $|A|$ and $|B|$ respectively such that $G = K \times L$. It is enough to show that $\mathrm{Ann}(B) \subset \mathrm{Ann}(L)$. Clearly $\mathrm{Ann}(L)$ consists of all characters χ of G for which $L \not\subset \mathrm{Ker}\chi$. Assume the contrary, that there is a character χ of G with $\chi \in \mathrm{Ann}(B)$ and $L \subset \mathrm{Ker}\chi$. Now

$$
\begin{aligned}
0 &= \chi(B) \\
&= \sum_{b \in B} \chi(b) \\
&= \sum_{b \in B} \chi\big(k(b)\big)\chi\big(l(b)\big) \\
&= \sum_{b \in B} \chi\big(k(b)\big),
\end{aligned}
$$

that is,

$$
0 = \sum_{b \in B} \chi\big(k(b)\big), \tag{1}
$$

where $k(b) \in K$, $l(b) \in L$ and $k(b)l(b) = b$. Let $|A| = |K|$ be a power of the prime p. There is a minimal nonnegative integer n such that each $\chi\big(k(b)\big)$ in (1) is a power of a fixed primitive (p^n)th root of unity. In the $n = 0$ case we get the contradiction

$$
0 = \chi(B) = \sum_{b \in B} \chi\big(k(b)\big) = |B|.
$$

Thus $n \geq 1$. Now by Lemma 3.1.1, it follows that $|B|$, the number of the terms in (1), is a multiple of p. This leads to the contradiction that p divides the relatively prime $|B|$ and $|A|$.

This completes the proof.

The next exercise concerns with an analogous result.

Exercise 11.2.6. Let $G = AB$ be a normalized factorization of the finite abelian group G, where each prime divisor of $|A|$ is greater than $|B|$. Show that then B

can be replaced by a subgroup of G. (Hint: Use the indirect assumption and ideas of the previous proof and consider the following equation.)

$$\begin{aligned}
0 &= \chi(B) \\
&= \sum_{b \in B} \chi(b) \\
&= \sum_{b \in B} \chi\big(k(b)\big)\chi\big(l(b)\big) \\
&= \sum_{b \in B} \chi\big(k(b)\big).
\end{aligned}$$

Theorem 11.2.2. *If $G = AB$ is a normalized factorization of the finite abelian group G, such that $|A| = m$ and $|B| = n$ are relatively prime and $2 \le m \le 2p - 1$, where p is the least prime divisor of n, then A can be replaced by a subgroup.*

Proof. There are subgroups K and L of G with the property that $G = K \times L$ and $|K| = m$, $|L| = n$. If $\mathrm{Ann}(A) \not\subset \mathrm{Ann}(K)$, then there is a character χ of G for which $\chi(A) = 0$ and $\chi(K) \ne 0$, that is, $\chi(k) = 1$ for each $k \in K$. Using this χ we get

$$\begin{aligned}
0 &= \chi(A) \\
&= \sum_{a \in A} \chi(a) \\
&= \sum_{a \in A} \chi\big(k(a)\big)\chi\big(l(a)\big) \\
&= \sum_{a \in A} \chi\big(l(a)\big),
\end{aligned}$$

that is,

$$0 = \sum_{a \in A} \chi\big(l(a)\big), \tag{2}$$

where $k(a) \in K$, $l(a) \in L$ and $k(a)l(a) = a$. There is a minimal nonnegative integer r such that each $\chi\big(l(a)\big)$, $a \in A$ is a power of a fixed primitive rth root of unity. In the case $r = 1$ we get the contradiction

$$0 = \chi(A) = \sum_{a \in A} \chi\big(l(a)\big) = |A|.$$

Thus we may assume that $r \ge 2$. From (2) by Lemma 3.1.1, it follows that r divides n. But r is a divisor of m and n that are relatively prime.

This contradiction completes the proof.

7. Hajós n-property. If a finite abelian group G admits a factorization into n factors $A_1, \ldots, A_n$ and if from the factorization $G = A_1 \cdots A_n$ it follows that at least one of the factors must be periodic, then we say that G has the Hajós

n-property. In order to avoid trivial cases we assume that $|A_i| \geq 2$ for each i, $1 \leq i \leq n$. If $|G| = 1$, then as each $|A_i| \geq 2$, G does not admit any factorization and so G does not have the Hajós n-property for any n. If $|G| = p$ is a prime, then G has the Hajós n-property only for $n = 1$. It is obvious that if $|G| \geq 2$, then G has the Hajós n-property for $n = 1$. A. D. Sands sets forth the next problem. Given a finite cyclic group G. Determine each n for which G has the Hajós n-property. The solution is the following. Let $|G|$ be the product of m not necessarily distinct primes. The possible n values for which G has the Hajós n-property can be described in terms of m.

If $1 \leq m \leq 4$, then $1 \leq n \leq m$.

If $5 \leq m$, then $n = 1$ or $m - 1 \leq n \leq m$.

The solution is also available for elementary p-groups. Let p be a prime and let G be an elementary p-group with $|G| = p^m$. Then G has the Hajós n-property for the following values of n described in terms of p and m.

If $p = 2$, $1 \leq m \leq 5$, then $1 \leq n \leq m$.

If $p = 2$, $6 \leq m$, then $n = 1$ or $\lfloor (m + 1)/2 \rfloor \leq n \leq m$.

If $p = 3$, $1 \leq m \leq 3$, then $1 \leq n \leq m$.

If $p = 3$, $4 \leq m$, then $n = 1$ or $m - 1 \leq n \leq m$.

If $p = 5$, $1 \leq m$, then $n = 1$ or $n = m$.

Here $\lfloor a \rfloor$ denotes the greatest integer less than or equal to a.

Appendix A

Factoring by simulated subsets

In this appendix we give three proofs for Theorem 1.2.1.

Theorem 1. *Let $G = A_1 \cdots A_n$ be a factorization of the finite abelian group G, where A_i is a simulated subset of G for each i, $1 \leq i \leq n$; then at least one of the factors must be a subgroup of G.*

The first proof exploits characters in an essential way. Another interesting aspect of the proof is the appearance of a directed graph as a bookkeeping tool.

First Proof. Let H_i be the unique subgroup associated with the simulated subset A_i, that is for which $|A_i \cap H_i| + 1 \geq |A_i| = |H_i| \geq 3$. If $A_i = H_i$ for some i, $1 \leq i \leq n$, then the result holds. So we may suppose that $|A_i \cap H_i| = |A_i| - 1$ for each i, $1 \leq i \leq n$. By Lemma 4.2.1, in the factorization $G = A_1 \cdots A_n$ we may replace A_i by H_i to get the factorization $G = H_1 \cdots H_n$. Thus G is a direct product of the subgroups $H_1, \ldots, H_n$.

Now there exist $a_i \in A_i$, $h_i \in H_i$ with $a_i \notin H_i$, $h_i \notin A_i$. The element a_i can be written in the form $a_i = h_i d_i$ with a suitable element $d_i \in G$. If $d_i = e$, then $A_i = H_i$ and so we are done. Thus assume that $d_i \neq e$. By Lemma 4.2.1, $\chi(A_i) = 0$ implies $\chi(H_i) = 0$ and $\chi(d_i) = 1$.

If $n = 1$, then the result is trivial. We assume that $n \geq 2$ and proceed by induction on n. In the factorization $G = A_1 \cdots A_n$ the factor A_i can be replaced by H_i to give the factorization

$$G = A_1 \cdots A_{i-1} H_i A_{i+1} \cdots A_n$$

which in turn gives the factorization

$$G/H_i = (A_1 H_i)/H_i \cdots (A_{i-1} H_i)/H_i \cdot (A_{i+1} H_i)/H_i \cdots (A_n H_i)/H_i$$

of the factor group G/H_i for each i, $1 \leq i \leq n$. By the inductive assumption some factor $(A_j H_i)/H_i$ is equal to $(H_j H_i)/H_i$. In other words $A_j H_i$ is equal to the subgroup $H_j H_i$. This implies that $d_j \in H_i$. Thus, for each i, there exists a j such that $j \neq i$, $1 \leq j \leq n$ and $d_j \in H_i$. We record this information using a graph Γ. The nodes of Γ are the subgroups $H_1, H_2, \ldots, H_n$. The directed edge (H_j, H_i) codes the fact that $d_j \in H_i$.

As from each node there is an edge out, there must be a directed cycle C in Γ. Say let $H_1, H_2, \ldots, H_r$ be the nodes of this cycle in order. Thus

$$(H_1, H_2), \; (H_2, H_3), \ldots, (H_{r-1}, H_r), \; (H_r, H_1)$$

are the edges of C. Now the product $A_1 \cdots A_r$ is a factorization of the subgroup $H = H_1 \cdots H_r$ of G. Indeed, as

$$d_1 \in H_2, \; d_2 \in H_3, \ldots, d_{r-1} \in H_r, \; d_r \in H_1,$$

it follows that $A_1 \subset H, \ldots, A_r \subset H$. Further $|A_i| = |H_i|$. In addition the product $A_1 \cdots A_r$ is a partial product of the direct product $A_1 \cdots A_n$.

If $r < n$, then the inductive assumption implies that some $A_i = H_i$. If $r = n$, then the elements d_i belong to n distinct subgroups $H_1, \ldots, H_n$ in a direct product giving G. It follows that there is a character χ of G such that $\chi(d_i) \neq 1$ for each i, $1 \leq i \leq n$. On the other hand, as χ is not the principal character of G from the factorization $G = A_1 \cdots A_n$, it follows that $0 = \chi(A_1) \cdots \chi(A_n)$ and hence $\chi(A_i) = 0$ for some i, $1 \leq i \leq n$. Therefore we get the contradiction that $\chi(d_i) = 1$ for some i.

This completes the proof.

Exercise 1. Show that if (H_j, H_i) and (H_j, H_k) are edges of Γ such that $i \neq k$, then A_j is a subgroup. (Hint: Observe that $d_j \in H_i \cap H_k = \{e\}$.)

In the second proof the partition lemma (Lemma 3.2.7) plays an essential part. Let A be a simulated subset of a finite abelian group G and let H be the subgroup of G associated with A. There are elements $a \in A$, $h \in H$ such that $a \notin H$, $h \notin A$ and there is a $d \in G$ with $a = hd$. We assign the subgroup $K = \langle d \rangle$ to A. Obviously $A = H$ if and only if $K = \{e\}$. By Lemma 4.2.1, $\chi(A) = 0$ implies $\chi(d) = 1$ and so $\chi(K) = |K| \neq 0$. In other words $\chi(K) = 0$ implies $\chi(A) \neq 0$.

Second Proof. Consider a factorization

$$G = A_1 \cdots A_n, \tag{1}$$

where A_i is a simulated subset of G and H_i is the subgroup of G corresponding to A_i with the elements a_i, h_i, d_i defined as in the first proof. The theorem holds when $n = 1$ and so we assume that $n \geq 2$ and proceed by induction on n. We assume that $d_i \neq e$ for each i, $1 \leq i \leq n$ since otherwise there is nothing to prove. In factorization (1) the factor A_1 can be replaced by the subgroup $L_1 = H_1$ to get the factorization $G = L_1 A_2 \cdots A_n$. Considering the factor group G/L_1 we get the factorization

$$G/L_1 = (A_2 L_1)/L_1 \cdots (A_n L_1)/L_1.$$

The inductive assumption gives that one of the factors, say

$$(A_2 L_1)/L_1,$$

is a subgroup of G/L_1 and so $A_2 L_1 = L_2$ is a subgroup of G. It follows that $d_2 \in L_1$ and then $K_2 \subset L_1$. Considering the factor group G/L_2 from the factorization $G = L_2 A_3 \cdots A_n$ we get the factorization

$$G/L_2 = (A_3 L_2)/L_2 \cdots (A_n L_2)/L_2.$$

By the inductive assumption one of the factors, say $(A_3 L_2)/L_2$, is a subgroup of G/L_2 and hence $A_3 L_2 = L_3$ is a subgroup of G. It follows that $d_3 \in L_2$, that is, $K_3 \subset L_2$. Continuing in this way we get that there is a permutation $B_1, \ldots, B_n$ of the factors $A_1, \ldots, A_n$ such that $B_1 = A_1$ and the partial products

$$L_1 = H_1, \ \ L_2 = H_1 B_2, \ldots, L_n = H_1 B_2 \cdots B_n$$

form an ascending chain of subgroups of G. We may assume that $B_1 = A_1, \ldots,$ $B_n = A_n$ since this is only a matter of indexing the factors in (1). Thus we get

$$K_2 \subset L_1, \ldots, K_n \subset L_{n-1}.$$

Let $C_i = A_{i+1} \cdots A_n$, $1 \le i \le n - 1$.

From the factorization $G = A_1 C_1$ we get that $\chi(C_1) = 0$ holds for each $\chi \in \mathrm{Ann}(K_1)$. In other words $\mathrm{Ann}(K_1) \subset \mathrm{Ann}(C_1)$. From $C_1 = A_2 C_2$ it follows that $0 = \chi(A_2)\chi(C_2)$ for each $\chi \in \mathrm{Ann}(K_1)$. Therefore

$$\mathrm{Ann}(K_1) \cap \mathrm{Ann}(K_2) \subset \mathrm{Ann}(C_2).$$

By the partition lemma there are subsets X_2, Y_2 of G for which

$$C_2 = X_2 K_1 \cup Y_2 K_2,$$

where the union is disjoint and the products are direct. Suppose $Y_2 \ne \emptyset$. There is a $y_2 \in Y_2$. Multiplying the factorization $G = L_2 C_2$ by y_2^{-1} gives the factorization $G = Gy_2^{-1} = L_2(C_2 y_2^{-1})$. Now $K_2 \subset L_1 \subset L_2$ and $K_2 \subset (C_2 y_2^{-1})$ contradict the factorization. Thus $Y_2 = \emptyset$ must hold and so $C_2 = X_2 K_1$. This is equivalent to $\mathrm{Ann}(K_1) \subset \mathrm{Ann}(C_2)$.

From $C_2 = A_3 C_3$ it follows that $0 = \chi(A_3)\chi(C_3)$ for each $\chi \in \mathrm{Ann}(K_1)$. Therefore

$$\mathrm{Ann}(K_1) \cap \mathrm{Ann}(K_3) \subset \mathrm{Ann}(C_3).$$

By the partition lemma there are subsets X_3, Y_3 of G for which

$$C_3 = X_3 K_1 \cup Y_3 K_3,$$

where the union is disjoint and the products are direct. If $Y_3 \ne \emptyset$, then we can choose a $y_3 \in Y_3$ and get the factorization $G = L_3(C_3 y_3^{-1})$. Here $K_3 \subset L_2 \subset L_3$

and $K_3 \subset (C_3 y_3^{-1})$ contradict the factorization. Thus $Y_3 = \emptyset$ must hold and so $C_3 = X_3 K_1$. Continuing in this way we get that $C_{n-1} = X_{n-1} K_1$. But $C_{n-1} = A_n$ and we end up with the contradiction that A_n is periodic.

This completes the proof.

The third proof does not rely on characters.

Third Proof. From the factorization (1) by the argument of the second proof it follows that we may assume that

$$L_1 = H_1, \ L_2 = H_1 A_2, \ldots, L_n = H_1 A_2 \cdots A_n$$

form an ascending chain of subgroups of G. From the factorization $L_3 = L_2 A_3$ it follows that the elements of A_3 are incongruent modulo L_2. Note that $A_2 \subset L_2$. Now Observation 2.3.1 is applicable and gives that $A_2 A_3$ is not periodic.

From the factorization $L_4 = L_3 A_4$ it follows that the elements of A_4 are incongruent modulo L_3. We can see that $A_2 A_3 \subset L_3$ holds. Observation 2.3.1 shows that $(A_2 A_3) A_4$ is not periodic. Continuing in this way finally we get that $(A_2 \cdots A_{n-1}) A_n$ is not periodic. From the factorization $G = A_1 (A_2 \cdots A_n)$, by Lemma 2.1.1, follows the contradiction that $A_2 \cdots A_n$ is periodic with period d_1.

This completes the proof.

Appendix B

Proofs for Hajós's theorem

After proving Hajós's theorem for p-groups there are several ways to extend the result for all finite abelian groups. We give here three proofs. The first one is based on the fact that a subset A always can be replaced by A^{-1}. This line of arguing is due to K. Corrádi.

Theorem 1. *Let G be a finite abelian non-p-group and let $G = A_1 \cdots A_n$ be a factorization of G, where each A_i is a cyclic subset of prime cardinality. Then at least one of the factors must be a subgroup of G.*

First Proof. Let G be a finite abelian non-p-group and assume the contrary, that

$$G = A_1 \cdots A_n \tag{1}$$

is a factorization of G, where

$$A_i = \{e, a_i, a_i^2, \ldots, a_i^{p(i)-1}\}$$

are nonsubgroup cyclic subsets of G with prime cardinalities. Call the quantity

$$h(A_1, \ldots, A_n) = \prod_{i=1}^{n} |a_i|$$

the height of the factorization (1). Suppose that the height of our counterexample is the smallest possible.

Choose a prime divisor p of $|G|$. If each cyclic subset A_i with cardinality p contains only p-elements, then the product of these cyclic subsets forms a factorization of the p-component of G. The p-component of G is a p-group and so by Lemma 1.3.3, at least one of the occurring factors must be a subgroup of G. This contradiction gives that there is a factor of cardinality p which contains not only p-elements. We assume that this factor is A_1. Hence $|a_1|$ has a prime divisor, say

r, which is distinct from $p(1)$. In factorization (1) the factor A_1 can be replaced by A_1^r to get the new factorization

$$G = A_1^r A_2 \cdots A_n. \tag{2}$$

Since $|a_1^r| < |a_1|$,

$$h(A_1^r, A_2, \ldots, A_n) < h(A_1, \ldots, A_n)$$

and so by the minimality of the height of the counterexample we get that at least one of the factors $A_1^r, A_2, \ldots, A_n$ is a subgroup of G. Clearly A_1^r is a subgroup of G. There is a permutation $B_2, \ldots, B_n$ of the factors $A_2, \ldots, A_n$ such that

$$A_1^r \subset A_1^r B_2 \subset \cdots \subset A_1^r B_2 \cdots B_n$$

are subgroups of G. We assume that this permutation is the identical one and so

$$A_1^r \subset A_1^r A_2 \subset \cdots \subset A_1^r A_2 \cdots A_n$$

are subgroups of G. In factorization (1) each factor A_i can be replaced by A_i^{-1}. In general for each

$$[\varepsilon(2), \ldots, \varepsilon(n)] = [\pm 1, \ldots, \pm 1]$$

we can consider the factorization

$$G = A_1 A_2^{\varepsilon(2)} \cdots A_n^{\varepsilon(n)}. \tag{3}$$

By Lemma 2.1.1, $A_2^{\varepsilon(2)} \cdots A_n^{\varepsilon(n)}$ is periodic with period $a_1^{p(1)}$. It follows that

$$a_1^{p(1)} \in A_2^{\varepsilon(2)} \cdots A_n^{\varepsilon(n)}$$

and hence

$$a_1^{p(1)} = a_2^{\alpha(2)\varepsilon(2)} \cdots a_n^{\alpha(n)\varepsilon(n)},$$

where

$$0 \leq \alpha(1) \leq p(1) - 1, \ldots, 0 \leq \alpha(n) \leq p(n) - 1.$$

If $[\alpha(2), \ldots, \alpha(n)] = [0, \ldots, 0]$ for some $[\varepsilon(2), \ldots, \varepsilon(n)]$, then $a_1^{p(1)} = e$ is a contradiction. Thus for each $[\varepsilon(2), \ldots, \varepsilon(n)]$ there is an $\alpha(i) \neq 0$ with a minimal i. Let k be the minimum of these indices and let $[\varepsilon(2), \ldots, \varepsilon(n)]$ be a system of exponents at which this k is attained. For notational simplicity we may assume that $[\varepsilon(2), \ldots, \varepsilon(n)] = [1, \ldots, 1]$. Now

$$a_1^{p(1)} = a_k^{\alpha(k)} \cdots a_n^{\alpha(n)},$$

where $k \geq 2$, $1 \leq \alpha(k) \leq p(k) - 1$ and

$$0 \leq \alpha(k+1) \leq p(k+1) - 1, \ldots, 0 \leq \alpha(n) \leq p(n) - 1.$$

Replace A_k by A_k^{-1} in (2). Using the corresponding factorization (3) the element $a_1^{p(1)}$ is expressible in the form

$$a_1^{p(1)} = a_k^{(-1)\beta(k)} a_{k+1}^{\beta(k+1)} \cdots a_n^{\beta(n)},$$

where

$$0 \le \beta(k) \le p(k) - 1, \ldots, 0 \le \beta(n) \le p(n) - 1.$$

The two expressions for $a_1^{p(1)}$ combined together give

$$e = a_k^{-(\beta(k)+\alpha(k))} a_{k+1}^{\beta(k+1)-\alpha(k+1)} \cdots a_n^{\beta(n)-\alpha(n)}, \tag{4}$$

where $0 \le |\beta(i) - \alpha(i)| \le p(i) - 1$ for each i, $k+1 \le i \le n$ and $1 \le \beta(k) + \alpha(k) \le 2p(k) - 2$. After introducing the notation $\gamma(i) = \beta(i) - \alpha(i)$ for each i, $k+1 \le i \le n$ (4) becomes

$$e = a_k^{-(\beta(k)+\alpha(k))} a_{k+1}^{\gamma(k+1)} \cdots a_n^{\gamma(n)}. \tag{5}$$

Set

$$\delta(i) = \begin{cases} 1, & \text{if } \gamma(i) = 0, \\ \gamma(i), & \text{if } \gamma(i) \ne 0, \end{cases}$$

for each i, $k+1 \le i \le n$. Clearly $\delta(i)$ is relatively prime to $p(i)$. If $\alpha(k) + \beta(k)$ is relatively prime to $p(k)$, then from the factorization (1) we can get the factorization

$$G = A_1 \cdots A_{k-1} A_k^{-(\beta(k)+\alpha(k))} A_{k+1}^{\delta(k+1)} \cdots A_n^{\delta(n)}$$

which violates the representation (5). Thus $\beta(k) + \alpha(k) = p_k$ and so

$$a_k^{p(k)} \in A_{k+1}^{\delta(k+1)} \cdots A_n^{\delta(n)} = C.$$

From factorization (2) we can construct the factorization

$$G = A_1^r A_2 \cdots A_k A_{k+1}^{\delta(k+1)} \cdots A_n^{\delta(n)}.$$

Here $a_k^{p(k)}$ is an element of the subgroup

$$A_1^r A_2 \cdots A_k = H$$

because $a_k \in A_k$. As $a_k^{p(k)} \in C \cap H = \{e\}$, the contradiction that $a_k^{p(k)} = e$ follows. This completes the proof.

The second proof is a standard application of the partition lemma (Lemma 3.2.7). If the cyclic subset A is in the form $A = \{e, a, a^2, \ldots, a^{p-1}\}$, then $\mathrm{Ann}(A)$ consists of each character χ of G with $\chi(a) \ne 1$ and $\chi(a^p) = 1$. To A we assign the subgroup $K = \langle a^p \rangle$. It is clear that A is a subgroup of G if and only if $K = \{e\}$. Putting this in another form A is periodic if and only if $K = \{e\}$.

Second Proof. Arguing in the same way as in the first proof we see that there is a prime r and there is a permutation $B_1, \ldots, B_n$ of the factors $A_1^r, A_2, \ldots, A_n$ such that $B_1 = A_1^r$ and the partial products

$$B_1, \ B_1 B_2, \ldots, B_1 B_2 \cdots B_n$$

are subgroups of G. We may assume that the permutation is the identity since this is only a matter of relabelling the factors. So the partial products

$$A_1^r, \ A_1^r A_2, \ldots, A_1^r A_2 \cdots A_n$$

are subgroups of G.

Let $H_i = A_1^r A_2 \cdots A_i$ and $C_i = A_{i+1} \cdots A_n$. Note that as $a_i \in H_i$, it follows that $K_i \subset H_i$.

From the factorization $G = A_1 C_1$ it follows that $0 = \chi(C_1)$ for each $\chi \in \mathrm{Ann}(K_1)$, that is, $\mathrm{Ann}(K_1) \subset \mathrm{Ann}(C_1)$. As $C_1 = A_2 C_2$ we get that $0 = \chi(A_2)\chi(C_2)$ for each $\chi \in \mathrm{Ann}(K_1) \subset \mathrm{Ann}(C_1)$. Hence

$$\mathrm{Ann}(K_1) \cap \mathrm{Ann}(K_2) \subset \mathrm{Ann}(C_2).$$

By the partition lemma there are subset X_2, Y_2 of G such that

$$C_2 = X_2 K_1 \cup Y_2 K_2,$$

where the union is disjoint and the products are direct. If there is an element y_2 of Y_2, then in the factorization $G = H_2 C_2$ the factor C_2 can be replaced by $y_2^{-1} C_2$ to get the factorization $G = H_2(y_2^{-1} C_2)$. But here $K_2 \subset H_2$ and $K_2 \subset y_2^{-1} C_2$ violates the factorization as $K_2 \neq \{e\}$. Thus $Y_2 = \emptyset$ and so $C_2 = X_2 K_1$. This is equivalent to $\mathrm{Ann}(K_1) \subset \mathrm{Ann}(C_2)$.

As $C_2 = A_3 C_3$, it follows that $0 = \chi(A_3)\chi(C_3)$ for each $\chi \in \mathrm{Ann}(K_1) \subset \mathrm{Ann}(C_2)$. Therefore

$$\mathrm{Ann}(K_1) \cap \mathrm{Ann}(K_3) \subset \mathrm{Ann}(C_3).$$

By the partition lemma there are subsets X_3, Y_3 of G such that

$$C_3 = X_3 K_1 \cup Y_3 K_3,$$

where the union is disjoint and the products are direct. If there is an element $y_3 \in Y_3$, then in the factorization $G = H_3 C_3$ the factor C_3 can be replaced by $y_3^{-1} C_3$ to get the factorization $G = H_2(y_3^{-1} C_3)$. But here $K_3 \subset H_3$ and $K_3 \subset y_3^{-1} C_3$ violates the factorization as $K_3 \neq \{e\}$. Thus $Y_3 = \emptyset$ and so $C_3 = X_3 K_1$. This is equivalent to $\mathrm{Ann}(K_1) \subset \mathrm{Ann}(C_3)$.

Using $C_3 = A_4 C_4$ it follows that $0 = \chi(A_4)\chi(C_4)$ for each $\chi \in \mathrm{Ann}(K_1) \subset \mathrm{Ann}(C_3)$. Therefore

$$\mathrm{Ann}(K_1) \cap \mathrm{Ann}(K_4) \subset \mathrm{Ann}(C_4).$$

By the partition lemma there are subsets X_4, Y_4 of G such that

$$C_4 = X_4 K_1 \cup Y_4 K_4,$$

where the union is disjoint and the products are direct. If there is an element $y_4 \in Y_4$, then the factorization $G = H_4(y_4^{-1} Y_4)$ leads to the contradiction $K_4 \subset H_4 \cap (y_4^{-1} Y_4) = \{e\}$. Thus $Y_4 = \emptyset$ and consequently $C_4 = X_4 K_1$.

Continuing in this way finally we get that $C_{n-1} = X_{n-1} K_1$. But $C_{n-1} = A_n$ and it follows that $A_n = K_1$.

This contradiction completes the proof.

Third Proof. Consider the factorization (1). The argument in the first proof leads to the ascending chain of subgroups

$$H_1 = A_1^r, \ H_2 = A_1^r A_2, \ldots, H_n = A_1^r A_2 \cdots A_n.$$

The factorization $H_3 = H_2 A_3$ implies that the elements of A_3 are incongruent modulo H_2. As $A_2 \subset H_2$, Observation 2.3.1 is applicable and gives that the product $A_2 A_3$ is not periodic. In a similar way step by step we can conclude that

$$A_2 A_3 A_4, \ldots, A_2 A_3 \cdots A_n$$

are not periodic. On the other hand from the factorization $G = A_1(A_2 \cdots A_n)$, by Lemma 2.2.1 , it follows that $A_2 \cdots A_n$ is periodic with period $a_1^{p(1)}$.

This contradiction completes the proof.

Appendix C

Special cases of Rédei's theorem

Rédei's theorem for groups of type (p, p) was proved in Section 1.4, where interpolation with polynomials over $\mathrm{GF}(p)$ played a role. Here we give another proof in which roots and divisibility of polynomials over $\mathrm{GF}(p)$ are applied. The interpolation idea is due to L. Lovász and A. Schrijver. The divisibility idea goes back to E. Wittmann.

Theorem 1. *Let p be a prime and let X be a subset of the affine plane $[\mathrm{GF}(p)]^2$ such that $|X| = p$. If X determines at most $(p+1)/2$ directions, then X is a straight line.*

Proof. Let X be a subset of $[\mathrm{GF}(p)]^2$ such that $|X| = p$ and X determines at most $(p+1)/2$ directions. The $p = 2$ case of the theorem is trivial so we will assume that $p \geq 3$. Note that X cannot determine all possible $p+1$ directions on $[\mathrm{GF}(p)]^2$ since $p + 1 > (p+1)/2$. Consequently there is a direction on the plane that is not determined by X. Using a direction determined by X as the direction of the first coordinate axis and using a direction not determined by X as the direction of the second coordinate axis, the points of X can be represented in the form

$$(0, a_0), \ (1, a_1), \ldots, (p - 1, a_{p-1}).$$

We may assume that $a_0 = a_1 = 0$. We wish to show that all the a_is are 0 modulo p. The way to do this will be to show that the polynomial

$$(x - a_0)(x - a_1) \cdots (x - a_{p-1})$$

is simply x^p viewed as a polynomial over $\mathrm{GF}(p)$.

The slopes of the directions determined by X are

$$U = \left\{ \frac{a_i - a_j}{i - j} : i, j \in \mathrm{GF}(p), i \neq j \right\}.$$

By the assumption of the theorem $|U| \le (p+1)/2$. Note that $y \notin U$ implies that

$$a_0 - 0y, a_1 - 1y, \ldots, a_{p-1} - (p-1)y$$

is only a rearrangement of $0, 1, \ldots, p-1$. There are at least $p-(p+1)/2 = (p-1)/2$ choices for y with $y \notin U$.

Let $c_0, c_1, \ldots, c_m$ be all the distinct values occurring among $a_0, a_1, \ldots, a_{p-1}$, taken modulo p. We define three polynomials with coefficients in $\mathrm{GF}(p)$:

$$D(x,y) = \prod_{i=0}^{p-1} \left(x - (a_i - iy)\right) = \sum_{i=0}^{p} d_i(y)x^i,$$

$$E(x) = D(x,0) = \prod_{i=0}^{p-1}(x - a_i) = \sum_{i=0}^{p} e_i x^i,$$

$$F(x) = \prod_{i=0}^{m}(x - c_i),$$

and we wish to prove that $E(x) = x^p$, or equivalently, that $E'(x) = 0$.

We first show that many of the coefficients of $E(x)$ are 0, by examining $D(x,y)$.

We have $d_p(y) = 1$ and each $d_{p-i}(y)$, $1 \le i \le p$, is an elementary symmetric function in the p expressions $a_j - jy$. Let $S_i(z_j)$ denote the ith elementary function of the expressions $z_0, \ldots, z_{p-1}$. Then we have

$$d_{p-i}(y) = (-1)^{(p-i)} S_i(a_j - jy).$$

Note that the degree of $d_{p-i}(y)$ is at most $i - 1$.

Now $D(x,y) = -x + x^p$ for each $y \notin Y$. Thus for $y \notin Y$ we have $d_{p-i}(y) = 0$ for $1 \le i \le p$. Therefore for those values of i the polynomial $d_{p-i}(y)$ has at least $(p + 1)/2$ roots. Since the degree of $d_{p-i}(y)$ is at most $i - 1$, we see that for $1 \le i \le (p - 1)/2$, $d_{p-i}(y)$ is the zero polynomial. That means that in $D(x,y)$ each coefficient $d_i(y)$ is zero $(p + 1)/2 \le i \le p - 1$. Thus in $E(x)$ each coefficient $e_i = 0$ for $(p + 1)/2 \le i \le p - 1$. Moreover, since $a_0 = a_1 = 0$ and are roots of $E(x)$, $e_0 = e_1 = 0$. All that remains is to show that $e_2 = e_3 = \cdots = e_{(p-1)/2} = 0$. To do that, we show that $E'(x) = 0$.

Introduce $G(x) = E(x) - (x^p - x)$, a polynomial of degree at most $(p-1)/2$. It is not the zero polynomial since it has a term of degree 1. Since $E'(x)$ has degree at most $(p-3)/2$, the product $G(x)E'(x)$ has degree at most $p - 2$ or is the zero polynomial. We show that $E(x)$, which has degree p, divides $G(x)E'(x)$.

Clearly, $E(x)/F(x)$ divides $E'(x)$ in the ring of polynomials with coefficients from $\mathrm{GF}(p)$. Hence there is a polynomial $Q(x)$ such that

$$E'(x) = \frac{Q(x)E(x)}{F(x)}.$$

Thus

$$G(x)E'(x) = \frac{G(x)}{F(x)}Q(x)E(x).$$

Since $F(x)$ divides $x^p - x$ and $E(x)$, it divides their difference, $G(x)$. Hence $E(x)$ divides $G(x)E'(x)$. Since the degree of

$$G(x)E'(x)$$

is smaller than the degree of $E(x)$, $G(x)E'(x)$ is the zero polynomial, hence $E'(x) = 0$. Therefore $E(x) = x^p$.

This completes the proof.

We prove Rédei's theorem in the special case of finite elementary p-groups. The proof we present here uses directed graphs and has a similar flavour to the first proof in Appendix A.

Theorem 2. *Let $G = A_1 \cdots A_n$ be a normalized factorization of the finite elementary p-group G, where each $|A_i| = p$. Then at least one of the factors $A_1, \ldots, A_n$ must be a subgroup of G.*

Proof. We proceed by induction on n. The $n = 1$ case is trivial and the $n = 2$ case is settled in Lemma 1.4.5 and in Theorem 1 of Appendix C. So we focus our attention on the case $n \geq 3$. We may assume that none of the factors $A_1, \ldots, A_n$ is a subgroup since otherwise there is nothing to prove. In the normalized factorization $G = A_1 \cdots A_n$ each factor A_i can be replaced by a subgroup $H_i = \langle a_i \rangle$ of G, where $a_i \in A_i \setminus \{e\}$. Note that $e, a_i \in H_i \cap A_i$ and consequently A_i and H_i differ only at most by $p - 2$ elements. As A_i is not a subgroup of G, the subgroup H_i can be selected more than one way. (See Exercise 1 after the proof.) Let us fix a choice of the subgroups $H_1, \ldots, H_n$ and let us work with them. From the normalized factorization

$$G = A_1 \cdots A_{i-1} H_i A_{i+1} \cdots A_n$$

we get the normalized factorization

$$G/H_i = (A_1 H_i)/H_i \cdots (A_{i-1} H_i)/H_i \cdot (A_{i+1} H_i)/H_i \cdots (A_n H_i)/H_i$$

of the factor group G/H_i. From this factorization, by the inductive assumption, it follows that there is a factor, say $(A_j H_i)/H_i$, which is a subgroup of G/H_i. Now $A_j H_i$ is a subgroup of G. Clearly this subgroup must be equal to $H_j H_i$. Thus for each i, $1 \leq i \leq n$ there exist a j such that $1 \leq j \leq n$ and $j \neq i$ further $A_j H_i = H_j H_i$. We record this information using a directed graph Γ. The nodes of Γ are the subgroups $H_1, \ldots, H_n$ and let (H_j, H_i) be a directed edge of Γ if $A_j H_i = H_j H_i$. Clearly, A_j can be written in the form

$$A_j = \{e, a_j, a_j^2 d_{j,2}, \ldots, a_j^{p-1} d_{j,p-1}\},$$

where $d_{j,2}, \ldots, d_{j,p-1} \in H_i$. As into each node joins an edge, the graph contains at least one directed cycle, say

$$(H_1, H_2), \ (H_2, H_3), \ldots, (H_{s-1}, H_s), \ (H_s, H_1).$$

The partial product $A_1 \cdots A_s$ is direct and gives the subgroup $H = H_1 \cdots H_s$ of G. If $s \leq n - 1$, then by the induction assumption we are done and so we may assume that $s = n$, that is, Γ consists of one single directed cycle, say

$$(H_1, H_2), \ (H_2, H_3), \ldots, (H_{n-1}, H_n), \ (H_n, H_1).$$

As there is at most one edge going into any node, Γ is a union of disjoint directed cycles. (See Exercise 2 after the proof.) Construct a new graph Γ^* using the subgroups $H_1^*, H_2, \ldots, H_n$ in place of the subgroups $H_1, H_2, \ldots, H_n$. The graph Γ^* is constructed from Γ by erasing the edges (H_n, H_1), (H_1, H_2) and adding the edges (H_n, H_1^*), (H_1^*, H_2). As $H_1 \cap H_1^* = \{e\}$ it follows that A_n is a subgroup of G.
　　This completes the proof.

Exercise 1. Show that if A_i is not a subgroup of G, then A_i can be replaced by at least two distinct subgroups of G.

Exercise 2. Show that if there is a node of Γ, say H_j, from which two edges go out, say (H_j, H_i) and (H_j, H_k) with $i \neq k$, then the factor A_j is a subgroup of G. (Hint: Note that the product $H_1 \cdots H_n$ is direct giving G and so the product $H_i H_j H_k$ is direct. Hence $A_j \subset H_i H_j \cap H_k H_j = H_j$.)

Appendix D

The background of some constructions

A number of multiple factorizations involving nonsubgroup cyclic subsets were described in Sections 4.3 and 4.4 without shedding any light on how they were constructed. We will illustrate how these multiple factorizations were constructed. In the second part we will describe how cube tilings guided factorization constructions.

Let $a_1, \ldots, a_n$ be elements of the finite abelian group G. Let $r(1), \ldots, r(n)$ be prime divisors of $|a_1|, \ldots, |a_n|$ respectively. Using a_i and $r(i)$ we can form the cyclic subsets

$$A_i = \{e, a_i, a_i^2, \ldots, a_i^{r(i)-1}\}$$

of G. By the character test, if for each nonprincipal character χ of G there is an i, $1 \leq i \leq n$ such that

$$\chi(a_i) \neq 1 \qquad \text{and} \qquad \chi\big(a_i^{r(i)}\big) = 1,$$

then the product $A_1 \cdots A_n$ is a multiple factorization of G. (The multiplicity of the factorization is $r(1) \cdots r(n)/|G|$.) Further if $a_i^{r(i)} \neq e$ for each i, $1 \leq i \leq n$, then none of the cyclic subsets $A_1, \ldots, A_n$ is a subgroup of G.

This suggests the following procedure. Compute the order of $\chi(a)$ for each character χ and for each element a of G and arrange the result into a table. Let the rows be labeled by the characters of G and let the columns be labeled by the elements of G. The intersection of the row and column labeled by χ and a respectively contains the order of $\chi(a)$. Let r be the order of $\chi(a)$. If $1 < r < |a|$, then $\chi(a) \neq 1$, $\chi(a^r) = 1$, and $a^r \neq e$. We refer to such a situation saying that the element a and the number r cover the character χ. We mark each r occurring in the column of a, say, by putting parentheses (brackets, braces etc.) around r. If the row of the character χ of G contains a marked entry, then this character is

000	1	+	+	+	+	+	+	+	+
001	3					+	+	+	+
002	3					+	+	+	+
010	2		+			+			
011	6					+			
012	6					+			
100	2			+			+		
101	6						+		
102	6						+		
110	2				+				+
111	6								+
112	6								+

Table 1

covered by some element a of G and integer r. A marked table can help to establish whether each nonprincipal character of G is covered and so helps to find a multiple factoring of G by nonsubgroup cyclic subsets. We may improve this plan by noting that if a and a' generate the same cyclic subgroup, then the column labeled by a and a' are the same. The observation that the rows labeled by χ and χ^{-1} are the same makes a further shortcut possible.

We describe the details of the procedure in connection with an example. Let G be a finite abelian group of type $(2,2,3)$. Let x, y, z be basis elements of G such that $|x| = |y| = 2$, $|z| = 3$ and $G = \langle x \rangle \times \langle y \rangle \times \langle z \rangle$. We record the element $x^i y^j z^k$ of G by the triplet (i, j, k). For the sake of brevity we suppress the parentheses and commas. For example the elements x, xyz, xz^2 of G will be recorded by the triplets $100, 111, 102$ respectively. As a first step we list all cyclic subgroups of G. The result is summarized in Table 1. The first column contains the elements of G. The second column gives the orders of the elements. The further columns correspond to cyclic subgroups of G. For instance the last column corresponds to the subset consisting of the elements $000, 001, 002, 110, 111, 112$, that is, the elements $e, z, z^2, xy, xyz, xyz^2$ that form the cyclic subgroup $\langle xyz \rangle$ of G.

We can see that G has 3 cyclic subgroups with composite order and we can choose $011, 101, 111$ as generators of these subgroups.

Now turn to the characters of G. Consider the character χ of G defined by

$$\chi(x) = \rho^i, \ \chi(y) = \rho^j, \ \chi(z) = \sigma^k,$$

where ρ and σ are primitive 2nd and 3rd roots of unity. This character will be

	011	101	111
001	(3)	(3)	(3)
011	6	(3)	6
101	(3)	6	6
111	6	6	(3)
010	[2]	1	2
100	1	[2]	2
110	[2]	[2]	1

Table 2

identified by the triplet (i, j, k). Again for the sake of brevity we suppress the parentheses and commas. Now we form Table 2 containing the orders of $\chi(a)$. The rows are labeled by the characters of G. But we use only one of χ and χ^{-1}. The columns are labeled by $011, 101, 111$.

In order to cover the characters $011, 101, 111$ we have to choose the elements $101, 011, 111$ paired with the numbers $3, 3, 3$ respectively. These choices give the bonus that the character 001 is covered free of charge. The last three rows remained uncovered. Choosing the elements 011 and 101 paired with the integers 2 and 2 we can cover these characters as well. From Table 2 with the marked entries we can read off that the product of the cyclic subsets

$$A_1 = \{e, yz, (yz)^2\}, \qquad A_2 = \{e, xz, (xz)^2\},$$
$$A_3 = \{e, xyz, (xyz)^2\}, \qquad A_4 = \{e, yz\},$$
$$A_5 = \{e, xz\}$$

is a 9-fold factorization of G. (This factorization is identical with the one in Exercise 4.3.1, only the order of the elements has changed.)

We carry out one more construction. Let G be a finite abelian group of type $(4, 4)$. From Table 1 we can see that G has six cyclic subgroups of order 4. The generators of these subgroups can be chosen to be $01, 10, 11, 12, 13, 21$.

Next we compute the orders of $\chi(a)$ and summarize the result in Table 4. In order to cover the first six rows the choices are unique. Marking the 2s occurring in the columns by parentheses we get the bonus that the last three rows are covered free of charge.

Finally from Table 4 with the marked entries we can read off that the product

00	1	+	+	+	+	+	+	+	+	+	+
01	4					+					
02	2		+			+					+
03	4					+					
10	4						+				
11	4							+			
12	4								+		
13	4									+	
20	2			+			+		+		
21	4										+
22	2				+			+		+	
23	4										+
30	4						+				
31	4									+	
32	4								+		
33	4							+			

Table 3

	01	10	11	12	13	21
01	4	1	4	(2)	4	4
10	1	4	4	4	4	(2)
11	4	4	(2)	4	1	4
12	(2)	4	4	4	4	1
13	4	4	1	4	(2)	4
21	4	(2)	4	1	4	4
02	(2)	1	(2)	1	(2)	(2)
20	1	(2)	(2)	(2)	(2)	1
22	(2)	(2)	1	(2)	1	(2)

Table 4

of the cyclic subsets

$$A_1 = \{e, x\}, \qquad A_2 = \{e, xy\},$$
$$A_3 = \{e, y\}, \qquad A_4 = \{e, xy^2\},$$
$$A_5 = \{e, xy^3\}, \qquad A_6 = \{e, x^2y\}$$

forms a 4-factorization of G without subgroup factors. (Note that this factorization is identical with the factorization in Exercise 4.3.2.)

In the remaining part, following the ideas of G. Hajós, we describe how cube tilings can help to construct certain factorizations. Consider the 3-dimensional Euclidean space with coordinate unit vectors e_1, e_2, e_3 filled with rectangular boxes of dimensions v_1, v_2, v_3 such that the boxes are abutting face to face and the edges are parallel to the coordinate axis. Divide the boxes into $v_1 v_2 v_3$ unit cubes then cut each unit cube into smaller cells of dimensions $1/u_1, 1/u_2, 1/u_3$. Here u_1, u_2, u_3, v_1, v_2, v_3 are given positive integers. The corners of the small cells form a lattice L_1. The corners of the unit cubes form a lattice L_2. The corners of the large cells form a lattice L_3. The lattice L_3 is spanned by $v_1 e_1, v_2 e_2, v_3 e_3$. The lattice L_2 is spanned by e_1, e_2, e_3. The lattice L_1 is spanned by $(1/u_1)e_1, (1/u_2)e_2, (1/u_3)e_3$. Clearly L_i is an abelian group with operation of addition of vectors and L_2, L_3 are subgroups of L_1. The factor group $G = L_1/L_3$ is a direct sum of cyclic groups of orders $u_1 v_1$, $u_2 v_2$, $u_3 v_3$ respectively. Let g_i denote the coset $(1/u_i)e_i + L_3$, $1 \le i \le 3$. Then g_1, g_2, g_3 are basis elements of G. The collection of $u_1 v_1 u_2 v_2 u_3 v_3$ small cells filling a large cell can be identified with the elements of G. Rearranging these cells leads to new decompositions of G. Let

$$A_i = \{0, g_i, 2g_i, \ldots, (u_i - 1)g_i\}, \quad H_i = \langle u_i g_i \rangle, \quad 1 \le i \le 3.$$

Then $A = A_1 + A_2 + A_3$ corresponds to $u_1 u_2 u_3$ small cells that form a unit cube and $H = H_1 + H_2 + H_3$ corresponds to the $v_1 v_2 v_3$ unit cubes that form a large cell. The factorization

$$G = (H_1 + H_2 + H_3) + A_1 + A_2 + A_3$$

corresponds to the fact that the cubes form a tiling. We construct a new set B from H by removing

$$(v_1 - 1)u_1 g_1 + H_2, \quad (v_2 - 1)u_2 g_2 + H_3, \quad (v_3 - 1)u_3 g_3 + H_1$$

from H and adding

$$(v_1 - 1)u_1 g_1 + H_2 + g_2, \quad (v_2 - 1)u_2 g_2 + H_3 + g_3, \quad (v_3 - 1)u_3 g_3 + H_1 + g_1$$

to H. Replacing $(v_1-1)u_1 g_1 + H_2$ by $(v_1-1)u_1 g_1 + H_2 + g_2$ corresponds to an infinite column of unit cubes parallel to the e_2 being shifted parallel to e_2. Replacing $(v_2 - 1)u_2 g_2 + H_3$ by $(v_2 - 1)u_2 g_2 + H_3 + g_3$ corresponds to an infinite column of unit cubes parallel to the e_3 being shifted parallel to e_3. Finally, replacing $(v_3 - 1)u_3g_3 + H_1$ by $(v_3 - 1)u_3g_3 + H_1 + g_1$ corresponds to an infinite column of unit cubes parallel to the e_1 being shifted parallel to e_1. The infinite columns we are shifting are not blocking each other's movements as they are disjoint and the resulting cube system is a tiling. The new cube tiling corresponds to a new factorization of G in the form

$$G = B + (A_1 + A_2 + A_3).$$

Of course we can carry out the described factorization construction without any reference to cube tilings and can check if the factorization has the desired properties in a purely algebraic manner.

Note that if the numbers u_1v_1, u_2v_2, u_3v_3 are pairwise relatively primes, then the group G is cyclic. Thus the precedure above can be used to construct factorization for finite cyclic groups. Two instances of this factorization procedure appeared in the book. At the first occasion, the first five of the numbers

$$u_1, \ v_1, \ u_2, \ v_2, \ u_3, \ v_3 \tag{1}$$

were chosen to be at least 2 and v_3 was set to be 1. In this case $A_3 = \{0\}$. This setting led in Section 6.3 to a factorization $G = B + A_1 + A_2$, where none of the sets B, A_1, A_2, $A_1 + A_2$ was periodic. (In Section 6.3 we used multiplicative notation.)

In the second case we chose all the numbers (1) to be at least 2. This setting led to a factorization $G = B + A_1 + A_2 + A_3$, where both B and $A = A_1 + A_2 + A_3$ spanned the whole of G. This construction is used in the proof of Theorem 9.2.3 construction (1). (Again note that in Section 9.2 we used multiplicative notation.)

Appendix E

Cyclotomic polynomials

We shall use $F_n(x)$ to denote the nth cyclotomic polynomial for each integer $n \geq 1$. It is the monic polynomial of least degree with rational coefficients satisfied by a primitive nth root of unity. We shall not prove the classical results here concerning these polynomials.

$$F_n(x) = \prod (x - \rho),$$

where ρ varies over all primitive nth roots of unity. Since

$$x^n - 1 = \prod (x - \sigma),$$

where σ varies over all nth roots of unity it is easy to see that

$$x^n - 1 = \prod_{d \mid n} F_d(x),$$

where d varies over all divisors of n. From this it follows that $F_n(x)$ has integer coefficients; $F_n(x)$ has degree $\phi(n)$, where ϕ is Euler's totient function. It is a result of Kronecker that, if m and n are relatively prime, then $F_n(x)$ is irreducible over the field of the mth roots of unity.

The nth cyclotomic polynomial is not irreducible over the nth cyclotomic field. In fact $F_n(x)$ factors into monic linear factors with coefficients from the nth cyclotomic field.

Exercise 1. Let ρ be a 9th primitive root of unity. Show that $F_9(x)$ factors into two cubic polynomials over $Q(\rho^3)$. Hint: $F_9(x)$ factors into linear factors over $Q(\rho)$,

$$F_9(x) = (x - \rho^1)(x - \rho^2)(x - \rho^4)(x - \rho^5)(x - \rho^7)(x - \rho^8).$$

Then regroup the factors to

$$\left[(x - \rho^1)(x - \rho^4)(x - \rho^7)\right]\left[(x - \rho^2)(x - \rho^5)(x - \rho^8)\right].$$

If μ is the Möbius function, then

$$F_n(x) = \prod_{d|n} (x^d - 1)^{\mu(n/d)}.$$

Other formulae involving these polynomials follow from this formula or may be proved by considering all the roots of appropriate equations.

If p is a prime, then

$$F_{p^e}(x) = \frac{x^{p^e} - 1}{x^{p^{e-1}} - 1} = 1 + x^{p^{e-1}} + x^{2p^{e-1}} + \cdots + x^{(p-1)p^{e-1}}.$$

The coefficients here are 0 or 1 and $F_{p^e}(1) = p$. Since

$$\frac{x^n - 1}{x - 1} = \prod_{\substack{d|n \\ d>1}} F_d(x),$$

it follows that, for all $d > 1$ such that d is not a prime power, $F_d(1) = 1$ and $F_d(x)$ involves both positive and negative coefficients. It may be checked that $n = 105$ is the smallest value such that $F_n(x)$ has coefficient greater than 1.

If n, m are relatively prime, then

$$F_n(x^m) = \prod_{d|n} F_{nd}(x).$$

If, on the other hand, every prime divisor of m also divides n, then

$$F_n(x^m) = F_{nm}(x).$$

In the general case we may write $m = m_1 m_2$, where every prime divisor of m_1 also divides n and m_2 is relatively prime to n. Then

$$F_n(x^m) = \prod_{d|m_2} F_{nm_1 d}(x).$$

For example with $n = 20$, $m = 45$ we have $m_1 = 5$, $m_2 = 9$ and

$$F_{20}(x^{45}) = F_{100}(x) F_{300}(x) F_{900}(x).$$

If $p_1, p_2, \ldots, p_k$ are the distinct prime factors of n and $n = ml$, where $m = p_1 p_2 \cdots p_k$, then $F_n(x) = F_m(x^l)$. So all cyclotomic polynomials may be determined from a knowledge of those corresponding to square free values.

G. Hajós proved that if $Z = A + B$ is a factorization of Z, where A is finite, then B is periodic. Let p be the smallest among the absolute values of the periods of B. We call p the period of B. Let d be the diameter of A. From the proof, presented in Section 6.3, we can read off that $p < 2^d$. Applying cyclotomic polynomials, M. N. Kolountzakis has given an asymptotically better upper bound.

Theorem 1. *Let $Z = A + B$ be a factorization of Z, where d is the diameter of A and p is the period of B. There is an integer d_1 and there are positive constants c_1, c_2 such that*

$$p < c_1 \exp(c_2 \sqrt{d \ln d} \sqrt{\ln \ln d})$$

for $d > d_1$.

Proof. Consider a factorization $Z = A + B$ such that d is the diameter of A and p is the period of B. We may assume that $A \subset \{0, \dots, d\}$. Let $H = \langle p \rangle$. From $Z = A + B$ we get the factorization

$$Z/H = (A + H)/H + (B + H)/H.$$

Clearly Z/H is a cyclic group of order p. We can identify Z/H with Z_p whose elements are $0, 1, \dots, p - 1$. The operation is the addition of integers modulo p. There is a factorization $Z_m = A_1 + B_1$. Here $m = p$ and $A_1 = A$. (We changed p to m because it reminds us of primes.) We assign polynomials to A_1 and B_1 by

$$A_1(x) = \sum_{a \in A_1} x^a, \quad B_1(x) = \sum_{b \in B_1} x^b.$$

Let χ be a character of Z_m. Note that if $\chi(A_1) = 0$, then $\chi(1)$ is a root of $A_1(x)$. If the order of $\chi(1)$ is s, then $F_s(x)$ divides $A_1(x)$ over the rationals. Conversely if $F_s(x)$ divides $A_1(x)$, then $\chi(A_1) = 0$ for each character χ of Z_m with $|\chi(1)| = s$.

For each divisor d of m, $d > 1$, the cyclotomic polynomial $F_d(x)$ divides $A_1(x)B_1(x)$. Let

$$F_{s(1)}(x), \dots, F_{s(k)}(x)$$

be all the cyclotomic polynomials among these that divide $A_1(x)$. Set $t = s(1) \cdots s(k)$.

We claim that if $t < m$, then t is a period of B_1.

In order to prove the claim assume that $t < m$. We want to show that $B_1 = B_1 + t$, that is,

$$\chi(B_1) = \chi(B_1)\chi(t) \tag{1}$$

holds for each character χ of Z_m. Clearly (1) holds if $\chi(B_1) = 0$ or $\chi(t) = 1$. We may assume that $\chi(B_1) \neq 0$ and $\chi(t) \neq 1$. Now χ is not the principal character of Z_m. Applying χ to $Z_m = A_1 + B_1$ gives that $0 = \chi(A_1)\chi(B_1)$. As $\chi(B_1) \neq 0$, it follows that $\chi(A_1) = 0$. Now $|\chi(1)|$ is one of $s(1), \dots, s(k)$, say $|\chi(1)| = s(i)$. This gives the

$$\chi(t) = [\chi(1)]^t = [[\chi(1)]^{s(i)}]^{t/s(i)} = 1$$

contradiction which proves the claim.

We know that $s(1), \ldots, s(k)$ are distinct divisors of m and so we may arrange them such that

$$1 < s(1) < s(2) < \cdots < s(k).$$

As $F_{s(i)}(x)$ divides $A_1(x)$ over the rationals and $\deg F_{s(i)}(x) = \phi(s(i))$, it follows that

$$\phi(s(1)) + \cdots + \phi(s(k)) \leq \deg A_1(x) \leq d. \tag{2}$$

It is known from analytic number theory that for each $\varepsilon > 0$ there is an integer n_0 such that

$$(e^{-\gamma} - \varepsilon)\frac{n}{\ln \ln n} \leq \phi(n) \tag{3}$$

for $n > n_0$. Here $\gamma = 0.57721\ldots$ is the Euler constant. Let us fix an ε, n_0 and set $c_3 = e^{-\gamma} - \varepsilon$. From (3) we get

$$\ln c_3 + \ln \ln n \leq \ln \phi(n) + \ln \ln \ln n.$$

There is an $n_1 > n_0$ such that

$$\ln \phi(n) \leq \ln n \leq 2 \ln \phi(n) \tag{4}$$

for each $n > n_1$.

We claim that there is an integer d_2 such that

$$s(1) + \cdots + s(k) \leq (e^{-\gamma} + 2\varepsilon) d \ln \ln d \tag{5}$$

for $d > d_2$.

To prove the claim we estimate the sum $s(1) + \cdots + s(k)$.

$$
\begin{aligned}
\sum_{i=1}^{k} s(i) \;&=\; \sum_{s(i) \leq n_1} s(i) + \sum_{s(i) > n_1} s(i) \\
&\leq\; n_1^2 + \sum_{s(i) > n_1} (e^{-\gamma} + \varepsilon)\phi(s(i)) \ln \ln(s(i)) \quad &\text{(see (3))} \\
&\leq\; n_1^2 + (e^{-\gamma} + \varepsilon)d \ln \ln(s(k)) \quad &\text{(see (2))} \\
&\leq\; n_1^2 + (e^{-\gamma} + \varepsilon)d(\ln 2 + \ln \ln d) \quad &\text{(see (4))} \\
&\leq\; (e^{-\gamma} + 2\varepsilon)d \ln \ln d \quad &\text{(for } d > d_1\text{)}
\end{aligned}
$$

as required.

From (5) it follows that there is a positive constant c_4 such that $k \leq c_4\sqrt{d \ln \ln d}$ for $d > d_2$. As B_1 is not periodic, $m \leq t$ holds. Finally there is an integer d_1 such that

$$
\begin{aligned}
m \leq t \leq \prod_{i=1}^{k} s(i) \;&\leq\; [(e^{-\gamma} + 2\varepsilon)d \ln \ln d]^k \\
&\leq\; c_1 \exp(c_2\sqrt{d \ln d}\sqrt{\ln \ln d})
\end{aligned}
$$

for $d > d_1$.

This completes the proof.

References

1. K. Amin, K. Corrádi, and A. D. Sands, The Hajós property for 2-groups, *Acta Math. Hungar.* **89** (2000), 189–198.

2. H. J. Bernstein, Extension of Hajós' factorization theorem to some non-abelian groups, *Comm. Pure Appl. Math.* **21** (1968), 289–311.

3. B. Bollobás and I. Leader, The number of k-sums modulo k, *Journal of Number Theory* **78** (1999), 27–35.

4. N. G. de Bruijn, On bases for the set of integers, *Publ. Math. Debrecen* **1** (1950), 232–242.

5. N. G. de Bruijn, On the factorization of finite abelian groups, *Indag. Math. Kon. Ned. Akad. Wetersch.* **15** (1953), 258–264.

6. N. G. de Bruijn, On the factorization of finite cyclic groups, *Indag. Math. Kon. Ned. Akad. Wetersch.* **15** (1953), 370–377.

7. N. G. de Bruijn, Some direct decompositions of the set of integers, *Math. of Computation* **18** (1964), 537–546.

8. B. Cipra, Disproving the obvious in higher dimensions, *What's happening in the mathematical sciences* Vol. 1 1993.

9. G. D. Cohen, S. Litsyn, A. Vardy and G. Zémor, Tiling of binary spaces, *SIAM J. Discrete Math.* **3** (1996), 393–412.

10. K. Corrádi, On direct product of cyclic subsets, *Acta Sci. Math.* **60** (1995), 119–129.

11. K. Corrádi, P. Z. Hermann, and S. Szabó, On factorizations of nonabelian groups, *Acta Sci. Math. (Szeged)* **67** (2001), 529–533.

12. K. Corrádi, A. D. Sands, and S. Szabó, Simulated factorizations, *Journal of Algebra* **151** (1992), 12–25.

13. K. Corrádi, A. D. Sands, and S. Szabó, Factoring by simulated subsets, *Journal of Algebra* **175** (1995), 320–331.

14. K. Corrádi, A. D. Sands and S. Szabó, Factoring by simulated subsets II, *Communications in Algebra* **27** (1999), 5367–5376.

15. K. Corrádi and S. Szabó, Some special cases of Keller's conjecture, *Journal of Pure and Appl. Alg.* **49** (1987), 247–252.

16. K. Corrádi and S. Szabó, Rédei László egy tételéről, *Mat. Lapok.* **34** (1987), 41–59.

17. K. Corrádi and S. Szabó, A Hajós-Minkowki-tétel és néhány ehhez kapcsolódó eredmény, *Mat. Lapok.* **34** (1987), 25–39.

18. K. Corrádi and S. Szabó, Keller's conjecture for certain p-groups, *Journal of Algebra* **119** (1988), 213–217.

19. K. Corrádi and S. Szabó, A new proof of Rédei's theorem, *Pacific Journal of Math.* **140** (1989), 53–61.

20. K. Corrádi and S. Szabó, On the algebraic form of Keller's conjecture, *Beiträge zur Algebra und Geometrie* **30** (1990), 17–34.

21. K. Corrádi and S. Szabó, A combinatorial approach for Keller's conjecture, *Periodica Math. Hung.* **21** (1990), 95–100.

22. K. Corrádi and S. Szabó, A Hajós-Keller type result on factorization of finite cyclic groups, *Acta Math. Acad. Sci. Hung.* **55** (1990), 311–313.

23. K. Corrádi and S. Szabó, Cube tilings and covering a complete graph, *Discrete Math.* **85** (1990), 319–321.

24. K. Corrádi and S. Szabó, Factorization of periodic subsets II, *Math. Japonica.* **36** (1991), 165–172.

25. K. Corrádi and S. Szabó, Factoring by subsets that are simulated or of cardinality of a prime power, *Bull. Soc. Math. Belg.* **43** (1991), 37–42.

26. K. Corrádi and S. Szabó, Factoring by subsets of cardinality of a prime or a power of a prime, *Communications in Algebra* **19(5)** (1991), 1585–1592.

27. K. Corrádi and S. Szabó, Factoring certain infinite groups by cyclic subsets, *Pure Math. and Appl. A* **2** (1992), 285–290.

28. K. Corrádi and S. Szabó, A contribution to Keller's conjecture, *Studia Sci. Math. Hung.* **27** (1992), 37–42.

29. K. Corrádi and S. Szabó, An extension for Hajós' theorem, *Journal of Pure and Applied Algebra* **79** (1992), 217–223.

30. K. Corrádi and S. Szabó, Factoring by subsets of cardinality of prime power, *Periodica Math. Hungarica* **26** (1993), 201–205.

31. K. Corrádi and S. Szabó, A generalized form of Hajós' theorem, *Communications in Algebra* **21(11)** (1993), 4119–4125.

32. K. Corrádi and S. Szabó, Factoring by nonsubgroup factors, *Beiträge zur Algebra und Geometrie, Contributions to Algebra and Geometry* **35** (1994), No 1, 85–89.

33. K. Corrádi and S. Szabó, Factoring by subsets of cardinality prime or four, *Journal of Algebra* **164** (1994), 91–100.

34. K. Corrádi and S. Szabó, An elementary proof for a result on simulated factoring, *Acta Mathematica Hungarica* **64** (1994), 139–142.

35. K. Corrádi and S. Szabó, Hajós' theorem for multiple factorizations, *Acta Mathematica Hungarica* **64** (1994), 305–308.

36. K. Corrádi and S. Szabó, A Hajós type result on factoring finite abelian groups by subsets, *Mathematica Pannonica* **5** (1994), 275–280.

37. K. Corrádi and S. Szabó, Direct product of weakly periodic subsets, *Rivista di Matematica Della Universita di Parma* **3** (1994), 295–299.

38. K. Corrádi and S. Szabó, Solution to a problem of A. D. Sands, *Communications in Algebra* **23** (1995), 1503–1510.

39. K. Corrádi and S. Szabó, The size of an annihilator in a factorization, *Mathematica Pannonica* **9** (1998), 195–204.

40. K. Corrádi and S. Szabó, Periodic factorization of a finite abelian 2-group, *Rend. Sem. Mat. Univ. Pol. Torino* **57** (1999), 303–308.

41. K. Corrádi and S. Szabó, A Rédei type factorization result for a special 2-group, *Mathematica Pannonica* **11** (2000), 279–282.

42. K. Corrádi and S. Szabó, Periodicity forcing factorization types for finite abelian 2-groups, *Atti Sem. Mat. Fis. Univ. Modena* **48** (2000), 481–494.

43. E. M. Coven and A. Meyerowitz, Tiling the integers with translates of one finite set, *Journal of Algebra* **212** (1999), 161–174.

44. T. Etzion and A. Vardy, On perfect codes and tilings, *SIAM J. Discrete Math.* **11** (1998), 205–223.

45. I. Fáry, Die Aequivalente des Minkowski-Hajósschen Satzes in der Theorie der Topologischen Gruppen, *Comm. Math. Helv.* **23** (1949), 283–287.

46. C. de Felice, Completing codes by Hajós factorization of groups, in: J. Almeida, G. M. S. Gomes, P. V. Silva (Editors), *Semigroups, Automata and Languages*, World Scientific, Porto, 1996, pp. 59–66.

47. C. de Felice, An application of Hajós factorization to variable length codes, *Theoret. Comp. Sci.* **164** (1996), 223–252.

48. O. Fraser and B. Gordon, Solution to a problem of L. Fuchs, *Quart. J. Math. Oxford* **25** (1974), 1–8.

49. O. Fraser and B. Gordon, Solution to a problem of A. D. Sands, *Glasgow Math. J.* **20** (1979), 115–118.

50. L. Fuchs, On the possibility of extension Hajós' theorem to infinite groups, *Publ. Math., Debrecen* **5** (1959), 338–347.

51. L. Fuchs, *Abelian Groups*, Akadémia Kiadó, Budapest 1958.

52. Ph. Furtwängler, Über Gitter konstanter Dichte, *Monatshefte für Mathematik u. Physik* **43** (1936), 281–288.

53. S. M. Gagola, S. C. Garisson, and M. R. Pettet, On a vanishing product in the integral group ring, *Amer. Math. Monthly* **88** (1981), 195–196.

54. B. Gordon, Multiple tilings of Euclidean space by unit cubes, *Computers and Mathematics with Applications* **39** (2000), 49–53.

55. G. Hajós, Többméretű terek befedése kockaráccsal, *Mat. Fiz. Lapok* **45** (1938), 171–190.

56. G. Hajós, Többméretű terek egyszeres befedése kockaráccsal, *Mat. Fiz. Lapok* **48** (1941), 37–64.

57. G. Hajós, Über einfache und mehrfache Bedeckung des n-dimensionalen Raumes mit einem Würfelgitter, *Math. Zeit.* **47** (1942), 427–467.

58. G. Hajós, Sur la factorisation des groupes abèliens, *Časopis Pěs. Mat. Fys.* **74** (1949), 157–162.

59. G. Hajós, Sur la problème de factorisation des groupes cycliques, *Acta Math. Acad. Sci. Hungar.* **1** (1950), 189–195.

60. G. Hajós, A ciklikus csoportok faktorizációjának problémájához, *MTA III. Oszt. Közl.* **3** (1953), 1–6.

61. W. Hamaker, Factoring groups and tiling space, *Aequationes Math.* **9** (1973), 145–149.

62. R. Hill and R. W. Irving, On group partitions associated with lower bounds for symmetric Ramsey numbers, *European Journal of Combinatorics* **3** (1982), 35–50. (This is an application)

63. J. M. Irwin and E. A. Walker, *Topics in Abelian groups*, Glenview, Illinois, 1963.

64. I. Kaplansky, *Infinite Abelian Groups*, University of Michigan Press, Ann Arbor, Michigan, 1969.

65. O. H. Keller, Über die lückenlose Einfüllung des Raumes Würfeln, *J. Reine Angew. Math.* **163** (1930), 231–248.

66. O. H. Keller, Ein Satz über die lückenlose Erfüllung des 5- und 6-dimensional Raumes mit Würfeln, *J. Reine Angew. Math.* **177** (1937), 61–64.

67. J. H. E. Kempermen, On small sumsets in an abelian group, *Acta Math.* **103** (1960), 63–88.

68. K. H. Kim and F. W. Roush, Robinson's conjecture on abelian groups, *Journal of Pure and Appl. Alg.* **25** (1982), 113–120.

69. M. N. Kolountzakis, Translational tilings of the integers with long periods, *The Electronic Journal of Combinatorics* **10** (2003), #R22.

70. J. P. S. Kung, M. R. Murty, and G. C. Rota, On the Rédei zeta function, *Journal of Number Theory* **12** (1980), 421–436.

71. J. C. Lagarias and P. W. Shor, Keller's cube tiling conjecture is false in high dimension, *Bulletin Amer. Math. Soc.* **27** (1992), 279–283.

72. J. C. Lagarias and P. W. Shor, Cube tilings and nonlinear codes, *Disc. and Comp. Geom.* **11** (1994), 359–391.

73. J. C. Lagarias and S. Szabó, Universal spectra and Tijdeman's conjecture on factorization of cyclic groups, *The Journal of Fourier Analysis and Applications* **7** (2001), 63–70.

74. J. C. Lagarias and Y. Wang, Tiling the line with translates of one tile, *Invent. Math.* **124** (1996), 341–365.

75. J. C. Lagarias and Y. Wang, Spectral sets and factorizations of finite abelian groups, *Journal of Funct. Anal.* **145** (1997), 73–98.

76. N. H. Lam, Hajós factorizations and completion of codes, *Theoretical Computer Science* **182** (1997), 245–256.

77. J. Lawrence, Tiling R^d by translates of the orthants, *Proc. of the Second Oklahoma Conf. Convexity and Related Combinatorial Geom.*, ed. D. C. Kay and M. Breen, 1982, 203–207.

78. B. Lindstrom, On group and nongroup perfect codes in q symbols, *Math. Scand.* **25** (1969), 149–158.

79. C. T. Long, Addition theorems for sets of integers, *Pacific Journal of Math.* **23** (1967), 107–112.

80. L. Lovász and A. Schrijver, Remarks on a theorem of Rédei, *Studia Sci. Math. Hungar.* **16** (1981), 449–454.

81. G. Lyubeznik, An equation in abelian groups and multiple lattice tilings of n-dimensional space, *Archiv der Math.* **38** (1982), 217–225.

82. J. Mackey, A cube tiling of dimension eight with no facesharing, *Discrete Comput. Geom.* **28** (2002), 275–279.

83. H. Minkowski, *Geometrie der Zahlen*, Teubner, Leipzig, 1896.

84. H. Minkowski, *Diophantische Approximationem*, Teubner, Leipzig, 1907.

85. H. Minkowski, *Gesammelte Abhandlungen von Hermann Minkowski*, Chelsea, New York, 1967.

86. E. Molnár, Sui mosaici dello spazio di dimensione n, *Atti della Academia Nazionale dei Lincei, Rend. Sc. Fis Mat. e Nat.* **51** (1971), 177–185.

87. E. Molnár, A Minkowski-sejtés Hajós-féle bizonyításáról, *Mat. Lapok* **34** (1987), 11–24.

88. D. Moews, Keller's conjecture for certain finite groups, *Journal of Algebra* **142** (1991), 435–440.

89. D. J. Newman, Tessellation of integers, *Journal of Number Theory* **9** (1977), 107–111.

90. E. E. Obaid, On a variation of Sands' method, *Internat. J. Math. Math. Sci.* **9** (1986), 597–604.

91. C. Okuda, The factorization of abelian groups, Ph.D. Thesis, The Pennsylvania State University, 1975.

92. J. E. Olson, An addition theorem for finite abelian groups, *Journal of Number Theory* **9** (1977), 63–70.

93. J. E. Olson, A problem of Erdős on abelian groups, *Combinatorica* **7** (1987), 285–289.

94. L. J. Paige, A note on finite abelian groups, *Bulletin Amer. Math. Soc.* **53** (1947), 590–593.

95. O. Perron, Über lückenlose Ausfüllung des n-dimensionalen Raumes durch kongruente Würfel, *Math. Zeit.* **46** (1940), 1–26, 161–180.

96. O. Perron, Modulartige lückenlose Ausfüllung des R^n mit kongruenten Würfeln, *Math. Ann.* **117** (1940), 415–447, 609–658.

97. L. Rédei, Jelentés az 1942. évi Kőnig Gyula jutalomról, Hajós György munkáinak ismertetése, *Mat. Fiz. Lapok* **49** (1942), 1–16.

98. L. Rédei, Zwei Lückensätze über Polynome in endlichen Primkörpern mit Anwendung auf die endlichen Abelschen Gruppen und die Gaussischen Summen, *Acta Math.* **79** (1947), 273–290.

99. L. Rédei, Vereinfachter Beweis des Satzes von Minkowski-Hajós, *Acta Sci. Math. Szeged* **13** (1949), 21–35.

100. L. Rédei, Die Reduction des gruppentheoretischen Satzes von Hajós auf den Fall von p-Gruppen, *Monatschefte für Math.* **53** (1949), 221–226.

101. L. Rédei, Kurzer Beweis des gruppentheoretischen Satzes von Hajós, *Comm. Math. Helv.* **23** (1949), 272–282.

102. L. Rédei, Ein Beitrag zum Problem der Factorisation von endlichen Abelschen Gruppen, *Acta Math. Acad. Sci. Hungar.* **1** (1950), 197–207.

103. L. Rédei, Zetafunctionen in der Algebra, *Acta Math. Acad. Sci. Hungar.* **6** (1955), 5–25.

104. L. Rédei, Die gruppentheoretischen Zetafunctionen und der Satz von Hajós, *Acta Math. Acad. Sci. Hungar.* **6** (1955), 271–279.

105. L. Rédei, Hazai vizsgálatok a véges csoportok elméletében, *MTA III. Oszt. Közl.* **5** (1955), 315–325.

106. L. Rédei, Hajós György 50. születésnapjára, *Mat. Lapok* **13** (1962), 217–227.

107. L. Rédei, Ein Überdeckungssatz für endlichen Abelschen Gruppen im Zusemmenhang mit dem Hauptsatz von Hajós, *Acta Sci. Math. Szeged* **26** (1965), 55–61.

108. L. Rédei, Logische Dualität der Frobenius-Stickelbergerschen und Hajósschen Hauptsätze der Theorie der Endlichen Abelschen Gruppen, *Acta Math. Sci. Hungar.* **16** (1965), 327.

109. L. Rédei, Die neue Theorie der endlichen abelschen Gruppen und Verallgemeinerung des Hauptsatzes von Hajós, *Acta Math. Acad. Sci. Hungar.* **16** (1965), 329–373.

110. L. Rédei, *Lückenhafte Polynome über endlichen Körpern*, Birkhäuser Verlag, Basel 1970, (English translation: *Lacunary Polynomials over Finite Fields*, North-Holland, Amsterdam, 1973.)

111. A. Restivo, S. Salemi and T. Sportelli, Completing codes, *Theoretical Informatics and Applications* **23** (1989), 135–147.

112. R. M. Robinson, Multiple tilings of n-dimensional space by unit cubes, *Math. Zeit.* **166** (1979), 225–264.

113. R. M. Robinson, Solution of an equation in abelian groups, *Amer. Math. Monthly* **86** (1979), 690.

114. R. M. Robinson, *Can cubes avoid meeting face to face?* The Mathematical Gardner, Prindle-Weber-Schmidt, Boston, 1980.

115. S. Saidi, Codes for perfectly correcting errors of limited size, *Discrete Mathematics* **118** (1993), 207–223.

116. A. D. Sands, On the factorisation of finite abelian groups, *Acta Math. Acad. Sci. Hungar.* **8** (1957), 65–86.

117. A. D. Sands, The factorization of abelian groups, *Quart. J. Math. Oxford* **10** (1959), 81–91.

118. A. D. Sands, On the factorisation of finite abelian groups II, *Acta Math. Acad. Sci. Hungar.* **13** (1962), 153–169.

119. A. D. Sands, The factorization of abelian groups II, *Quart. J. Math. Oxford* **13** (1962), 45–54.

120. A. D. Sands, On a problem of L. Fuchs, *J. London Math. Soc.* **37** (1962), 277–284.

121. A. D. Sands, Factorization of cyclic groups, *Proc. Coll. on Abelian Groups*, Tihany, Hungary, 1963, 139–146.

122. A. D. Sands, On the factorisation of finite abelian groups III, *Acta Math. Acad. Sci. Hungar.* **25** (1974), 279–284.

123. A. D. Sands, On a conjecture of G. Hajós, *Glasgow Math. J.* **15** (1974), 88–89.

124. A. D. Sands, On the factorization of finite groups, *J. London Math. Soc.* **7** (1974), 627–631.

125. A. D. Sands, On Keller's conjecture for certain cyclic groups, *Proc. Edinburgh Math. Soc.* **22** (1979), 17–21.

126. A. D. Sands, Simulated factorizations II, *Aequationes Math.* **44** (1992), 48–59.

127. A. D. Sands, Simulated factorizations III, *Algebra Colloq.* **6** (1999), 177–185.

128. A. D. Sands, Replacement of factors by subgroups in the factorization of abelian groups, *Bull. London Math. Soc.* **32** (2000), 297–304.

129. A. D. Sands, Factoring finite abelian groups, *Journal of Algebra* **274** (2004), 540–549.

130. A. D. Sands and S. Szabó, Factorization of periodic subsets, *Acta Math. Acad. Sci. Hungar.* **57** (1991), 159–167.

131. T. Schmidt, Über die Zerlegung des n-dimensionalen Raumes mit gitterförmig angeordnete Würfeln, *Schr. math. Semin. u. Inst. angew. Math. Univ. Berlin* **1** (1933), 186–212.

132. K. Seitz, Vizsgálatok a véges Abel-csoportok Hajós-féle elmélete köréből, *MTA III. Oszt. Közl.* **21** (1972), 119–127.

133. K. Seitz, Investigations in the Hajós-Rédei theory of finite abelian groups, *Karl Marx University, Budapest*, 1975 (MR 53 no. 655 (1977)).

134. S. K. Stein, Factoring by subsets, *Pacific J. Math.* **22** (1967), 523–541.

135. S. K. Stein, A symmetric star body that tiles but not as a lattice, *Proc. Amer. Math. Soc.* **36** (1972), 543–548.

136. S. K. Stein, Algebraic tiling, *Amer. Math. Monthly* **81** (1974), 445–462.

137. S. K. Stein, Tiling packing, and covering by clusters, *Rocky Mount. Journal of Mathematics* **16** (1986), 277–321.

138. S. K. Stein and S. Szabó, *Algebra and Tiling: Homomorphisms in the Service of Geometry*, Mathematical Association of America, 1994. (Carus mathematical monographs, 25)

139. S. Szabó, Multiple tilings by cubes with no shared faces, *Aequationes Math.* **25** (1982), 83–89.

140. S. Szabó, A type of factorization of finite abelian groups, *Discrete Math.* **54** (1985), 121–125.

141. S. Szabó, A reduction of Keller's conjecture, *Periodica Math. Hung.* **17** (1986), 265–277.

142. S. Szabó, A stronger form of Rédei's theorem for p-groups, *K. Marx Univ. Dept. of Math.* 1988, 15–21.

143. S. Szabó, On a problem of K. Seitz, *Bull. Soc. Math. Belg. Math.* **41** (1989), 95–100.

144. S. Szabó, On a problem of A. D. Sands, *Aequationes Math.* **38** (1989), 186–191.

145. S. Szabó, Multiple factorizations by cyclic subset, *Journal of Algebra* **134** (1990), 28–35.

146. S. Szabó, Cube tilings as contributions of algebra to geometry, *Beiträge zur Algebra und Geometrie, Contributions to Algebra and Geometry* **34** (1993), No 1, 63–75.

147. S. Szabó, An elementary proof of Hajós' theorem through a generalization, *Mathematica Japonica* **40** (1994), 1–9.

148. S. Szabó, Products of lacunary subsets of a finite abelian group, *Glasnik Matematički* **29** (1994), 235–238.

149. S. Szabó, Multiple simulated factorizations, *Studia Sci. Math. Hungar.* **32** (1996), 23–29.

150. S. Szabó, Factoring a certain type of 2-groups by subsets, *Rivista di Matematica* **6** (1997), 25–29.

151. S. Szabó, Groups with the Rédei property, *Le Matematiche* **52** (1997), 357–364.

152. S. Szabó, Factoring elementary p-groups, *Journal of Algebra* **206** (1998), 170–182.

153. S. Szabó, Elementary 3-groups with Hajós n-property, *Demostratio Mathematica* **31** (1998), 627–631.

154. S. Szabó, Factoring by cyclic and simulated subsets, *Acta Math. Hung.* **85** (1999), 123–133.

155. S. Szabó, Periodicity forcing factorizations for odd abelian p-groups, *Universitatera din Timişoara Anale* **37** (1999), 139–147.

156. S. Szabó, Factoring an infinite abelian group by subsets, *Period. Math. Hungar.* **48** (2000), 481–494.

157. S. Szabó, An extension of a result of A. D. Sands, *Demostratio Mathematica* **33** No. 3 (2000), 459–465.

158. S. Szabó and K. Amin, Factoring abelian groups of order p^4, *Mathematica Pannonica* **7** (1996), 197–207.

159. S. Szabó and C. Ward, Factoring abelian groups and tiling binary spaces, *Pure Math. and Appl.* **8** (1997), 111–115.

160. S. Szabó and C. Ward, Factoring elementary groups of prime cube order into subsets, *Mathematics of Computation* **67** (1998), 1199–1206.

161. S. Szabó and C. Ward, Factoring groups having periodic maximal subgroups, *Bol. Soc. Mat. Mexicana* **3** (1999), 327–333.

162. T. Szele, Neuer vereinfacher Beweis des gruppentheoretischen Satzes von Hajós, *Publ. Math., Debrecen* **1** (1949), 56–62.

163. T. Szőnyi, Around Rédei's theorem, *Discrete Math.* **208/209** (1999), 557–575.

164. C. B. Swenson, Direct sum subset decompositions of abelian groups, Ph.D. Thesis, Washington State University, 1972.

165. C. B. Swenson, Direct sum subset decompositions of Z, *Pacific Journal of Mathematics* **53** (1974), 629–633.

166. R. Tijdeman, Decomposition of the integers as a direct sum of two subsets, Number Theory (Paris 1992–1993) London Math. Soc. Lecture Note Ser., Vol. 215, Cambridge Univ. Press, Cambridge 1995, 261–276.

167. E. Wittmann, Einfachter Beweis des Hauptsatzes von Hajós-Rédei für elementare Gruppen von Primzahlquadratordnung, *Acta Math. Acad. Sci. Hungar.* **20** (1969), 227–230.

168. E. Wittmann, Über verschwindende Summen von Einheitswurzeln, *Elem. der Math.* **26** (1971), 42–43.

Index

affine plane, 22
affine space, 7
algebraic integer, 133
alphabet, 1
    binary, 282
Amin, K., *321*, *330*
annihilator, 5, 68
antiisomorphism, 123

Bernstein, H. J., 289, *321*
block design, 1
Bollobás, B., 2, *321*
Bruijn, N. G. de, 164, *321*

cardinal number, 270
cardinality, 15
character, 5, 66
    faithful, 102
    principal, 66
character group, 123
Cipra, B., *321*
clique, 28
code, 1
    binary, 283
    error correcting, 283, 284
    linear, 285
    maximal, 282
    subgroup, 283
    variable length, 282
    words, 282
Cohen, G. D., *321*
complete graph, 280
complete map, 2, 235
complex plane, 65
computer search, 238, 254, 281

concatenation, 282
conjecture
    Corrádi's, 123
    Furtwängler's, 4, 99, 108
    Keller's, 4, 26, 107
    Minkowski's, 3, 99
    Rédei's, 233
    Restivo, Salemi, Sportelli, 283
    Tijdeman's, 7, 240
Corrádi trick, 12, 19, 25, 137
Corrádi, K., 290, *321*
Coven, E. M., 2, *323*
covering system, 130
cross, 286
cube tiling, 3
cyclotomic field, 133, 317
cyclotomic polynomial, 57, 317

Daddel, A., 2
difference set, 1
dihedral group, 289
Diophantine approximation, 3
directed graph, 297
Dirichlet correspondence, 123
distortion element, 8
distortion place, 8
divisibility, 307

Erdős, P., 2
Etzion, T., *323*
Euclidean space, 3
Euler constant, 320
Euler's function, 317

Fáry, I., *323*

factorization, 3
    multiple, 4, 94, 311
    normalized, 5, 20
    quasi-periodic, 290
Felice, C. de, 282, *324*
Fourier analysis, 7
Fraser, O., *324*
Frattini subgroup, 126
Fuchs, L., 33, 289, *324*
Furtwängler, Ph., 4, *324*

Gagola, S. M., *324*
Galois field, 280
Garisson, S. C., *324*
Ginsburg, A., 2
Gordon, B., 4, 279, *324*
group
    Z-bad, 288
    Z-good, 288
    cyclic, 13
    elementary, 9
    infinite, 7
    Prüferian, 288
group ring, 5, 78

Hadamard matrix, 1
Hajós, G., 3, 164, *324*
Hamaker, W., *324*
Hamiltonian group, 289
Hamming distance, 283
Hamming sphere, 284
Hermann, P. Z., 290, *321*
Hilbert, D., 286
Hill, R., 281, *325*

interpolation, 307
invariant, 4
Irving, R. W., 281, *325*
Irwin, J. M., *325*

Kaplansky, I., *325*
Keller, O. H., 4, *325*
Kempermen, J. H. E., *325*
kernel of a character, 154

Kim, K. H., *325*
Kolountzakis, M. N., 319, *325*
Kronecker, L., 317
Kung, J. P. S., *325*

Lagarias, J. C., 4, 29, *325*
Lam, N. H., 283, *325*
Latin square, 2, 234
lattice, 3
Lawrence, J., *325*
Leader, I., 2, *321*
lemma
    partition, 75, 298, 303
    power replacement, 21
    zero divisor, 79
    Zorn, 282
Lindstrom, B., 283, *326*
Long, C. T., *326*
Lovász, L., *326*
Lyubeznik, G., *326*

Möbius function, 318
Mackey, J., 4, 31, *326*
Meyerowitz, A., 2, *323*
Minkowski, H., 3, *326*
Moews, D., *326*
Molnár, E., *326*
multiset, 66
Murty, M. R., *325*

Newman, D. J., 2, *326*

Obaid, E. E., *326*
Okuda, C., 288, *326*
Olson, J. E., *326*

Paige, L. J., 2, *326*
partition, 10, 22
period, 8
permutation, 9
Perron, O., *327*
Pettet, M. R., *324*
pigeon hole principle, 99, 129
Prüferian, 288
primitive product, 112

projective plane, 1
property
    Hajós, 6, 215
    P, 33
    Rédei, 7, 239

Rédei, L., 5, *327*
Ramsey number, 280
rank of a group, 288
Restivo, A., 283, *328*
Robinson, R. M., 4, *328*
root of unity, 5, 57
Rota, G. C., *325*
Roush, F. W., *325*

S. Litsyn, S., *321*
Saidi, S., *328*
Salemi, S., 283, *328*
Sands, A. D., 40, 288, *321*, *328*
Schmidt, T., *329*
Schrijver, A., *326*
Seitz, K., *329*
semicross, 2
Shor, P. W., 4, 29, *325*
Sportelli, T., 283, *328*
square-free, 318
star body, 286
Stein, S. K., 3, 289, *329*
subset
    Z-subset, 288
    r-free, 280
    cyclic, 3, 13
    distorted cyclic, 34
    distorted periodic, 34
    lacunary cyclic, 134
    normalized, 5, 21
    periodic, 5, 8
    replaceable, 9
    simulated, 7, 138
    symmetric, 280
    weakly periodic, 33
Swenson, C. B., *331*
Sylow subgroup, 288
Szőnyi, T., *331*

Szabó, S., 290, *321*, *329*
Szele, T., *331*

tensor product, 30
theorem
    Chinese remainder, 277
    Hajós's, 14, 301
    Rédei's, 13, 21, 307
Tijdeman, R., 7, *331*
transversal, 2
type
    of a factorization, 5, 47
    of a group, 4, 9

Vardy, A., *321*, *323*

Walker, E. A., *325*
Wang, Y., *325*
Ward, C., 238, *331*
Wittmann, E., *331*

Zémor, G., *321*
zero divisor, 5
zero polynomial, 140
Ziv, A., 2